Understanding Complex Systems

Future scientific and technological developments in many fields will necessarily depend upon coming to grips with complex systems. Such systems are complex in both their composition – typically many different kinds of components interacting simultaneously and nonlinearly with each other and their environments on multiple levels – and in the rich diversity of behavior of which they are capable.

The Springer Series in Understanding Complex Systems series (UCS) promotes new strategies and paradigms for understanding and realizing applications of complex systems research in a wide variety of fields and endeavors. UCS is explicitly transdisciplinary. It has three main goals: First, to elaborate the concepts, methods and tools of complex systems at all levels of description and in all scientific fields, especially newly emerging areas within the life, social, behavioral, economic, neuro- and cognitive sciences (and derivatives thereof); second, to encourage novel applications of these ideas in various fields of engineering and computation such as robotics, nano-technology and informatics; third, to provide a single forum within which commonalities and differences in the workings of complex systems may be discerned, hence leading to deeper insight and understanding.

UCS will publish monographs, lecture notes and selected edited contributions aimed at communicating new findings to a large multidisciplinary audience.

More information about this series at http://www.springer.com/series/5394

Dean S. Hartley III

An Ontology of Modern Conflict

Including Conventional Combat and Unconventional Conflict

Springer

Dean S. Hartley III
Hartley Consulting
Oak Ridge, TN, USA

ISSN 1860-0832 ISSN 1860-0840 (electronic)
Understanding Complex Systems
ISBN 978-3-030-53213-0 ISBN 978-3-030-53214-7 (eBook)
https://doi.org/10.1007/978-3-030-53214-7

This Springer imprint is published by the registered company Springer Nature Switzerland AG
The registered company address is: Gewerbestrasse 11, 6330 Cham, Switzerland

Foreword

This book represents a certain amount of work in its compilation; however, for me, it was more of a pleasure than a chore. Those I asked to review it and make comments are the ones who did the real work.

I thank Paul Works and the U.S. Army's Training and Doctrine Command (TRADOC) Analysis Center (TRAC) for providing the opportunity to develop an ontology that applies to unconventional conflict. Thanks are also due to Lee Lacy who taught me about ontologies and was my co-worker on the TRAC projects. Wayne P. Hughes, Jr. (Captain, USN, Retired) and Dr. Ken Jobson both provided valuable references to books and documents that I used to improve my work. Finally, I thank Mike McCurdy who reviewed this book and made comments. *Mahalo nui loa*!

Errors and omissions are my responsibility, as I do not always accept the advice I am given.

Oak Ridge, TN, USA

Dean S. Hartley III

Preface

Conventional combat has been studied for centuries. And just when we think we understand it, someone does something unexpected, changing the nature of war. One recent development is the problem posed by asymmetrical war, in which we discover that it is not enough to have the best conventional military possible when the enemy refuses to fight by the "rules." (This is not actually new; however, not everyone studies/remembers/understands history.) Another development is the realization that conventional wars do not always end with one side capitulating absolutely. Conventional combat can morph into unconventional conflict (and unconventional conflict can morph into conventional combat). Further, in the "bad old days" if you defeated someone, you were allowed the "spoils of conflict." That is, you could take what you wanted and leave the area devastated. Today we have a moral impulse to accept the "if you broke it, you own it" philosophy, which leads to efforts to "fix" what is broken, where "fixing" can include changing political and cultural systems along with repairing physical damage. We include things other than "pure conventional combat" in something called "unconventional conflict."

The simplest thing that can be said about unconventional conflict is that it is not conventional war. It may include combat operations. It may include multiple conflicting parties. It may have a time span measured in years. It may be confined to a single country or span a continent. It certainly includes social and cultural behavior issues and the parties to the conflict may not all play by the same set of rules. Regardless of the details, unconventional conflict is real and messy and appears to be here to stay. Some unconventional conflicts may be optional. That is, your country may choose not to be engaged. However, some are not optional, at least not for all parties. The current crop of Islamist terrorists has chosen to include a large part of the globe in their conflict, waging terror campaigns in the USA, in Europe, in Africa, and in Asia. Whatever it is, unconventional conflict is important.

We use the term "modern conflict" to be inclusive of both conventional combat and unconventional conflict. A particular modern conflict will have its own configuration of parts of conventional combat and unconventional conflict. Not only may the configuration of the conflict be unknown beforehand, the configuration may evolve during the conflict itself. Some "mission creep," decried by many, may be due to this evolution. In any case we cannot expect the conflict at hand to be the last that ever occurs. Therefore, it is important to understand the entire domain of modern conflict.

Modern conflict is a complex adaptive system. Complex adaptive systems exhibit emergent properties—that is, they have properties that are not predictable from simple descriptions of the system. For example, high school physics is complicated, but seemed relatively straightforward, with mass and momentum, levers and inclined planes, potential and kinetic energy, and so forth. Even the addition of atomic physics, with neutrons, protons, and electrons, did not seem to be much of a stretch. Chemistry was a separate field that was fun because you could create color changes in liquids and explosions. Then we discovered that all of chemistry is an emergent property of physics—in particular those electrons. Some atoms like to connect to other atoms because of the configuration of the electrons. How could you predict that? Then we find that biology is an emergent property of chemistry and that social systems are an emergent property of biology and conflict is an emergent property of social systems. What emerges from conflict?

We are undertaking a description of modern conflict. How do you describe something that is likely to have unforeseen ramifications? The answer is that you do the best you can. An ontology is an organizational tool for describing a domain of knowledge. At its best, it captures all of the important parts, such as atoms, neutrons, protons, electrons, mass, potential and kinetic energy, etc. And it captures all of the important relationships among the parts. It supports the statement of what is known. If it leaves out things that are not yet known, at least it should have a structure that allows them to be added once they are discovered. Typically, an ontology is imperfect and needs revisions as it is reviewed and used. In some cases, the domain of knowledge itself is so imperfect that an ontology of that knowledge can have errors in its structure and content. Modern conflict is definitely a candidate for that "imperfect knowledge" category. Humans have engaged in conflict since before the dawn of recorded history and have thought and written about it ever since.

This volume describes **an** ontology of modern conflict—the Modern Conflict Ontology (MCO). The point of an ontology is to create a structure to contain the knowledge of a domain and then to fill it with that knowledge. It may be that having this ontology will help us to deal with conflicts. Perhaps, more importantly, it may be that the ontology will help us understand the things we don't know that we need to know to succeed.

The initial ontology work was performed for the U.S. Army Training and Doctrine Command (TRADOC) Analysis Center (TRAC) and resulted in the construction of the Irregular War Ontology (IWO). These results were published in two TRAC documents (Hartley and Lacy, Irregular Warfare (IW) Metrics Ontology Final Report, TRAC-H-TR-13-020, 2011, IW Ontology Final Report, 2013b).

Subsequently, the author expanded the IWO into the Unconventional Conflict Ontology (UCO). The first Springer publication concerning the UCO concentrated on the value of the ontology in modeling unconventional conflict (Hartley D. S., Unconventional Conflict: A Modeling Perspective, 2017). That book described the advantages of having a comprehensive, holistic ontology of the unconventional conflict domain in creating a conflict model and in verifying, validating, and accrediting (VV&A) the model. In 2018, the author created a book that described the details of the ontology (Hartley D. S., An Ontology for Unconventional Conflict, 2018). This book on the Modern Conflict Ontology is the third in a series of books about ontologies of conflict. Each of these ontologies has been constructed to be as general as possible; however, they begin from an essentially US viewpoint of warfare and may retain some unconscious assumptions because of that. Further, an ontology, like any other model of reality, abstracts those parts of reality that are relevant to its purpose and omits those that are irrelevant. Ontologies constructed for different purposes will differ from one another. There is also a measure of art to the process of constructing an ontology. If something can be divided into parts differently, different authors may choose dissimilar divisions, yielding equally valid but superficially distinctive ontologies.

The first three chapters of the book begin the discussion of the MCO. The first chapter introduces the foundational concepts. The second chapter discusses modern conflict in detail. The third chapter provides an overview of ontologies in sufficient detail to make the rest of the book understandable, but without covering the minutia of the subject.

The next ten chapters describe the parts of the MCO. Each part is a sub-ontology and is discussed in detail, including connections to the other parts. Instances are used very liberally to ensure that the concepts are made concrete.

- The central elements of the MCO are Actor classes (representing autonomous things, such as people), Action classes (representing the actions that Actors can initiate), Object classes (representing passive things), and Metric or State Variable classes (representing the conditions of the other elements).
- The MCO contains two supplementary sub-ontologies, one describing the instruments of national power and the other the state of the world. Diplomatic, Information, Military, and Economic (DIME) actions have been named the instruments of national power. Where the DIME paradigm describes how the state of the world can be changed, the PMESII paradigm describes the state of the world. State variables (also called metrics) are categorized as Political, Military, Economic, Societal, Information, and Infrastructure (PMESII) variables. These two ontologies are described as supplementary because much of their information content is replicated by the Action Ontology and the Metric Ontology.
- There are two Concept Ontologies that contain connections among the elements based on the related meanings and operations among the elements, relations that are obvious to humans because they know the meanings of words, but not obvious to a computer.
- The MCO also contains a Theories Ontology, which comprises a structure for theories of cause and effect salient to modern conflict.

- There are five Scenario Ontologies that provide the structure for describing a scenario. These include scenario identification, agendas, rules of engagement, connections (such as kinship, ownership, and co-location) among entities in a group or network, and sets connecting actions to results.
- In addition to *things*, the MCO is constructed of *relationships* between things. The basic relationships in the MCO are the object relationships that connect classes and instances. These relationships range from structural ontology relationships such as *is-a* and *hasInstance* to domain-specific relationships such as *hasTaskGoalPair* and *owns*.

The final chapter consolidates the descriptions of the ontology into a discussion of "what we can know." It describes the implementation history and changes from the UCO to the MCO, plus potential future enhancements. It also discusses some uses of the ontology and finishes with some concluding thoughts.

The front matter of the book includes a listing of the acronyms used in this book and their definitions. The end matter includes a bibliography of the citations in the text and an index of important terms.

Oak Ridge, TN, USA Dean S. Hartley III

Contents

Acronyms

Acronym	Definition
AAR	Actor-Action-Result
AO	Area of Operations
APSO	Aggravated Peace Support Operations
BFO	Basic Formal Ontology
C^2	Command and Control
C4I	Command, Control, Communications, Computers, and Intelligence
CAS	Complex Adaptive System
CD	Counter-Drug
CI	Counterinsurgency
CO	Company
COG	Center of Gravity
COIN	Counter Insurgency
DEXES	Deployable Exercise Support system
DIA	Defense Intelligence Agency
DIME	Diplomatic, Informational, Military, Economic
DIME+	DIME plus Other
DIMEFIL	DIME plus Financial, Intelligence, Law Enforcement
DM2	DoDAF Meta-Model
DoD	Department of Defense
DoDAF	DoD Architecture Framework
DOLCE	Descriptive Ontology for Linguistic and Cognitive Engineering
DP	Dimensional Parameter
DR	Disaster Relief
EA	Enterprise Architecture
EMP	Electromagnetic Pulse
FASP	Foreign Assistance Standardized Program
FAST	Flexible Asymmetric Simulation Technologies
FID	Foreign Internal Defense
FON	Freedom of Navigation
GFO	General Formal Ontology

GTO	Goal-Task-Owner
HA	Humanitarian Assistance
HA/DR	Humanitarian Assistance and Disaster Relief
HHC	Headquarters and Headquarters Company
HN	Host Nation
HQ	Headquarters
HSCB	Human, Social, Cultural Behavior
ICBM	Intercontinental Ballistic Missile
ID	Identification
IDP	Internally Displaced Population or Person
IED	Improvised Explosive Device
IFOMIS	Institute for Formal Ontology and Medical Information Science
IGO	Intergovernmental Organization
IHRL	International Human Rights Law
IO	International Organization
ISIS	Islamic State of Iraq and Syria
ISSM	Interim Semi-static Stability Model
IW	Irregular War or Warfare
IWO	Irregular War Ontology
JUORS	Japan–US Operations Research Seminar
LIC	Low Intensity Conflict
LOE	Line of Effort
MCO	Military Contingency Operations
MCO	Modern Conflict Ontology
MIO	Maritime Intercept Operation
MIS	Management Information System
MLRS	Multiple Launch Rocket System
MOE	Measure of Effectiveness
MoFE	Measure of Force Effectiveness
MoM	Measure of Merit
MoP	Measure of Performance
MoPE	Measure of Political Effectiveness
MSCA	Military Support to (Domestic) Civil Authorities
NEO	Noncombatant Evacuation Operation
NGO	Non-Governmental Organization
NI	National Integrity
OE	Operational Environment
OIF	Operations Iraqi Freedom
OOTW	Operations Other than War
ORSBM	Oak Ridge Spreadsheet Battle Model
OWL	Web Ontology Language
PE	Peace Enforcement
PK	Peacekeeping
PMESII	Political, Military, Economic, Social, Information, Infrastructure
PMESII+	PMESII-KE

PMESII-KE	PMESII plus Kinetics, Environmental
PMESII-PT	PMESII plus Physical Environment, Time
PO	Peace Operation
POL	Petroleum, Oil, and Lubricants
PSYOPS	Psychological Operations
QCA	Qualitative Comparative Analysis
QJM	Quantified Judgment Model
QJMA	Quantified Judgment Method of Analysis
ROE	Rules of Engagement
SaF	Stocks-and-Flows
SAR	Search and Rescue
SNA	Social Network Analysis
SSTR	Stability, Security, Transition, and Reconstruction
STEM	Science, Technology, Engineering, and Mathematics
SUMO	Standard Upper Merged Ontology
SWAG	Scientific Wild-Assed Guess
TOE	Table of Organization and Equipment
TRAC	TRADOC Analysis Center
TRADOC	Training and Doctrine Command
TWG	Tactical War Game
UCO	Unconventional Conflict Ontology
UN	United Nations
US	United States
USPACOM	US Pacific Command
V&V	Verification and Validation
VBA	Visual Basic for Applications
VV&A	Verification, Validation and Accreditation
WAG	Wild-Assed Guess
WMD	Weapons of Mass Destruction
XML	Extensible Markup Language

List of Figures

List of Tables

Chapter 1
Introduction

Modern conflict is a combination of conventional combat and unconventional conflict. Each is a complex adaptive system, and the combination is a complex adaptive system with a larger domain. It is more **complicated** because there are additional interactions; however, it is not obviously more complex in the sense of becoming some sort of extra-complex adaptive system. We may view a conflict as a "thing" that takes place or doesn't, which doesn't seem to fit with the concept of a system. Even if we consider a conflict as a process, we can have a problem calling it a system. However, if we look at the internals of a conflict or consider the generality of conflicts, we see the interplay of a multitude of processes within conflicts. It is a system. Within a conflict we often see changes of tactics due to changing situations and even changing of strategies and goals ("mission creep"). This is adaptation at work. When we look from one conflict to another, later, conflict, we see adaptation at work. The complexity of conflict is almost a given. However, the complexity of the ontology necessary to describe conflict will give evidence to the complexity of the subject. We have the elements: complex, adaptive, and system. However, we expect complex adaptive systems to exhibit "emergence," the generation "of a new quality of macroscopic collective behavior the manifestations of which are the spontaneous formation of distinctive temporal, spatial or functional structures (Fellman et al. 2015)." Conflict can certainly do that! Despite the plans of the participants of a conflict, unexpected changes frequently occur. New countries are formed; old countries disappear; and new internal political systems are generated, sometimes in the "winning" entity.

In an earlier book, I described an ontology of unconventional conflict (Hartley D. S., An Ontology for Unconventional Conflict, 2018). I also mentioned possible future enhancements: aligning the ontology with a more formal ontology framework and adding combat to the domain. This book describes the results of these two enhancements.

D. S. Hartley III, *An Ontology of Modern Conflict*, Understanding Complex Systems, https://doi.org/10.1007/978-3-030-53214-7_1

It is not surprising that the addition of conventional combat requires changes to the ontology for unconventional conflict. The changes are significant, but not overwhelming. However, I have taken this opportunity to bring the ontology into closer agreement with the organizational conventions advocated by many professional ontologists.

The point of the ontology is to describe what we can know about conflicts in general and to support recording what we know about a particular conflict. To those ends, we need to be able to describe the actors, objects, and actions of the conflict. We need to talk about the relationships among them. We need to talk about what drives them to do what they do.

What is "modern conflict"? That is the key question that this book begins to answer. The approach is to describe an ontology of modern conflict. This ontology contains the structure of the things that are known about modern conflict – the defined concepts and the relationships among the concepts. This structure illuminates the knowledge and serves up a holistic view of modern conflict. In turn, this holistic view provides a background that allows detailed examinations of the concepts and relations, keeping these examinations in perspective to the whole.

1.1 Introduction to Modern Conflict

The current standard paradigm for understanding conflict involves consideration of the instruments of power, that is, the means for effecting change in a situation, and the state variables that describe the situation at any given time. The simplest listing of the instruments of power is Diplomatic, Informational, Military, and Economic (DIME). The simplest listing of the state variable categories is Political, Military, Economic, Social, Information, and Infrastructure (PMESII). Each of the DIME and PMESII categories can be decomposed into observable and measurable elements. This paradigm has been very important in advancing the understanding of unconventional conflict; however, it misses two important components of the situation, the actors that effect the changes and suffer the effects of those changes and the objects – passive elements (such as infrastructure) – that are also affected. The actors and objects are implied by the PMESII state variables, but not explicitly defined. The new paradigm of this ontology explicitly defines the actors, actions, objects, and state variables. It includes ontological characterizations of each, which are separate from the DIME/PMESII decomposition. However, the ontology also includes the links to the DIME/PMESII decomposition, thus supporting two different tools for understanding unconventional conflict. Because modern conflict contains unconventional conflict as an integral part and because the additional combat elements fit seamlessly into the new paradigm, the ontology for modern conflict has the same structure as the one for unconventional conflict.

PMESII

The traditional thinking about conventional conflict focused on military actions and military metrics. Clearly, more is needed to organize thinking about unconventional conflict. Figure 1.1, derived from *Joint Operations* (Chairman of the Joint Chiefs of Staff 2011), illustrates this point. The six domains, political, military, economic, social, information, and infrastructure (**PMESII**), are shown as parts of the operational environment. Each has its own link and node structure; however, all of the domains are interconnected. The figure emphasizes that the operational center of gravity (COG) may differ from the strategic COG. In the military, the COG of an organization is the source of its strength. The concept was developed by von Clausewitz to describe the location of the cohesive strength of a military force – the point at which an attack should be aimed (Clausewitz 1993).

The categories of PMESII are expanded as follows:

Political: The primary components of the political domain are governance (policies, personnel, organizations, freedom, etc.), the rule of law (judiciary, law

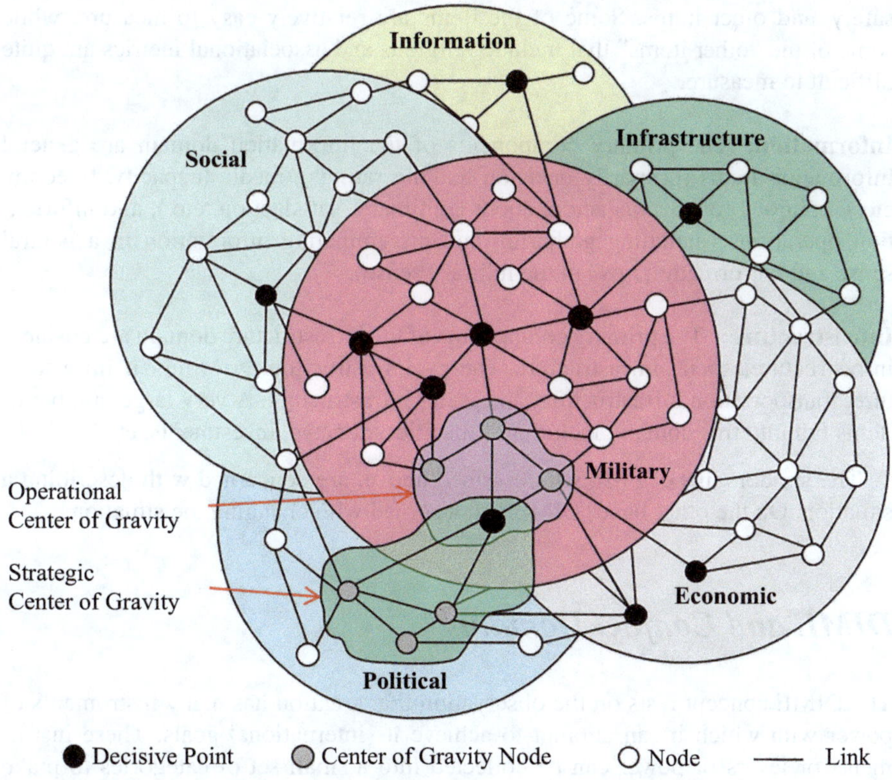

Fig. 1.1 PMESII diagram

enforcement, crime, etc.), and politics (leadership, factions, etc.), as well as some miscellaneous items (intervenor status, stability/peace and legitimacy ratings, etc.). Measuring the status of some of these items is difficult, but obviously necessary.

Military: The primary components of the military domain are conflict, government (the relation with, intelligence services, organizational sizes, etc.), and security (provisions), as well as other items (insurgents, terrorists, paramilitary forces, capacities, etc.). Measuring the status of these items is easier than measuring the political status. (The US military traditionally measures itself in terms of capacity, capability, and readiness.)

Economic: The primary components of the economic domain are agriculture, crime, energy, finance, governmental economic actions, jobs, and "other" things (including various industries other than agriculture). Measuring the status of economic items is a relatively well-defined process compared to some other PMESII items.

Social: The primary components of the social domain are basic needs (food, water, shelter, etc.), education, health, movement (freedom of, restrictions on, forced, etc.), safety, and other items. Some of the items are relatively easy to measure, while some of the "other items" that include religious and associational metrics are quite difficult to measure.

Information: The primary components of the information domain are general information items (primarily decision-making ratings), media (capacity, freedom, etc.), opinions (of various actors about legitimacy, satisfaction, etc.), and information operations (including gathering and disseminating information in a neutral sense and "information ops" in the military sense).

Infrastructure: The primary components of the infrastructure domain are business infrastructure, social infrastructure, energy infrastructure, government infrastructure, transportation infrastructure, and water infrastructure. A very large number of items fall into this domain, including capacities, damage, investments, etc.

The six domains of PMESII, described above, are concerned with describing a situation. On the other hand, DIME is concerned with changing the situation.

DIME and Conflict Domains

The DIME concept rests on the observation that a nation has many instruments of power with which it can attempt to achieve its international goals. These instruments or levers of power can be collected into a small set of categories to make remembering them easier. The most popular set is DIME – Diplomatic, Informational, Military, and Economic (Fig. 1.2).

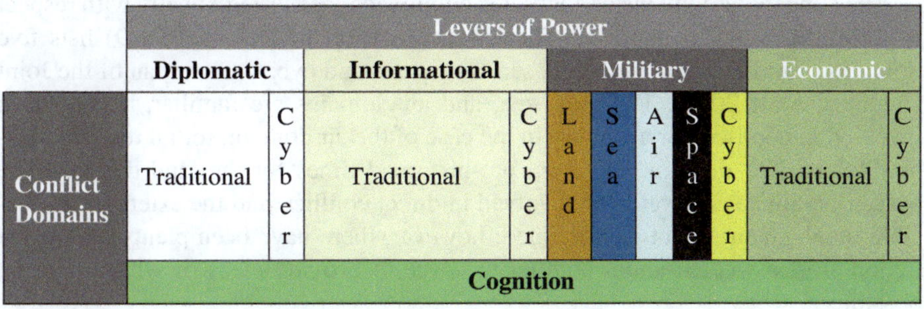

Fig. 1.2 Levers of power and domains of conflict

Diplomatic power rests in negotiations and agreements. To the extent that a nation regards itself as bound to honor its agreements, diplomacy can result in changes in the actions of one or more nations. However, there are advantages to be had short of binding agreements. Offering to provide or withhold favors among countries can also result in changes of actions.

Informational power lies in gaining information from others and in controlling the information desired by others. With cyberwarfare, remotely controlled sabotage is also possible (Lubold and Volz 2018; Taylor and Germano 2018). Differences in what is known between contending parties can be decisive, whether the domain is economic, military, or diplomatic.

Military power is an obvious component. However, the fact that there are limits to its effectiveness and on its use should serve warning that the other levers of power also have limits (both individually and collectively).

Economic power is also an obvious component. Nearly immediate effects can be seen from such actions as freezing bank accounts. Long-term strategies may involve the stronger economic power causing the weaker power to spend itself into defeat. Particular sectors of the economy can be significant in their own right. For example, the need to import oil and fuels (energy) costs the importing nation directly and may have less direct costs in the form of obligations and possible coercion. Energy independence allows a nation to avoid these problems and becoming a net exporter adds to the nation's influence (Olson 2018; Wall Street Journal Editorial Staff 2018). "When China wants to compel the United States to do something, it doesn't turn to force. … China threatens the sale of US debt, pressures US businesses invested in China's market, uses boycotts, restricts critical imports and exports, and employs predatory practices (McFate 2019)."

Figure 1.2 also displays the concept of domains of conflict. In the pre-computer world, the diplomatic lever of power was limited to the realm of nation-states (and polities that desire to become nation-states). The informational lever of power was important between and among nation-states and corporations, with individuals participating as minor players. The economic lever of power was similar, with individuals in the role of consumers.

The concept of domains of conflict is employed most systematically with respect to the military lever of power. The US Department of Defense (DoD) lists five domains of military conflict: land, sea, air, space, and cyber (Chairman of the Joint Chiefs of Staff 2017). The land, sea, and air domains are familiar, having been involved in conflicts for a century in the case of the air domain, for all recorded history for the sea domain, and from pre-historical times for the land domain. The space domain has not yet been involved in direct conflict, and the extent of its possible involvement is not well-defined; however, there have been plenty of science fiction stories written about conflict in space to provide ideas of what might be possible.

As shown in Fig. 1.2, the cyber domain is effectively part of each of the other levers of power. The cyber domain is definitely new. It is so new that much of the applicable terminology either is unfamiliar to many or has varying definitions, depending on the source. The definitions below are provided by Lucas Kello, who approaches the problem from an international relations perspective as opposed to a military perspective (Kello, The Virtual Weapon and International Order, 2017).

Cyberspace is all of the computers and networks in existence, including computers that are isolated from networks (air-gapped). Note that this includes embedded computers, such as those in cars (at least 1 in each modern car, with some having more than 30 computers). Cyberspace is manipulable through code (and on/off switches). (Note that the cyberspace domain in Fig. 1.2 actually refers to the cyber domain (defined next) due to differences between Kello's and DoD definitions.)

The ***cyber domain*** includes cyberspace and all of the human and institutional actors who operate and control cyberspace. The cyber domain also includes machinery that is controlled by cyberspace. The cyber domain affects and is affected by cyberspace, and the human part is responsive to human inputs of a psychological, social, and political nature.

"***Cybersecurity*** consists of measures to protect cyberspace from hostile action." It also includes measures to "protect the cyber domain from threats emanating from" cyberspace. When cybersecurity involves the military, it is called ***cyber defense***.

Information security is used to describe the control of information flows. This can be the suppression of subversive information in an autocratic state or efforts to control the exchange of child pornography. Information security is often conflated with cybersecurity.

"***Malware*** denotes software designed to interfere with the function of a computer or degrade the integrity of its data." Malware may directly affect cyberspace or may do so indirectly by affecting human operators who then install malware into cyberspace (usually inadvertently, although conceivably the threat of harm to the operator might lead to the operator knowingly installing the malware).

The term ***cyberweapon*** is restricted to malware that is capable of and intended to do sufficient damage that the effects would be classified as the effects of a weapon in the realm of international relations.

"***Cybercrime*** entails the use of a computer for an illicit purpose under the existing penal code of a nation."

"***Cyberattack*** refers to the use of code to interfere with the functionality of a computer system for a political or strategic purpose." "Neither the goal nor the effects of a cyberattack need to be contained in cyberspace." "If the effects of a cyberattack produce significant physical destruction or loss of life, the action can be labelled ***cyberwar***, a term that should be used sparingly."

"***Cyber exploitation*** refers to the penetration of an adversary's computer system for the purpose of exfiltrating (but not defiling) data." When the goal is to steal military or industrial secrets, the exploitation is known as ***cyber espionage***. If the goal is to obtain sensitive information about a public official or organization to be released at an opportune moment to influence government actions or undermine public confidence, then the exploitation is known as [***cyber***] ***kompromat***.

Information conflict takes place among nation-states, corporations, and individuals. It took place before the advent of computers and continues after their creation. The informational lever of power originally could be characterized as spying (obtaining information from the enemy or competitor), deception (ensuring the incorrectness of the information that the other obtains), and counterintelligence (thwarting the attempts of the other to obtain information). The addition of the cyberspace domain has not removed these operations, but added to them. Computers can aid in the spying, deception, and counterintelligence, and they can be used to corrupt the computers and cognified systems of the other side. (***Cognified systems*** are systems with computational power, often connected to other systems through networks.)

The cognition domain is the "sixth domain" (counting among the military domains) because we now have the ability to affect the cognitive abilities of humans (and augmented humans). We have practiced the art of persuasion (or rhetoric) since before the time of Aristotle, but only recently have been creating an additional science of persuasion, which vastly improves the effectiveness of persuasive actions. We have developed new learning science interventions that affect cognition. In addition, we have developed tailored pharmaceuticals that affect cognition (beyond the unfocussed effects of naturally occurring drugs such as peyote and opium and the volcanic gases of the Greek oracles). We now need to pursue cognitive superiority in addition to the traditional air, land, and sea superiority and diplomatic and economic superiority and the newer information, space, and cyber superiority requirements (Hartley and Jobson, Cognitive Superiority: Information to Power, the Road to Winning in the Sixth Domain, 2020).

The DIME paradigm was a description of the *national* levers of power. However, today we have to contend with non-nation-state actors, including individuals, who have the power of terrorism, asymmetric warfare, biological warfare, and cyber warfare to act against nation-states (or other organizations and individuals) as only nation-states could in the past. Accordingly, we have changed the label to be just "levers of power."

DIME/PMESII

Variations to PMESII have been proposed, such as PMESII-PT, which adds physical environment and time to the package. Instead, we add the Kinetic and Environmental categories (PMESII-KE or PMESII+).

Kinetic: The primary components of the kinetic domain are logistics, damage and attrition, and command, control, communications, computers, and intelligence (C4I).

Environmental: The primary components of the environmental domain are atmospheric, earth, water, and fire effects.

Variations to DIME have also been proposed, such as DIMEFIL, which adds Financial, Intelligence, and Law Enforcement to the DIME package. Instead, we add an "Other" category.

Other Power: This includes criminal and environmental (e.g., natural disasters) forces.

As mentioned above, there have been variations of DIME/PMESII that have been argued (including my offerings); however, for our purposes, we will use PMESII to refer to all state variables, regardless of taxonomy details. The acronym DIME refers to the levers of power that a (nation) state has to influence the PMESII situation. As with PMESII, we will use DIME to refer to all such interventions, regardless of taxonomy details. Collectively, these will be referred to as DIME/PMESII+ or DIME/PMESII or simply as PMESII for brevity (Hartley D. S., DIME/PMESII Models, 2015).

Contemplations on Conflict

This isn't a capstone work on modern conflict, but actually a foundation for increasing our understanding of modern conflict.

The conceptual description of combat that I chose is drawn from several well-respected military thinkers. Having spent several decades working with combat models, I am familiar with the choices that many modelers have made in what to represent and how to represent it in simulations of combat. However, this is like being familiar with many different forests, without having a grasp of a good conceptual model of a forest. (I can say with assurance that many of the modelers who created those combat models also lacked that conceptual model – many of them were just "making it up as they went along.") The Oak Ridge Spreadsheet Battle Model (ORSBM) was preceded by the work of many and was written to express such a conceptual model (Hartley D. S., Predicting Combat Effects, 2001). It has parts that are not germane to the MCO and does not cover unconventional conflict.

This is not the only effort to consider an extension of traditional views of conflict. The conflict modeling community is involved in defining conflict because it is tasked with creating models that embody the actions that the operational and doctrinal community defines. Gallagher, Caswell, Hanlon, and Hill discuss a need to rethink the traditional hierarchy of military models (Gallagher et al. 2014). The traditional hierarchy begins with System/Engineering models, followed by Engagement models and Mission models, and ending with Campaign models at the most aggregated, lowest resolution level. The authors advocate adding two levels, Defense Enterprise and Government, Non-government, and Coalition Instruments of Power, at the highest level.

The principal descriptions of conflict that are used here are Sun Tzu (1963); Machiavelli (1966); Clausewitz (1993); Liddell Hart (1967); Wylie (2014); DuBois, Hughes and Low (1997); the US Marine Corps (1997); and a modern Chinese book by Qiao and Wang (1999). These sources are described in Chap. 2; however, Fig. 1.3

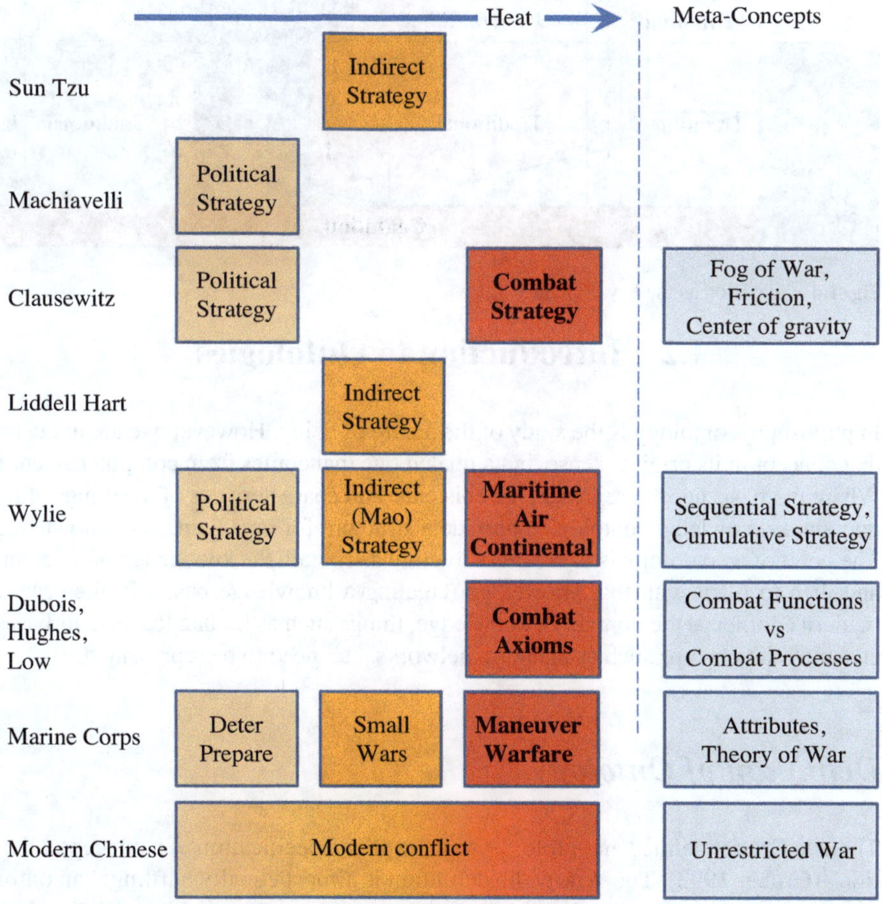

Fig. 1.3 Themes in theories of war

provides an overview of the themes that each developed. The themes are divided (roughly) into the political/pre-war, unconventional conflict, and war stages. These stages represent increasing kinetic conflict, metaphorically, increasing heat. Some of the sources also provided important meta-concepts.

As the battle moves toward increasingly kinetic conflict, the tactics of control move from persuasion to coercion. This is true whether the lever being used is diplomatic, informational, economic, or military. As Carl von Clausewitz said, "*war is only a branch of political activity; that it is in no sense autonomous* (Clausewitz 1993)." [The emphasis is contained in the source.] In Fig. 1.4, we take a step back and derive all actions from the cognitive domain, with the purpose of the actions being to impact the cognitive domain of the opponent. Naturally, the opponent is doing likewise, with the positions reversed. Achieving cognitive superiority is required for winning.

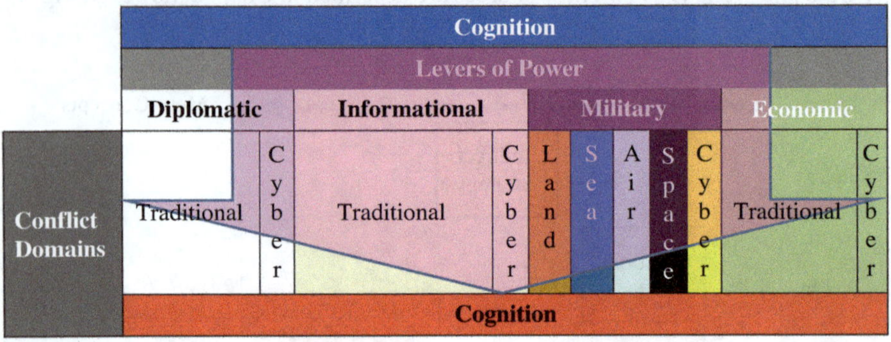

Fig. 1.4 Cognition as the key domain

1.2 Introduction to Ontologies

In philosophy, ontology is the study of the nature of being. However, we are not using that concept in its original sense, but a related one that comes from computer science. Where there is a need to represent the existence and characteristics of the things of the domain of knowledge, ontologies provide a structure for organizing that knowledge. The point of an ontology is to create a structure to contain the knowledge of a domain and then to fill it with that knowledge (creating a knowledge base). In the case of modern conflict as the domain of knowledge, things such as human leaders, air bases, etc. and relationships such as agendas, networks, etc. need to be represented.

Definition of Ontology

Thomas Gruber defined an ontology as an "explicit specification of a conceptualization" (Gruber 1993). The Wikipedia definition is almost equally baffling: "an ontology is a formal naming and definition of the types, properties, and interrelationships of entities that really or fundamentally exist for a particular domain of discourse" (Wikipedia 2016). We will use a less formal definition: an ontology is a usable description of what is known about a domain of knowledge.

Ontologies support sharable and reusable knowledge representations. These representations specify the concepts within a domain of knowledge, the attributes of the concepts, and the relationships among these concepts. Ontologies can also support reasoning – making inferences – about the data, which may be particularly important when the classes have been instantiated in a knowledge base.

Representing Knowledge in an Ontology

Other than tutorial ontologies used in explaining what an ontology is and the kinds of relationships that may be represented, ontologies are too large to be expressed pictorially or in human languages and require computer language expression. The basic idea is that there are *things* (an undefined term) to be represented. These *things* are divided into *classes* and *relationships* (with, perhaps, other elements). The basic relationship *is-a* is included to connect these. That is, a *class is-a thing* and a *relationship is-a thing*. This relationship is a fundamental statement that one thing has all of the characteristics of the other, but is restricted in some way. Thus, a *pie is-a dessert* and a *cake is-a dessert*, but a *pie* is not the same thing as a *cake*. Different domains of knowledge require different classes and different relationships, so their ontologies will differ; however, there are commonalities of structure.

Figure 1.5 illustrates the pictorial representation of a taxonomy – an ontology with only one relationship and only one parent class for each child class. The boxes with three horizontal subdivisions represent classes. The connector with an open triangle head points from a subclass to a class and represents the *is-a* relationship.

The part of the diagram to the left of the vertical line is a standard expanded taxonomy – that is, every class is displayed. The part of the diagram to the right of the vertical line is a short-hand representation of the same taxonomy. The sole level-0 class, *thing*, is included. The two level-1 classes, *subThing1* and *subThing2*, are collapsed into a representation class, here named *category*. The four level-2 classes, *subSubThing11*, *subSubThing12*, *subSubThing21*, and *subSubThing22*, are collapsed into a representation class named *subcategory*. For large taxonomies, this short-hand representation is necessary to show structures without filling a diagram with tiny, unintelligible class boxes. Often the top levels are shown expanded and the lower levels are collapsed. For ontologies with multiple relationships or multiple parents allowed, collapsing may not make sense; however, if there are sections that are pure taxonomies, those sections may be collapsed, with "category" and "subcategory" suffixes or prefixes to indicate what is taking place.

Eventually an ontology reaches its lowest level, shown in Fig. 1.6 as the *element* class. Usually (although not universally) only the lowest-level classes can be instantiated. Here, if the lowest-level class were to represent a person, an instantiation would represent a particular person, say *Frank*. The conventions on types of connectors vary as to type of line and type of arrowhead. As long as there is text saying the relationship is an instantiation or that one end of the relationship is an

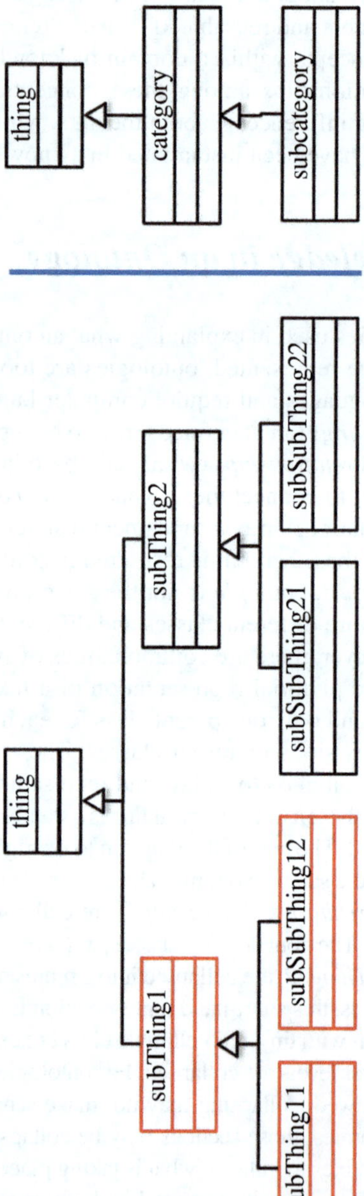

Fig. 1.5 Two representations of a taxonomy

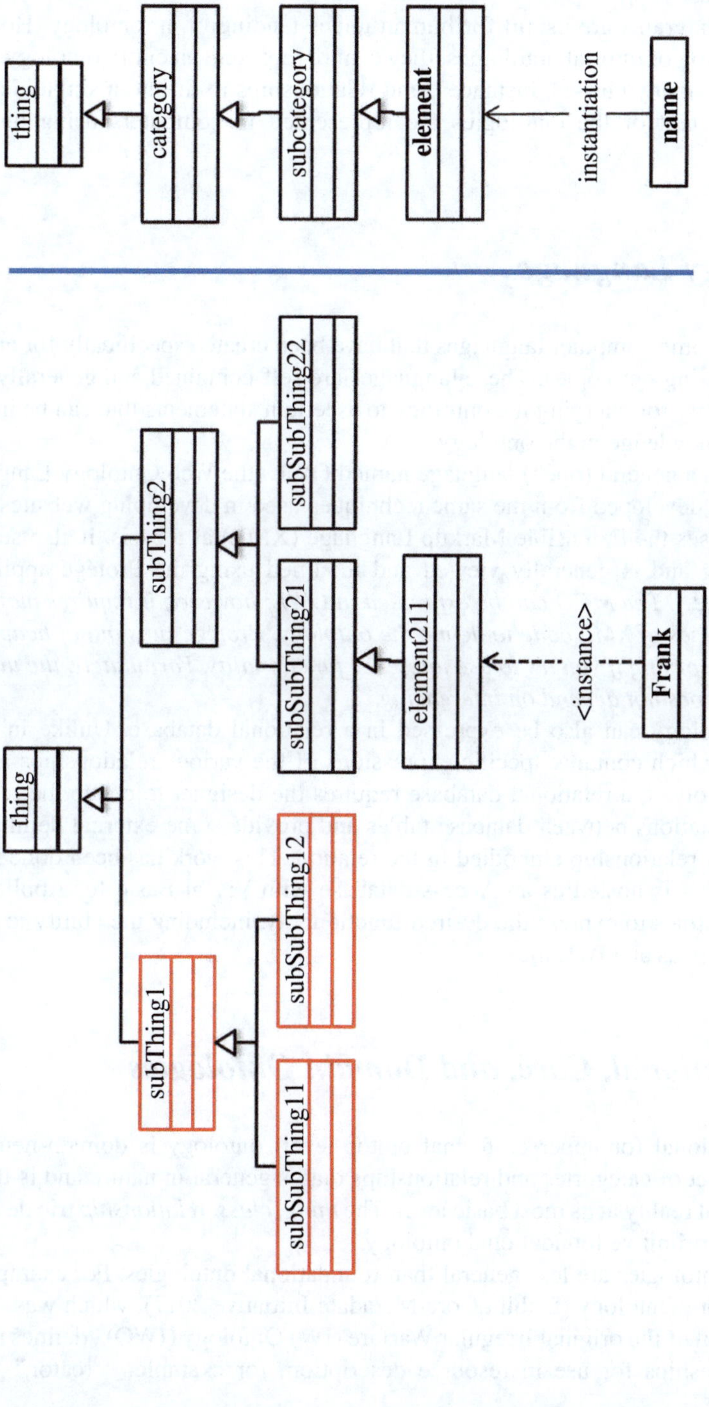

Fig. 1.6 Representing instances

instance, the meaning will be clear. An instance may also be represented as a box with only one or two horizontal subdivisions. The point is to distinguish instances from classes.

These diagrams are useful for human understanding of an ontology. However, except for toy or tutorial ontologies, they cannot represent an entire ontology. There are just too many classes, instances, and relationships to fit into a single diagram. All of the rest of the ontologies are represented in computers using ontology languages.

Ontology Languages

There are some computer languages that have been created specifically for creating and expressing ontologies. These languages are self-contained and generally focus on the facility for querying the ontology to ascertain statements that can be inferred from the knowledge in the ontology.

There is a general (open) language named OWL [the Web Ontology Language], which was developed from the same techniques used in developing websites (Lacy 2005). It uses the Extensible Markup Language (XML) as a basis. It also supports inferencing and is generally viewed and modified using the Protégé application (Protégé n.d.). *The MCO can be expressed in OWL; however, it requires more than a million lines of XML code to define the ontology. Protégé runs out of heap space and quits working if you try to use inference functionality. Fortunately, the utility of the MCO does not depend on inferencing.*

An ontology can also be expressed in a relational database. Unlike in OWL, however, which contains specific expressions of the various relationships and the classes involved, a relational database requires the designer to create the relationships as relations between database tables and provide some external definition of the type of relationship embodied in the relation. This work has been done for the MCO, and it is hosted as an Access database with Visual Basic for Applications (VBA) routines to express the desired functionality, including the ability to export the ontology as an OWL file.

Foundational, Core, and Domain Ontologies

A foundational (or upper or formal or top-level) ontology is domain-neutral. It defines a set of categories and relationships that is general in nature and is thought to represent reality at its most basic level. The *thing*, *class*, *relationship* trio described above is a primitive foundational ontology.

Core ontologies are less general than foundational ontologies. For example, the Dublin Core Ontology (Dublin Core Metadata Initiative 2017), which was used in the creation of the original Irregular Warfare (IW) Ontology (IWO) , defines a set of 15 relationships for use in resource description, for example, "creator," "date," "description," "format," and "language." Other core ontologies introduce relationships that are useful for extending a foundational ontology in its application to domain ontologies.

Domain or material ontologies (such as the one depicted in this book) describe a particular domain of knowledge. A domain ontology may import a core ontology and a foundational ontology to aid in interoperability with other domain ontologies and to avoid the problem of defining the contents of the core and foundational ontologies within the domain ontology.

1.3 Previous Work

The initial ontology work was performed for the US Army Training and Doctrine Command (TRADOC) Analysis Center (TRAC) and resulted in the construction of the Irregular War Ontology (IWO). These results were published in two TRAC documents (Hartley and Lacy, Irregular Warfare (IW) Metrics Ontology Final Report, TRAC-H-TR-13-020, 2011, IW Ontology Final Report, 2013b).

Subsequently, the author expanded the IWO into the Unconventional Conflict Ontology (UCO). The first Springer publication concerning the UCO concentrated on the value of the ontology in modeling unconventional conflict (Hartley D. S., Unconventional Conflict: A Modeling Perspective, 2017). That book described the advantages of having a comprehensive, holistic ontology of the unconventional conflict domain in creating a conflict model and in verifying, validating, and accrediting (VV&A) the model. In 2018, the author created a book that described the details of the ontology (Hartley D. S., An Ontology for Unconventional Conflict, 2018). This book is the third in a series of books about ontologies of conflict.

1.4 Organization of the Book

Figure 1.7 maps the 14 chapters of the book. The first mention in each chapter will spell out Modern Conflict Ontology (MCO). Most future references will be to "MCO," unless there seems to be a good reason to spell it out.

Chapter 1 is this chapter, in which we introduce the concepts of modern conflict and ontologies. In this chapter, modern conflict is introduced philosophically: How does one go about understanding it? What are its major components? What theories have been proposed? What are major sources of information about it? This chapter also introduces ontologies philosophically: What is an ontology? How is knowledge represented in an ontology? What languages are used in ontologies? What are the basic varieties of ontologies? In Chaps. 2 and 3, we move from philosophy to systematic discussions of conflicts and ontologies, particularly the Modern Conflict Ontology.

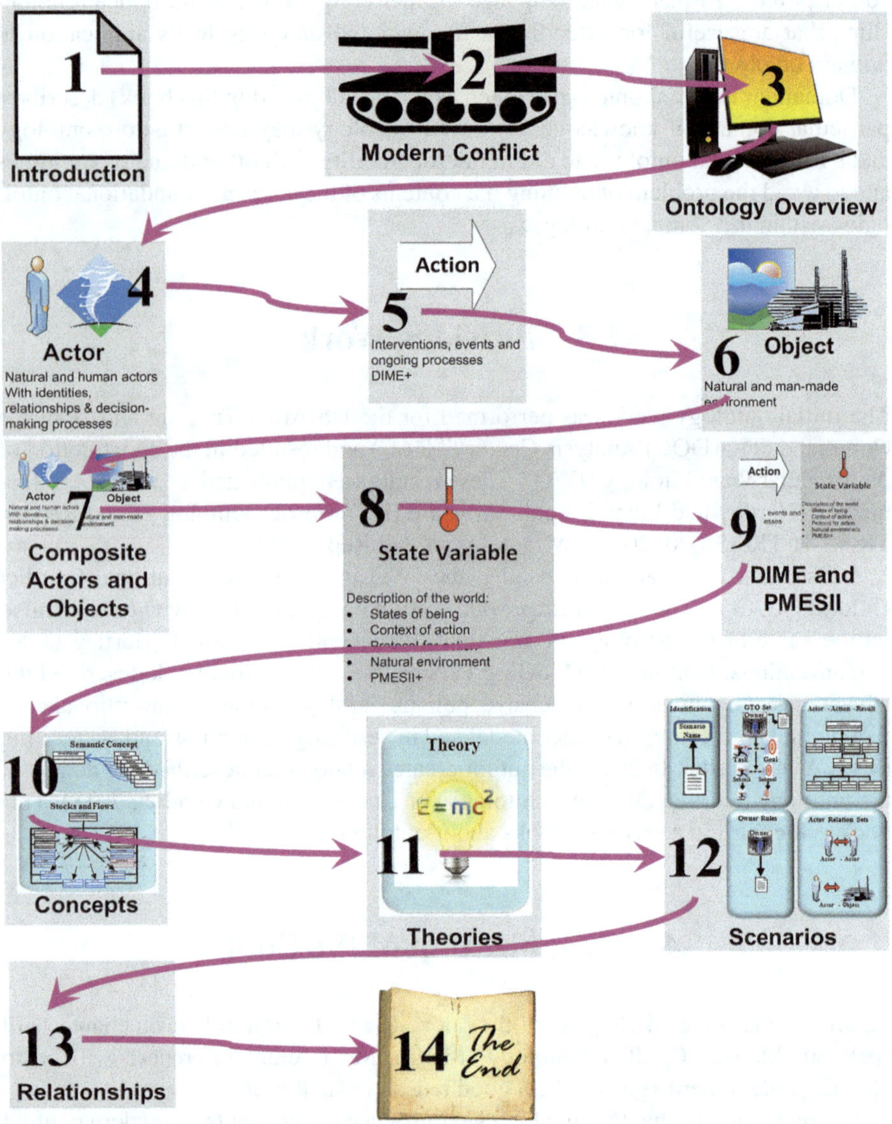

Fig. 1.7 Chapter topics

Chapter 2 provides a more detailed discussion of modern conflict. Multiple books can be and have been written on the various aspects of modern conflict. By comparison, the description presented in the chapter is superficial; however, it is sufficient to provide the background understanding of the domain that is the topic of the MCO. This chapter discusses modern conflict in detail. It defines the components of modern conflict, considers the relative likelihood of those components occurring, and examines the status of modern conflict as a complex adaptive system. In this chapter, we also review the various theories of conflict that have been advanced and reviewed over the course of years and centuries. We also present several illustrations of different kinds of conflict. We conclude the chapter with a characterization of modern conflict. The contents of this chapter provide a solid understanding of the modern conflict domain, which is necessary for creating and understanding the MCO.

Chapter 3 provides an overview of ontologies and the concepts involved in creating and understanding them. This chapter only provides enough information to understand the application of ontologies to modern conflict in the MCO. We begin by discussing the choice of the best foundational ontology for the creation of an ontology for modern conflict. In this chapter we also present an overview of the structure of the Modern Conflict Ontology (MCO). The central elements of the MCO are Actor classes (representing autonomous things, such as people), Action classes (representing the actions that Actors can initiate), Object classes (representing passive things), and Metric or State Variable classes (representing the conditions of the other elements). It is also useful to compare the MCO with a related ontology and we do so. The MCO is derived from the Unconventional Conflict Ontology (UCO) (Hartley D. S., An Ontology for Unconventional Conflict, 2018), so a discussion of the changes from the UCO to the MCO is in order. The contents of this chapter provide a solid understanding of ontologies and the particular constructs that are needed to model the modern conflict domain, which are necessary for creating and understanding the MCO.

Chapter 4 describes the Actor Ontology, a sub-ontology of the MCO. Actors are the motivating elements in the MCO: they cause things to happen. Generally, Actors are people, but not always. Chapter 4 expands the discussion of Actor classes, introduced in Chap. 3. The chapter begins with an overview of Actors, including changes from the UCO and insights into Actor classes. It continues with a discussion of each of the Actor categories and subcategories and a listing of all of the Actor classes in each. (The amplification of the definitions of the composite Actor classes is postponed until Chap. 7.) The chapter concludes with a discussion of the relationships used in the chapter and a review of Actors.

Chapter 5 describes the Action Ontology. Actions are the elements that cause changes to the Objects and Actors in the MCO. Chapter 5 expands the discussion of Action classes, introduced in Chap. 3. The chapter begins with an overview of Actions, including changes from the UCO and insights into Action classes. It continues with a discussion of each of the Action categories and subcategories and a listing of all of the Action classes and their definitions in each. The Action classes also are the basis for the DIME paradigm; however, this discussion is reserved for Chap. 9. The chapter concludes with a discussion of the relationships used in the chapter and a review of Actions.

Chapter 6 describes the Object Ontology. Objects are the passive elements that are only the recipients of Actions. Chapter 6 expands the discussion of Object classes, introduced in Chap. 3. The chapter begins with an overview of Objects, including changes from the UCO and insights into Object classes. It continues with a discussion of each of the Object categories and subcategories and a listing of all of the Object classes in each. (The amplification of the definitions of the composite Object classes is postponed until Chap. 7.) The chapter concludes with a discussion of the relationships used in the chapter and a review of Objects.

Chapter 7 discusses the new composite Actor, Object, and Object plus Actor classes. The complicated possibilities introduced by these classes are not immediately evident from their definitions in Chaps. 4 and 6. The exposition is designed to build on the concepts, introducing one or two new facets at each stage, to make the complete picture more easily understood. Instances are used very liberally to ensure that the concepts are made concrete.

Chapter 8 describes the State Variable or MType Metric Ontology, with particular emphasis on the meaning of Metrics. It expands the discussion of Metric classes, introduced in Chap. 3. In this chapter we discuss the types of Metrics; however, we only provide samples of the names of the Metric classes because of their very large number (almost 6000). The Metric classes also are the basis for the PMESII paradigm; however, this discussion is reserved for Chap. 9.

Chapter 9 describes two supplementary ontologies, one describing the instruments of national power and the other the state of the world. The DIME Ontology classifies the Actions according to the Diplomatic, Information, Military, and Economic (DIME) paradigm, plus an Other category. DIME is the paradigm for instruments of national power. The PMESII Metric Ontology classifies the Metrics according to the Political, Military, Economic, Societal, Information, and Infrastructure (PMESII), plus Kinetic and Environmental categories and a Null category. The PMESII paradigm describes the state of the world. Chapter 9 also illustrates how the Action classes are distributed across the DIME categories and how the Actor, Action, and Object classes are distributed across the PMESII categories through their Metric classes. These two ontologies are described as supplementary because much of their information content is replicated by the Action Ontology and the MType Ontology.

Chapter 10 describes the two Concept Ontologies. The Semantic Concept Ontology contains connections among the elements based on the related meanings of the elements, relationships that are obvious to humans because they know the meanings of words, but not obvious to a computer. The Stocks-and-Flows Ontology also contains connections among elements; however, these connections are operational in nature. The name is derived from the simplest operational concept, illustrated by a water tank (a stock of water) and two verbs, "fill" and "empty," that describe flows into and out of the water tank. This ontology provides the explicit connections, of which a computer would be ignorant.

Chapter 11 describes the Theories Ontology. In this chapter we describe the structure of the ontology, its contents and their validity, and its uses with respect to model validity. The Theories Ontology is a work in progress. It contains a structure and a set of classes. However, the contents require review, emendation, and additions by experts in the various disciplines (such as psychology, sociology, and anthropology) before it can be regarded as nearly complete.

Chapter 12 describes the five Scenario Ontologies. These ontologies require instantiations for full understanding. That is, the structures are defined in the MCO; however, their value lies in describing a particular scenario, and examples are required (and provided) to see how they work. The Scenario Identification Ontology identifies a scenario. The Goal-Task-Owner (GTO) Sets Ontology defines the important parties and their agendas in a scenario. The Scenario Rules Ontology defines the restrictions the parties apply to their own actions in a scenario. The Actor-Action-Result (AAR) Sets Ontology defines the connections of the parties' actions to results in a scenario. And the Scenario Relations Ontology defines the relationships between Actors (such as kinship), between Actors and Objects (such as ownership), between Objects (such as co-location), among entities in a group, and among entities in a network.

Chapter 13 explores the relationships that connect the Actors, Objects, and Actions into the complex structures that express the modern conflict domain. The basic relationships in the MCO are the object relationships that connect classes and instances. These relationships range from structural ontology relationships such as *is-a* and *hasInstance* to domain-specific relationships such as *hasTaskGoalPair* and *owns*. The data relationships are technical artifacts of the need to specify connections to data in an ontological environment in which most of the elements refer to things, not numbers or rankings. The annotation relationships are artifacts of the OWL language that support connecting additional information to a central element such as a class or instance without having to include the information in the name. Possible future enhancements are suggested.

Chapter 14 consolidates the descriptions of the ontology into a discussion of "what we can know." It describes the implementation history and changes from the UCO to the MCO, plus potential future enhancements. It also discusses some uses of the ontology and presents some concluding thoughts.

Chapter 2
Modern Conflict

A review of the modern conflict domain that is to be captured in the Modern Conflict Ontology (MCO) is required to ensure that all relevant aspects are captured.

The term, "modern conflict," is not standard usage. It is introduced here to emphasize that our expectations for conflict in the future should not be set by previous experience with combat in World War II or Korea. Nor should it be defined by what was expected in Europe if the Cold War had turned hot. In each of those cases, the primary measure of success was territory gained or lost. Attrition (losses of personnel and warfighting equipment) was considered to be either a secondary measure or a causally related co-variant measure with territorial control. Further, modern conflict is not defined exclusively by the unconventional conflicts that have occupied the United States over the past several decades.

2.1 Conflict

The noted historian John Keegan begins the conclusion of his book, *A History of Warfare*, saying, "'What is war?' was the question with which I began this book. Now that I have finished it, and if the reader has followed me to the end, I hope I have called into doubt the belief that there is a simple answer to that question or that war has any one nature (Keegan 1994)." In his description of warfare from earliest known records and anthropological studies implying the nature of prehistorical conflict to modern times, we see this great variety and see that our "conventional war" is of recent vintage.

We start with the understanding that conflict can range from disagreements between two individuals all the way to global thermonuclear war. Rather than using a spectrum ranging from cooperation, then competition, to conflict, with the implication that conflict involves fighting, we allow "conflict" to include a large part of competition, which allows us to include inter-state competition that may evolve into

D. S. Hartley III, *An Ontology of Modern Conflict*, Understanding Complex Systems, https://doi.org/10.1007/978-3-030-53214-7_2

fighting. The terminology differs somewhat from that used in the Competition Continuum (Joint Force Development 2019); however, this document maintains that not only do situations morph from one point on the continuum (spectrum) to another but that, for a given pair of parties, they are usually engaging in simultaneous activities at different points on the continuum. One implication of this statement is that the conflict is between or among human beings; however, a broader definition allows for situations in which one of the parties of the conflict is a force of nature, perhaps an erupting volcano, which engenders many of the same actions that would take place in some purely human conflicts.

In this book, we will narrow the domain somewhat, omitting nuclear warfare (although not omitting dirty bombs and stolen nuclear weapons) and omitting conflicts that are not of interest to nation-states. In this domain, military actions of all types often form a large part of the total activity of the conflict. National militaries may or may not be in charge (often the national diplomats, such as the US Department of State, are in charge of the national operations); however, in many cases, the "heavy lifting" is accomplished by the military, whether in combat or logistical roles. Accordingly, we will often use military terminology to describe the operations. This book has been written from a US viewpoint. Thus, references to the Departments of Defense and State refer to the US departments, and "domestic" refers to US domestic issues.

Unconventional Conflict

The interest in the unconventional part of conflict partially lies in the prevalence and significance of terrorism in the current world situation. We must understand this form of conflict if we want to achieve a desirable outcome. The fact that we have been engaged in terrorism-related conflicts for two decades (at the time of this writing, depending on the choice of dates for its beginning) implies that we have not yet figured out how to conclude them satisfactorily. We also have two ongoing unconventional conflicts that are simmering: one with North Korea and one with Iran. The Korean War (1950–1953) (technically a "police action") did not resolve the Korean conflict. It only marked the beginning of what has proved to be a long unconventional conflict with North Korea, which periodically threatens to become a hot war, whether conventional or nuclear remaining unknown. When the Shah of Iran was overthrown in late 1979, the new Iranian government began its long unconventional conflict with the United States and with other nations in the region. This conflict also has the potential to become a large conventional or nuclear conflict. In both of these latter conflicts, the emphasis has been on avoiding general war. Even absent these unconventional conflicts, there is cause to believe that unconventional conflict, rather than conventional war, will be prevalent in the future. Any party with a serious dispute with a party having unmatchable conventional military force will seek unconventional means to gain its own ends. Further, two nations with near-parity forces, both having nuclear forces, may engage in proxy conflicts to avoid the catastrophic effects of nuclear war. In the past many similar proxy conflicts have been unconventional conflicts and may be so in the future.

Conventional Conflict

Sean McFate, a professor of strategy at the National Defense University, provides "a very short history of conventional war" in his book, *The New Rules of War*. He describes it as a result of the Thirty Years' War and the Peace of Westphalia (1648) at its end. He said, "The 'Westphalian Order' made states the sheriff, outlawing mercenaries and those who hired them. States invested in their own standing armies and began their ascendancy. The relationship among force, power, and world order is stark. Those who control the means of violence get to make the rules that others must obey – or die. Nonstate rivals became defenseless without mercenaries and were easily defeated. Old medieval powerhouses such as the church had no choice but to kowtow to state rulers. Soon, nation-states reigned above all others." Under the Westphalian Order, only states are allowed to have militaries and wage war. This allowed the creation of various "laws of war" and the outlawing of all other forms of war – thus defining conventional war (McFate 2019).

Despite the recent emphasis on unconventional conflict, the United States has been involved in several nearly conventional conflicts since World War II. The Korean War itself was certainly fought as a conventional war. Large parts of the Vietnam War (1954–1975, with US entry in 1965) were fought as a conventional war, although with major unconventional aspects. The first Gulf War (August 1990 invasion of Kuwait by Iraq, Operation Desert Storm from January 1991 to February 1991) was an entirely conventional war operation. The second Gulf War (March 2003–May 2003) was an entirely conventional war operation. However, unlike the first Gulf War, the end of conventional conflict was followed by unconventional conflict.

New Rules

Sean McFate believes conventional war is obsolete. In his book, *The New Rules of War*, he describes how war as actually practiced differs from conventional war. For example, he says, "there is no such thing as war or peace – both coexist, always." By this he means that actors are working to push the limits to obtain their objectives, using methods that fall short of acts that provoke adversaries into declaring war. "Hearts and minds do not matter," as brute force is sufficient and is being used in various places now and will be used in the future. He claims, "technology will not save us," and "the best weapons do not fire bullets." Purchasing more sophisticated and expensive weapons will not win wars, but persuasion will (where brute force is not desirable). Shadow wars will dominate, in which forces wage war but use narrative warfare to obscure the identity of the principal actor and even that a war is taking place (McFate 2019).

Current wars are using mercenaries (we just call them contractors) and this trend will grow. The concept of nation-states as the sole actors capable and allowed (by international law) to wage war has become obsolete. Many of the approximately 194 nations recognized today are nations in name only. Many are narco-states, run by fabulously wealthy drug organizations, or states run by warlords as their personal adjuncts, or essentially lawless areas, denoted by names and boundaries for reference purposes and historical reasons, but not acting as nation-states serving a populace. Corporations and individual billionaires can afford to hire armies and wage wars as they see fit (McFate 2019).

Hybrid War

Hybrid war can be loosely defined as "conflicts involving regular and irregular forces engaged in both symmetric and asymmetric combat (Mansoor 2012)." The point of the book, *Hybrid Warfare*, edited by Murray and Mansoor, is that while the name may be new, the conflicts described by the name are not new, going back at least to the Peloponnesian War of the fifth century B.C. The book includes chapters on the combat of Rome against Germania in the first century B.C., the English attempts to subjugate the Irish around 1600, the American Revolution, the Peninsular War in Spain against Napoleon of 1807–1814, the Union's counterguerrilla war in 1861–1865, the Franco-Prussian War of 1870–1871, the British Empire small wars from 1790 to 1970, the Japanese experience in North China in 1937–1945, and the Vietnam War (Murray and Mansoor 2012).

According to Volodymyr Horbulin, Academician of the National Academy of Sciences of Ukraine, "The hybrid war as a form of Russia's aggressive solution to its geopolitical issues continues developing in every possible way, becomes more sophisticated and spreads out to the new battlegrounds." He continues, "Nowadays we are witnessing at least three ongoing large-scale 'hybrid' operations, each of which represents different possible types of 'hybrid wars' activities: traditional military operation in Syria (Turkey), non-military activity in the EU and the mix of all three types in Ukraine (Horbulin 2018)."

Cyber War

Naturally, cyber war did not exist before computers. Before the Internet, components of cyber war were parts of spying and counterintelligence. That has changed. Richard Clarke and Robert Knake, writing in *The Fifth Domain*, say, "As we write this in 2019, we see a pattern of malicious activity in cyberspace that suggest we are already engaged in a low-grade, simmering cyber conflict with Russia, China, and Iran (Clarke and Knake 2019)." The components of a cyber conflict include spying and counterintelligence using computers and network information flows. It also includes denial of service attacks and attacks on hardware (such as centrifuges) that are linked to computers. The broader cyber conflict includes attacks on both military and non-military infrastructure, conducted by nations, corporations, non-nation-states, and individuals.

The prospect is that future conflicts will always involve some aspects of unconventional conflict and often aspects of conventional war. This unpredictable mix of conflict types drives the definition of modern conflict.

2.2 Definition of Modern Conflict

Figure 2.1 illustrates the relations among the elements of modern conflict. The three dimensions are military *operations*, *variables*, and *theories*. Each axis shows categories for its dimension.

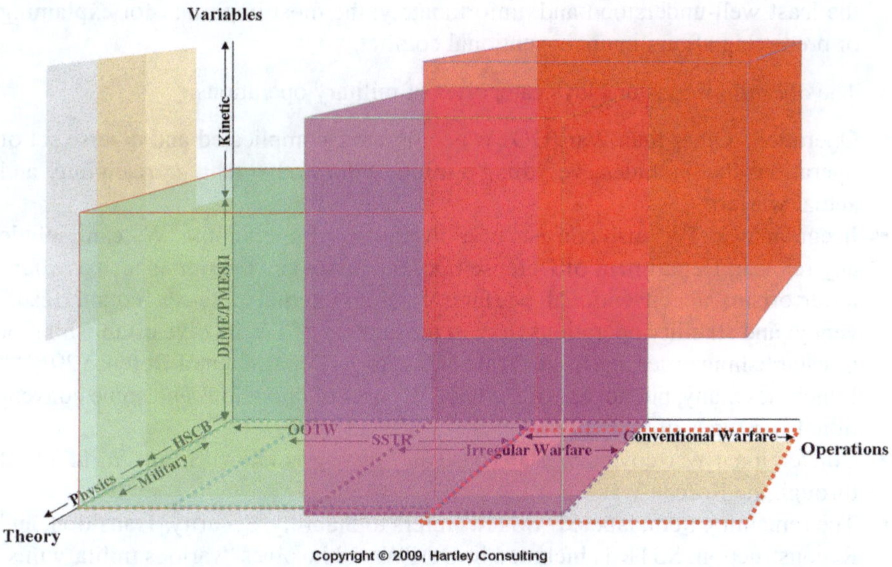

Fig. 2.1 Conflict dimensions

The *variables* axis shows the categories of variables that may be used to describe the situation. In this figure, they are divided into kinetic and DIME/PMESII variables.

- Kinetic variables include most of the standard variables used in describing conventional warfare, such as probability of kill, the physical characteristics of vehicles, the numbers of personnel and equipment, logistics descriptors, and physical environment descriptors.
- DIME variables are Diplomatic, Informational, Military, and Economic actions that can influence the situation. PMESII variables are Political, Military, Economic, Social, Information, and Infrastructure state variables that describe the state of the situation.

The *theories* axis displays the categories of theories that might be used to explain or predict the results of actions.

- The category labeled "physics" includes all that is often called the hard sciences and mathematics (or what educators refers to as STEM – Science, Technology, Engineering, and Mathematics). The theories in this category are generally testable, with fairly well-defined domains of applicability and known degrees of accuracy.
- The category labeled "military," which overlaps the other two categories, includes the various fields taught as military science, which include some hard science and some "softer" subjects, such as military history and leadership.
- The category labeled "HSCB," which stands for Human, Social, and Cultural Behavior, contains the fields of social science. The theories in this category are the least well-understood and, unfortunately, the most important for explaining or predicting affairs in unconventional conflict.

The *operations* axis displays categories of military operations.

- Operations Other than War (OOTW) comprise a complicated and diverse set of operations that includes everything a military does other than garrison duty and actual warfare.
- Irregular War (IW) also consists of a diverse set of operations. IW "can include any relevant Department of Defense (DoD) activity and operation such as counterterrorism; unconventional warfare; foreign internal defense; counterinsurgency; and stability operations that, in the context of IW, involve establishing or re-establishing order in a fragile state or territory (Department of Defense 2014)." It includes many, but not all, of the OOTW types of operations and some conventional war types of operations.
- Conventional war covers warfare basically as it was known from World War I through the Korean War.
- The remaining item, labeled "SSTR," refers to Stability, Security, Transition, and Reconstruction. SSTR is included in OOTW and involves "various military missions, tasks, and activities conducted outside the United States in coordination with other instruments of national power to maintain or reestablish a safe and secure environment, provide essential governmental services, emergency infrastructure reconstruction, and humanitarian relief (Department of Defense 2009)."

The first three-dimensional element in Fig. 2.1 is OOTW. It is represented by the green rectangular solid in the lower left of the figure. It spans all of the DIME/PMESII variables (described in more detail in Chap. 9) and some kinetic variables. On the *theory* axis, it spans the HSBC theories (described in Chap. 11), the military science theories, and some of the physical science theories. Obviously, it spans the OOTW portion of the *operations* axis.

The second element in Fig. 2.1 is IW. It is represented by the magenta rectangular solid in the middle of the figure, overlapping OOTW. On the *variables* axis, it includes all of the DIME/PMESII variables and more of the kinetic variables than does OOTW. On the *theories* axis, it adds more physical science theories than does OOTW.

The third element in Fig. 2.1 is conventional war. It is represented by the red rectangular solid in the upper right of the figure, overlapping IW. On the *variables* axis, it spans the kinetic variables and some of the DIME/PMESII variables. On the *theories* axis, it spans all of the physical science and military science variables and many of the HSCB theories.

Modern conflict is roughly the union of OOTW, IW, and conventional warfare.

We are focusing on modern conflict at the theater level. Generally, this means that the geographical area under consideration is roughly equivalent to one or more countries. Given a theater perspective, the granularity or level of resolution is restricted by practicality: a too fine granularity requires too many items (from a data population point of view) to enumerate in the ontology, and representing too many things at a too fine granularity can obscure useful insights. For example, the Host Nation's legislature (whether in existence or nascent) provides a fairly good standard of granularity. Another example is provided by non-governmental organizations (NGOs), such as the International Red Cross, which often play major roles in unconventional conflicts. In conventional wars, brigade or battalion equivalents probably represent the appropriate granularity. Any other group or person with approximately equivalent impact would have the same level of resolution or granularity. The amount of time that will have to be addressed is biased toward periods in years, rather than days or months. Modern conflicts that primarily involved natural disasters have tended to be resolved in months; however, unconventional conflicts that primarily involved cultural conflict among multiple sides have often taken years and sometimes decades to resolve. Conventional wars have ranged in time span from days to years.

Operations Conducted in Modern Conflict

In this section we expand on the definition of modern conflict given above by enumerating the many types of operations that are contained in this domain. Modern conflict is also difficult to grasp because of the large number of types of operations that can be undertaken. It is important to remember the following definitions are for types of operations, not types of situations. A single situation might require several types of operations. For example, a noncombatant evacuation might be required during a peacekeeping operation that also involves humanitarian assistance. On the other hand, a particular situation may only involve one external (e.g., US) operation, such as a noncombatant evacuation conducted to extract civilians during an insurgency. Despite this difference between operations and situations, an understanding of the possible operations helps in understanding the elements of possible situations.

Conventional Warfare

In previous centuries, war was easily divided into land warfare and naval warfare. In modern conflicts, there are more categories and the distinctions are less clear-cut. In Fig. 2.2, these are joined by five other major categories, aerial, space, information, special operations (ops), and support.

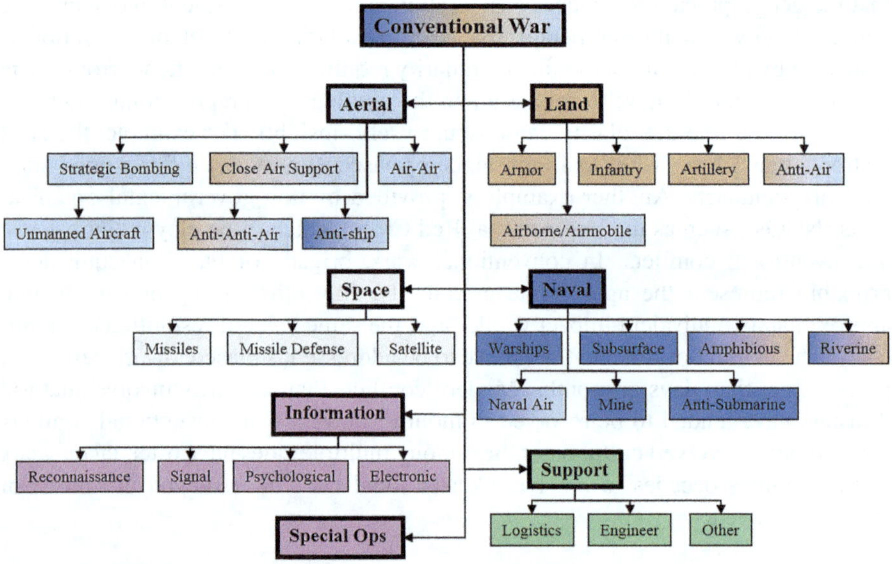

Fig. 2.2 Conventional war type-taxonomy

Land Operations are divided by capabilities or modes of operation. In previous times there were cavalry, mounted infantry, infantry, and engineers. Now there are armor, infantry, artillery, anti-air, and airborne/airmobile (which overlaps with aerial).

Aerial Operations are divided into strategic bombing, close air support, air-to-air, unmanned aircraft, anti-anti-air, and anti-ship, only one of which is purely aerial.

Space Operations are divided into missiles, missile defense, and satellites. In the future, manned space operations might be included.

Naval Operations are divided into (surface) warships, subsurface craft, amphibious operations, riverine operations, naval air, mine (and counter-mine) operations, and anti-submarine operations. The overlaps among air, land, and naval categories are obvious.

Information Operations take place within each of the previous categories and also are included in unconventional conflict (below), but are separated because of their cross-cutting importance. They are divided into reconnaissance, signal, psychological, and electronic (warfare) operations.

Special Operations are also included in unconventional conflict and overlap with naval and land operations. Special Ops are carried out by dedicated Special Forces units using unconventional methods and resources.

Support Operations include logistics, engineer, and many other operations. These operations are critical to the prosecution of all of the other operations.

Unconventional Warfare

Unconventional warfare is a catch-all category for warfare that doesn't follow mid-twentieth-century rules. Figure 2.3 shows the taxonomy of types of unconventional warfare. All of these are included in modern conflict.

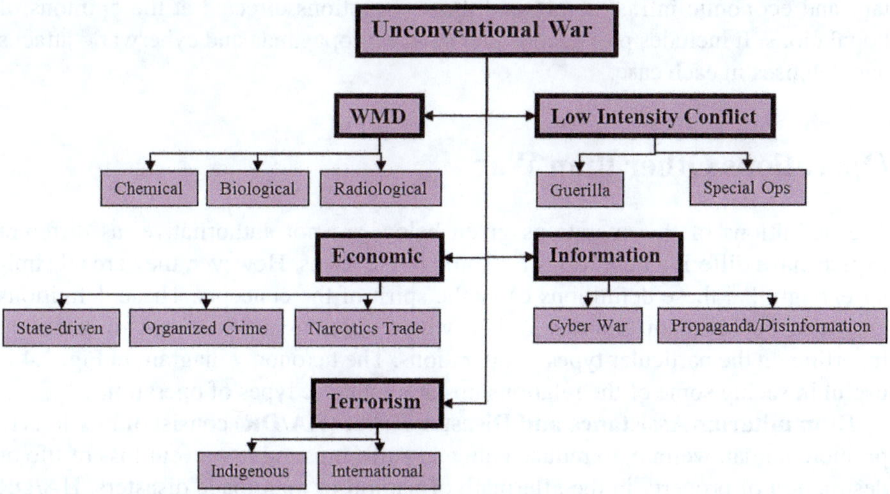

Fig. 2.3 Unconventional war type-taxonomy

Weapons of Mass Destruction (WMD) includes radiological warfare, chemical warfare, and biological warfare. These are defined as the use of radioactive materials, toxic chemicals (including toxins of biological origin), and infectious agents, respectively, with the intent to damage an organization or nation by killing or incapacitating humans, animals, or plants. Notice that nuclear warfare is purposefully omitted. It is "unconventional," but belongs in a category of its own and is not part of the MCO.

Economic Warfare is the use of any means that have damage to the economy of an opponent as their primary purposes and effects. It can involve physical actions such as blockades, economic actions such as freezing bank accounts, social actions such as supporting organized crime or narcotics trade, etc.

Low-Intensity Conflict (LIC) was coined to describe operations like those early in the Vietnam conflict in which small unit combat took place occasionally, interspersed with periods of tense waiting and maneuvering. It may include **Guerrilla Warfare**: warfare in which a small group of combatants such as paramilitary personnel, armed civilians, or irregulars use military tactics including ambushes, sabotage, raids, petty warfare, hit-and-run tactics, and mobility to fight a larger and less-mobile traditional military. LIC may also include **Special Operations**: military operations that are "special" or unconventional and carried out by dedicated Special Forces units using unconventional methods and resources. These include drone operations.

Terrorism is the threat or use of violence to achieve a political, religious, or ideological aim. It is considered a war crime under the laws of war when used to target non-combatants, such as civilians, neutral military personnel, or enemy prisoners of war.

Information Warfare and Information Operations are the use and management of information and communications techniques and technology in pursuit of a competitive advantage over an opponent. This includes operations directed at military and economic infrastructure and those operations directed at the opinions of populations. It includes propaganda and counterpropaganda and cyberwar – attacks and defenses in each case.

Operations Other than War

The definitions of the operations given below are not authoritative, as different experts have differing definitions for some of the terms. However, they are all similar enough that these definitions carry the spirit of the concepts. These definitions may also be found in (Hartley 2017), where citations are included to represent reporting on the particular types of operations. The taxonomy diagram in Fig. 2.4 is useful in seeing some of the relationships among these types of operations.

Humanitarian Assistance and Disaster Relief (HA/DR) consist of missions to promote human welfare, to reduce pain and suffering, and to prevent loss of life or destruction of property in the aftermath of natural or man-made disasters. HA/DR includes refugee problems and consequence management that result from the use of weapons of mass destruction (WMD).

Peace Operations (PO) are military operations to support diplomatic efforts to reach a long-term political settlement.

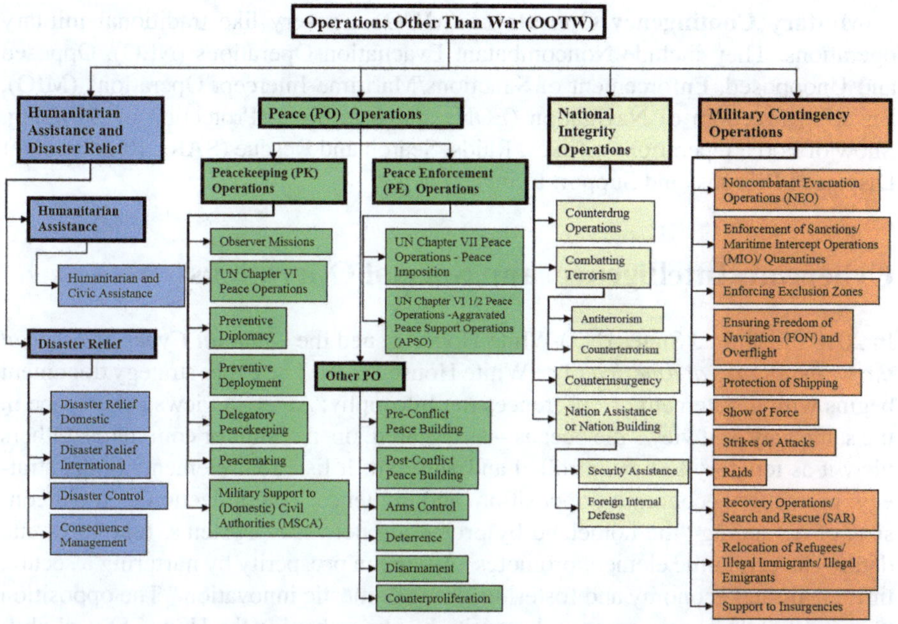

Fig. 2.4 OOTW operations type-taxonomy

- Peacekeeping Operations (PK): Military operations undertaken with the consent of all major parties to a dispute, designed to monitor and facilitate implementation of an agreement and support diplomatic efforts to reach a long-term political settlement. This also known as United Nations (UN) Chapter VI Peace Operations.
- Peace Enforcement Operations (PE): The authorized application of military force to compel compliance with resolutions or sanctions designed to maintain or restore peace and order.
- UN Chapter VII Peace Operations: Part of peace enforcement – operations, short of war, and requiring use of force to impose peace – also known as peace imposition.
- UN Chapter VI ½ Peace Operations: Operations having characteristics of both UN Chapter VI and UN Chapter VII operations and thus often referred to as Chapter VI ½.
- Operations requiring a show of force, or small tactical operations, to enforce peace – part of peace enforcement, also known as Aggravated Peace Support Operations (APSO).
- Other peace operations include Arms Control and Counterproliferation efforts.

National Integrity (NI) Operations are operations to promote national integrity. They include Counter-Drug (CD) Operations, Combatting Terrorism, Counterinsurgency (CI or COIN), Nation Assistance or Nation Building: Foreign Internal Defense (FID), and Stability Operations.

Military Contingency Operations (MCO) are very like traditional military operations. They include Noncombatant Evacuation Operations (NEO), Opposed and Unopposed, Enforcement of Sanctions/Maritime Intercept Operations (MIO), Ensuring Freedom of Navigation (FON) and Overflight, Protection of Shipping, Show of Force Operations, Strikes, Raids, Search and Rescue (SAR), Relocation of Displaced Persons, and Support to Insurgency.

Cyberwar, Intelligence, and Special Operations

In 2018 the United States (US) White House issued the *National Cyber Strategy of the United States of America* (The White House 2018). The cyber strategy document begins with a statement of differences in philosophy: America views cyberspace in the same way as it views the oceans – as a shared, open realm for commerce. Others view it as territory to be controlled and annexed. It lists four elements in the strategy: protection, economics, opposition, and influence. The protection element consists of defending "the homeland by protecting networks, systems, functions, and data." The economic element promotes "American prosperity by nurturing a secure, thriving digital economy and fostering strong domestic innovation." The opposition element will "preserve peace and security by strengthening the United States' ability – in concert with allies and partners – to deter and, if necessary, punish those who use cyber tools for malicious purposes." The influence element has the purpose of extending "the key tenets of an open, interoperable, reliable, and secure internet."

The protection and opposition elements confront conflict most directly. The protection element consists of defensive actions to secure federal networks and information against attack, secure the nation's critical infrastructure against attack, and combat cybercrime, in part by improving its reporting. The opposition element includes the active measures to "identify, counter, disrupt, degrade, and deter behavior in cyberspace that is destabilizing and contrary to national interests, while preserving United States overmatch in and through cyberspace."

There is no denying the importance of intelligence operations, special operations, or the new cyberwar operations. The problem with including them in the ontology has to do with consistency of scale. In each case, an individual operation will involve a very few people. The value premise is that the result will be useful at the operational or strategic level. It is easy to imagine a situation in which blowing up a single bridge will prevent the enemy from launching a major attack; however, not every bridge will enjoy this operational or strategic centrality. Including information about every bridge because some bridge may become important increases the content of the knowledge base, without guaranteeing an equal increase in the value of the information. Similarly, the cyber operation known as "Olympic Games," which unleashed the Stuxnet code on the Iranian centrifuges, had strategic effects; yet every cyber operation cannot be expected to rise to that level (Kello 2013). Accordingly, these operations are included as generic operations of their respective types, leaving the results to be included as the individual circumstances warrant. Hartley and Jobson provide a more detailed view of information war and related topics (Hartley and Jobson 2020).

Relative Likelihood of Overall Conflict Types

In 1994 in an INFORMS panel discussion, Dave Haut, then Chief of the Research and Analysis Division for the Department of Defense's US Pacific Command (USPACOM), asked the question, "Suppose there are problems in country *X* and the Ambassador has the choice of having a carrier battle group sail down the country's coastline in a show of force or playing golf with the country's Prime Minister; how does he decide which will be more effective?" (Haut 1994). The combat models of the time provided no way to frame such a question, much less any hope of illuminating the pros and cons of the alternatives. We still are unable to answer this question; however, we are getting closer.

In a presentation to the Japan-US Operations Research Seminar (JUORS) in 1995, Haut presented an estimate of the likelihood of various types of operations across the continuum from peace to nuclear war (Haut 1995; Haut and McCurdy 1994). Figure 2.5 displays a version of this figure, which shows that the various types of OOTW are much more likely to occur than are conventional or nuclear combat operations. This makes sense, as unless the sides in a conflict have approximate parity in military size and quality, conventional war is not likely to be the chosen conflict type.

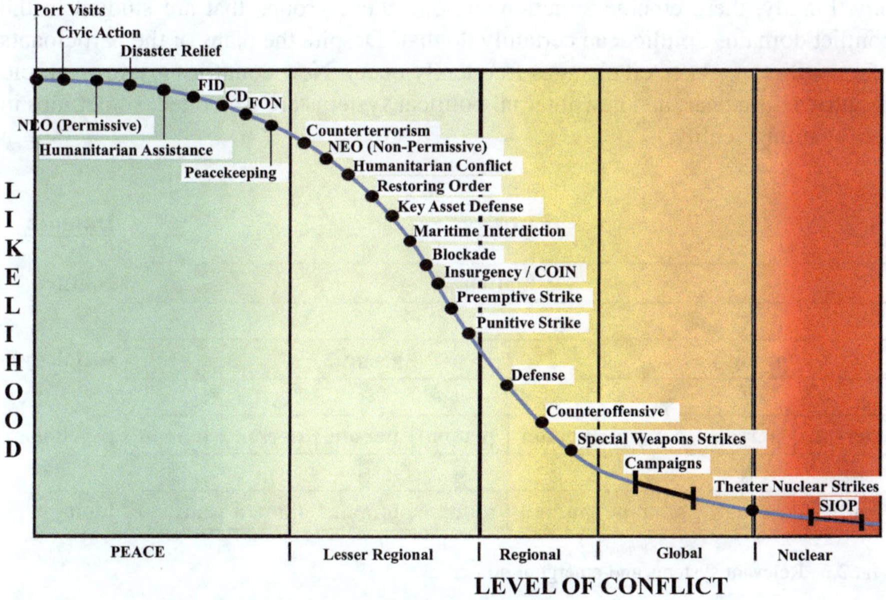

Fig. 2.5 Haut's continuum of operations

The Fund for Peace is an organization that, among other things, is interested in forecasting conflicts. It has developed a model that uses computer-based content analysis of communications to help in forecasting and assessing conflicts (Haken et al. 2010). The assessment part provides data on the actual types of conflicts.

The US Army is also interested in the forecasting and assessment problem. The Center for Army Analysis has improved its model by recognizing patterns of nation-state instability that lead to conflict (Shearer and Marvin 2010).

Conflict as a Complex Adaptive System

The first page of each Springer Complexity series book says, "Complex Systems are systems that comprise many interacting parts with the ability to generate a new quality of macroscopic collective behavior the manifestations of which are the spontaneous formation of distinctive temporal, spatial or functional structures" (Fellman et al. 2015). The macroscopic collective behavior referred to is also called an "emergent property."

Figure 2.6 is a sketch of the relevant emergent properties that lead to considering conflict as a complex system. It is a sketch in that there may be additional steps that are omitted in the figure; however, it portrays the essentials of the argument. The base of the figure shows humans as biological systems, studied in the domain of biology. From these biological systems, there somehow emerges a cognitive person for each. These "person" systems are studied in the domain of psychology. From these systems, there emerge group systems that are studied in the domain of sociology. Finally, there emerge situations among these groups that are studied in the conflict domain. Conflict can certainly do that! Despite the plans of the participants of a conflict, unexpected changes frequently occur. New countries are formed; old countries disappear; and new internal political systems are generated, sometimes in the "winning" entity.

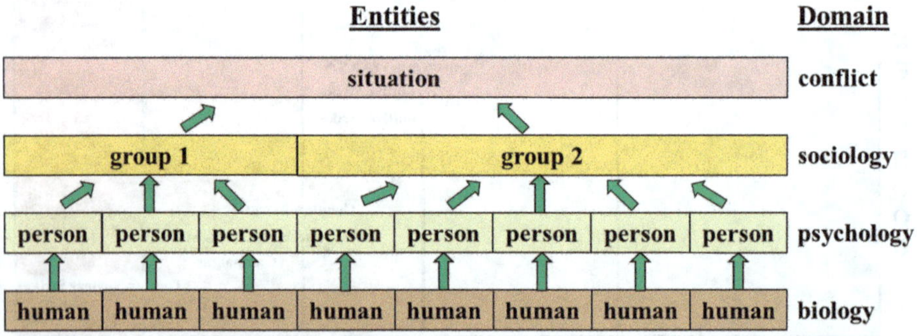

Fig. 2.6 Relevant systems and emergent properties

Thus, we see that conflict situations are complex systems that are related to other human complex systems. The figure does not show the adaptive nature of the systems. However, humans, as biologic systems, modify their behaviors based on the situation and have evolved (developed transformed biologic structures) over time to meet situational challenges. Persons also show adaptive behavioral changes (over a much shorter time frame). In sociology, it is known that groups adapt to their situations, changing internal structures and behaviors to meet differing challenges. Similarly, as the groups entangled in a conflict situation change, the situation changes. Further, new situations differ from old situations through group and personal learning.

2.3 Theories of Conflict

Stories of conflict are contained in the origin stories of most cultures. Our earliest literature includes stories of historical combat at various levels, e.g., *The Iliad* (Homer 1950) and *The Odyssey* (Homer 1946). Our earliest histories include chronicles of wars, e.g., Thucydides' *History of the Peloponnesian War* (Thucydides 1954/1972). Therefore, it is not surprising that theories about war also began to be composed early in our history.

The definition of modern conflict in the first part of this chapter is reductionist in nature, identifying the pieces of modern conflict. Theories of conflict are synthetic in approach, identifying the emergent nature of a domain. From our point of view, differences in theories are as important as commonalities, as we have no guarantee of the essential validity of any of them. Three classical theories of conflict and five modern theories provide essential insights into the nature of conflict.

Sun Tzu

Sun Tzu's *The Art of War* provides our earliest (*circa* 400–320 B.C.) example of theorizing about conflict (Sun-Tzu 1963). This work consists of short chapters containing precepts on war such as:

> All warfare is based on deception. Therefore, when capable, feign incapacity; when active, inactivity. When near, make it appear that you are far away; when far away, that you are near. Offer the enemy a bait to lure him; feign disorder and strike him. When he concentrates, prepare against him; where he is strong, avoid him. Anger his general and confuse him. Pretend inferiority and encourage his arrogance. Keep him under a strain and wear him down. When he is united, divide him. Attack where he is unprepared; sally out when he does not expect you.

Sun Tzu introduced the important concept of *shi* (势), which is translated in various ways. Griffith renders it as "energy," "force," "influence," and "authority"; however, he says commentators also include "situation" as a meaning. *Shi* is the title of one of the chapters in *The Art of War* (Sun-Tzu 1963). It embodies a strategy of present actions to create influence in the future and involves "normal" (conventional) and "extraordinary" (unconventional) forces to deceive the enemy as to where his attention should be focused.

Sun Tzu's work is credited with being a major influence on Mao Tse-tung and thus on many modern unconventional conflicts.

Machiavelli

Niccolò Machiavelli's *The Prince* (1513) is a second major work on conflict (Machiavelli 1966). This work also consists of short chapters containing precepts, although Machiavelli's precepts deal more with political machinations than with war. In his chapter titled, "What a Prince must do to be Esteemed," he begins:

> Nothing wins so much esteem for a prince as embarking on great enterprises and giving rare proofs of his ability. In our own times, for example, the present King of Spain, who may almost be called a new prince; setting out as a weak monarch, he proceeded to win so much renown and glory that he has now become the greatest king in Christendom.

Machiavelli's work is credited with influencing politicians (and statesmen) ever since. His influence has been so great that "Machiavellian" has become a common adjective (usually pejorative) used to describe many who have been successful.

Clausewitz

Carl von Clausewitz's *On War* (1832) is the third major work on conflict (Clausewitz 1993). Clausewitz undertook the task of a scientific analysis of war. He wanted to discover how war is conducted, why it is conducted that way, and how it should be conducted best. He wanted to create a true theory of war. His approach was similar to working a grade-school math problem where the teacher said, "show your work." Machiavelli and especially Sun Tzu would have had their grades reduced for only giving the answers without "showing their work." This work is much longer that the first two and is organized in sections (called "books") and chapters. A list of the section names is illustrative:

- On the Nature of War
- On the Theory of War
- On Strategy in General
- The Engagement
- Military Forces
- Defense
- The Attack
- War Plans

Clausewitz introduced such concepts as the ***fog of war*** (difficulty of knowing what is happening), ***friction*** (difficulty of getting things done), ***center of gravity*** (critical element without which things fall apart), ***absolute war*** (war pursued to the complete defeat of the opposition with no external interferences), and the concept of war as an extension of politics. Up until the final chapters of the book, he has described absolute war as the ideal, with real war (with its fits and starts and political dealings) as if it were a second-class effort. Actually, he was setting up for his final arguments. In the subchapter "War Is an Instrument of Policy," he writes

> Up to now we have considered the incompatibility between war and every other human interest, individual or social – a difference that derives from human nature, and that therefore no philosophy can resolve. We have examined this incompatibility from various angles so that none of its conflicting elements should be missed. Now we must seek out the unity into which these contradictory elements combine in real life, which they do by partly neutralizing one another. We might have posited that unity to begin with, if it had not been necessary to emphasize the contradictions with all possible clarity and to consider the different elements separately. This unity lies in *the concept that war is only a branch of political activity; that it is in no sense autonomous*. [The emphasis is contained in the source.]

This final concept lies at the heart of modern conflict.

Liddell Hart

B. H. Liddell Hart published his book on strategy in 1954 and published a revised edition, titled *Strategy*, in 1967 (Liddell Hart 1967). In his preface to the second edition, Liddell Hart points out that the development of nuclear weapons did not "change the basis or practice of strategy and would not free us from dependence on what are called 'conventional weapons'." Further, he said they would lead "to the increasing use of a guerrilla-type strategy." In his preface to the first edition, he remarked that in his study of "a long series of military campaigns, I first came to perceive the superiority of the indirect over the direct approach." He said that later he began to see that this superiority also applied more generally to conflicts in politics, business, and other spheres of interest.

Wylie

Rear Admiral J. C. Wylie published *Military Strategy: A General Theory of Power Control* in 1967 and added a postscript *circa* 1987 (Wylie 2014). Wylie wrote from a naval perspective; however, his intent was more general than that.

Strategy "A plan of action designed in order to achieve some end; a purpose together with a system of measures for its accomplishment." (Compare this to the concept of Goal-Task-Owner (GTO) Sets in the chapter on Scenarios, Chap. 12.) Wylie divided strategies into two types:

- A **sequential strategy** is a "series of discrete steps or actions, with each one of this series of actions growing naturally out of, and dependent on, the one that preceded it."

- A **cumulative strategy** is one in which "the entire pattern is made up of a collection of lesser actions, but these lesser or individual actions are not sequentially interdependent. Each individual one is no more than a single statistic, an isolated plus or minus, in arriving at the final result."

Wylie observed that there were a number of tacit theories of war among military leaders.

The Maritime Theory

Part 1 Control This part emphasizes the degree of knowledge and control of everything on the sea required for success – absolute is preferred. Encounters are episodic – fight only when "mutually agreeable."

Part 2 Exploitation Prior to World War II, exploitation of sea control to control of the land was mainly through economic, political, and social means, mainly by blockading of various sorts. During World War II, the technology supported landing forces in both contested and uncontested situations. With naval missile launchers, the added dimension of direct destruction from sea to land became possible.

The Air Theory

Part 1 Control This part emphasizes the degree of knowledge and control of everything in the air required for success – absolute is preferred. Encounters are episodic – fight only when "mutually agreeable."

Part 2 Exploitation Strategic bombing is one method of exploiting air control to achieve land control.

Part 3 Close air support Close air support is the second method of exploiting air control to achieve land control and is given third priority.

The Continental Theory

Part 1 Control The terrain acts as a dominant force over combat actions, constraining and controlling all actions.

Part 2 Continuity Combat is nearly continuous as loss of contact can be dangerous.

Part 3 Military destruction The ultimate goal is to destroy the enemy's military force and will to fight. The bigger the battle, the more conclusive will be the result.

Part 4 Support The rationale for naval forces is to deliver the ground forces to their theater and to provide supplies. The rationale for air forces is the same, plus to supply close air support.

The Mao Theory

Part 1 Power "[T]he rural peasant is the base of power."

Part 2 Fluidity In guerilla war, there is "no such thing as a decisive battle"; centralized control is "not only undesirable but impossible"; and the "tactics of defense have no place."

Wylie divided the Mao Theory into four phases:

- Phase 1 is the "political indoctrination of an expanding core of believers in the country."
- Phase 2 is "expanding guerrilla warfare combined with political, social, and economic warfare, with all of them directed against the incumbent government and its forces."
- Phase 3 is the use of an optional "organization of enlarged guerilla cadres and more orthodox armies."
- Phase 4 is the replacement of the existing government with a communist apparatus.

Comparison of the Theories

Wylie points out that the Maritime, Air, and Continental Strategies address particular situations and thus cannot be general strategies for warfare. Similarly, he regards the Mao Theory as "proven within the limits of its assumptions," implying that it, too, is not general enough.

DuBois, Hughes, and Low

DuBois, Hughes, and Low in *A Concise Theory of Combat* (1997) provide another important entry in works on conflict (DuBois et al. 1997). While this book does not have the centuries of analysis and criticism that have been accorded to other works, it does have a distinguished history. The authors had considerable military experience prior to putting it together (Brigadier General, US Army; Captain, US Navy; and Commander, US Navy, respectively). Further, they were compiling the results of almost 20 years of preparing and presenting research papers on many aspects of military conflict by a distinguished group of operations analysts who were experienced in quantitative historical study and military operations research. These activities were hosted by the US Naval Postgraduate School, the US Army War College, SAIC, the Institute for Defense Analysis, SRI International, and the Center for Naval Analyses.

Wayne Hughes made the following comment about the book in a personal message (Hughes 2018):

> Military combat covers a spectrum of intensities. The authors believe that the book is relevant to the *tactics* of all combat, from big to irregular warfare. At the operational, logistical, and strategy levels, there are major differences in the processes between nuclear, worldwide, regional, and irregular warfare.
>
> Combat results are not just attrition and exchange of territory. Though difficult or impossible to measure, equally important are the effects on the mind (demoralization) and spirit (willpower). That is why attrition models often are poor predictors.

He also said that on the ground, defense is the stronger form of war. However, at sea and for air attacks against the ground, the ability to attack effectively first makes offense stronger than defense.

Axioms

The six axioms below and their descriptions are quoted from Chapter 2 of their book (DuBois et al. 1997). (The italics are added to separate the axiom statement from the explanation.)

> **Axiom 1** *Military combat involves deadly interaction between military forces.* A military force is defined as a set of elements that are [sic] activated for the purpose of engaging in combat. Deadly interaction includes the direct use of deadly means and the threat of using deadly means. Insofar as nonlethal weapons may be used, they are always backed up by deadly means should the latter be needed.
>
> **Axiom 2** *In combat each side seeks to achieve a goal, called its mission, which has perceived value.* The mission of a force is the specific task assigned it by higher authority or presumed by the commander of the force based on guidance from higher authority. Combat is not undertaken for its own sake in isolation from other activities; it is conducted as a purposeful activity within the broader goals of entities external to combat.
>
> **Axiom 3** *Combat potential is embodied in military forces.* **Combat potential** is the latent capacity of a military force to achieve results in combat.
>
> **Axiom 4** *The commander on each side in combat activates combat potential to create combat power in furtherance of the mission.* **Combat power** is the realized capability of a military force at any instant of time to achieve results in combat.
>
> **Axiom 5** *Domination of the opposing military force is the ultimate means of accomplishing a mission.* Domination is the result of imposing the will of one force on the opposing force through all the interactions of combat, especially those affecting the will and spirit.
>
> **Axiom 6** *Uncertainty is inherent in combat.* **Uncertainty** is a state of doubt about the combat situation, including its outcome. Uncertainty is distinguished from chance, which concerns unpredictable events that happen without discernable human or other cause.

Components of Combat

DuBois, Hughes, and Low (DHL) identify several components of combat.

- An **element** is "a material or intangible thing of any kind, whether animate or inanimate, that exists in combat and can change the state of another element or itself." They divide elements into **agent** elements and **object** elements, depending on which is performing an action and which is being acted on. (This is similar to our Actor and Object division. However, the DHL definition implies that the agent and object labels are strictly situational, whereas the Actor classes and Object classes are disjoint in the MCO.) Elements may be single entities or aggregated entities, such as a platoon.

- An **action** is "an act performed by a single or aggregated element to change the state of one or more other elements, its own state, or both." Single actions have only a small duration. Actions may be combined into aggregated actions. Actions cannot be acted on; however, an action against an element may have the indirect effect by modifying some action of the object element. (This concept is identical to our Action.)
- An **attribute** is "a qualitative or quantitative modifier of a combat element [or action]." (Attributes are identical to our state variables.)
- A **state** is the combination of an element and the values of its attributes at a particular time. (This is the reason for calling our state variables "state" variables.)
- The **parties** to a combat are the two opposing forces and the geophysical combat environment.
- The **geophysical combat environment** is an aggregate element, "which acts as a non-adversarial third party."

Combat Functions and Processes

The authors also distinguish between combat functions and combat processes.

- A **primary combat function** is "a generic category of like actions taken by elements of either adversary in combat to achieve an intended result."
- A **primary combat process** is "combat activity of any kind that produces a common generic result."

At the lowest level, primary combat functions are categories of like actions taken by single elements. For aggregate elements, the primary combat functions are categories of aggregate actions. The authors considered lists of combat functions from various sources and derived the list of primary combat functions shown in Table 2.1.

Part of the reason for distinguishing functions from processes is that, although a combat function is intended to achieve a result, there is no guarantee that it will actually achieve that result. On the other hand, combat processes are descriptions of what happens, regardless of the intent. The authors separated the processes into externally and internally directed processes and derived the list shown in Table 2.2.

While two of these have the same names as two of the primary combat functions, the differences are instructive. Command-control, as a function, includes the intent to successfully issue commands and control the party's own forces. Command-control, as a process, describes the actual results of all activities (including chance events, adversary actions, and imperfect execution of own actions) on the command and control of the party's own forces.

Table 2.1 Primary combat functions

Command-control
Fighting
Support
Information acquisition

Table 2.2 Primary combat processes

Externally directed processes	Internally directed processes
Demoralization	Motivation
Destruction	Command-control
Suppression	Information acquisition
Neutralization	Communication
Disruption	Movement
Deception	Protection
	Sustainment

Another reason for distinguishing between functions and processes lies in the time scale. Functions, based on actions, have short duration, while processes may have longer durations, longer by an order or two of magnitude.

Figure 2.7 illustrates the differences between combat functions and combat processes. The left side of the figure starts with a combat situation at time t for Side 1 and for Side 2 (shown smaller because we are focusing on Side 1). The primary combat functions for Side 1 are shown, along with dashed arrows pointing to a shifted situation at time $t + \Delta t$ (the shift to the left indicating a desired change).

The right side of Fig. 2.7 shows the actual results. Side 1 performs a large number of elemental (DHL terminology for small-grained) activities (indicated by the numerous initial vertical arrows), as does Side 2 (indicated by the slanted arrows). Because we are only illustrating the effects on Side 1, we show that all of these arrows contribute to the primary combat processes of Side 1 (they also contribute to the primary combat processes of Side 2). The Side 1 primary combat processes are shown as parts of the large blue arrow, which is longer to indicate a greater duration than that of the elemental activities. (Note that some Side 1 primary combat processes, such as demoralization, are aimed at Side 2. The results of these processes on Side 2 are not shown. However, these processes also affect Side 1. The result of the demoralization process aimed at Side 2 might have null effect on Side 1, increase the morale of Side 1, or have a perverse demoralization effect on Side 1.) At some point, each elemental activity has a result, indicated by the lighter band within the large arrow. These elemental results may then produce aggregated results for each primary combat process. Finally, all of the combat process results are blended to produce the actual situation at time $t + \Delta t$. This is shown as un-shifted to illustrate that the actual results may differ from the desired (left-shifted) results. Note also that the actual results depend on both the Side 1 activities and the Side 2 activities because this is the case for each of the combat processes.

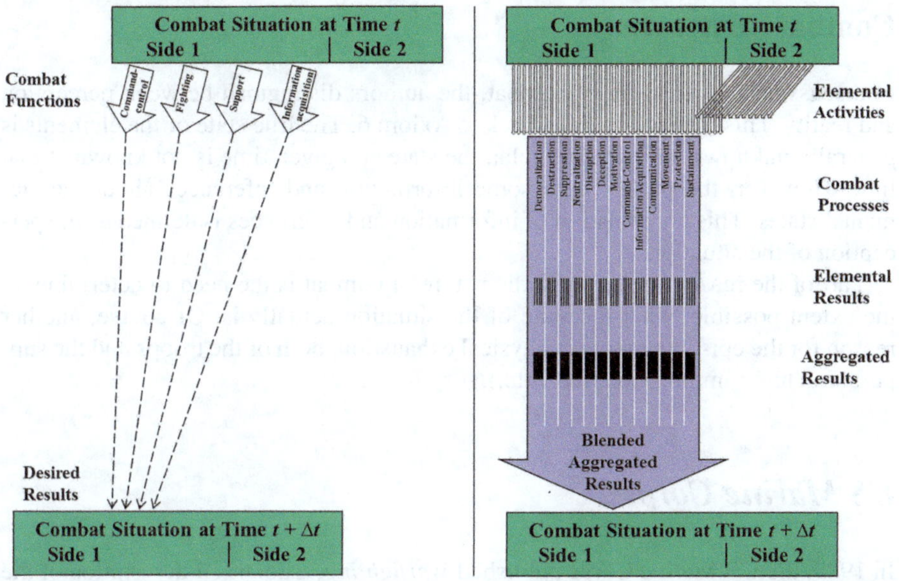

Fig. 2.7 Combat functions and processes

Combat Potential and Combat Power

As defined in Axiom 3, **combat potential** "is the latent capacity to achieve results in combat." As defined in Axiom 4, **combat power** "is the realized capability of a military force at any instant of time to achieve results in combat." The distinction between these two is similar to the distinction between combat functions and combat processes: one is based solely on the intention of a single side and the other is "determined by the combination of the actions of a force and the interactions with the opposing force and the combat environment, rather than by the unilateral actions of the force."

Combat potential includes the manpower available and its training, cohesion, and morale, the numbers of available weapons' types and their qualities, the force structure, etc. Combat power exists only, while combat is in progress: before and after combat, there is only combat potential. Combat power is granular, existing as the aggregation of smaller, elemental bits of combat power. It is directed toward achieving a mission (whether successful or not) and impacts elements, not actions. Combat power is the realization of effects on an enemy, which are not just destruction but also demoralization and instillation of fear. Further, combat power affects the states of the elements on both sides – over time. These changes in state are things that impact subsequent actions. That is to say, actions affect actions only indirectly.

Combat Structure

In discussing the structure of combat, the authors distinguish between perception and reality. This distinction links back to Axiom 6. The true state of the elements is generally unknown, and, in particular, the state at a given time is not known at that time. However, the parties have some information and inferences about the elements' states. This combination of information and inferences is defined as the perception of the situation.

Part of the reason for the episodic nature of combat is the need to determine, to the extent possible, what the state of the situation actually is. Of course, another reason for the episodic nature is physical exhaustion, both of the troops and the support structure (ammunition, fuel, etc.).

US Marine Corps

In 1989, the US Marine Corps published *Warfighting*, a doctrinal description of the nature of war and how it is fought by the Marines. This doctrine was revised in 1997 (U.S. Marine Corps 1997). All of the quotations in this section are from the latter source. There are many such quotations because this document is so well-written that rephrasing its contents would be counterproductive. Where the phrase "the doctrine" is used, the referent is the "doctrinal description" of this document.

The Nature of War

"War is a violent clash of interests between or among organized groups characterized by the use of military force." "The object in war is to impose our will on our enemy."

Attributes of war

- **Friction**, as defined by Clausewitz, is a constant of war.
- **Uncertainty** is also a constant of war. Uncertainty is based on lack of knowledge of the true state of things, what the causal effects of actions will be, and what may occur by chance.
- **Fluidity** as an attribute of war describes the relationship between episodes. An episode may seem to be a self-contained thing; however, it is the result of previous episodes and will flow into new episodes.
- **Disorder** has always been an attribute of war. Even when battles were fought in linear formations, once the battle started, the order began to decrease. The term "melee" was the official name for the muddle of combatants participating in hand-to-hand combat once all order was lost. In modern war, the order is conceptual and not visible and can be lost just as in the times of linear formations.

- **Complexity** is an attribute of warfare. "[W]ar is not governed by the actions or decisions of a single individual in any one place but emerges from the collective behavior of all the individual parts in the system interacting locally in response to local conditions and incomplete information."
- The **human dimension** is central to war. "Any doctrine which attempts to reduce warfare to ratios of forces, weapons, and equipment neglects the impact of the human will on the conduct of war and is therefore inherently flawed."
- **Violence** and **danger** are basic attributes of war. This means that **fear**, which is the natural human reaction, is a basic attribute.
- **Physical**, **moral**, and **mental** forces are the components that interact in war. The physical components are the most easily quantified. The moral or psychological forces, such as conscience, resolve, fear, morale, and leadership, all have significant importance. The mental components "provide the ability to grasp complex battlefield situations; to make effective estimates, calculations, and decisions; to devise tactics and strategies; and to develop plans." "Although material factors are more easily quantified, the moral and mental forces exert a greater influence on the nature and outcome of war."
- **Evolution** is an attribute of war. War's basic nature is constant; however, the means and methods change over time. The change is driven by circumstances and technology.
- **Science**, **Art**, and **Dynamics**: "*We thus conclude that the conduct of war is fundamentally a dynamic process of human competition requiring both the knowledge of science and the creativity of art but driven ultimately by the power of human will.*" [emphasis in the source]

These attributes affect all participants. Marines should understand the negatives, operate despite the problems, and attempt to make use of them to defeat the opponent.

The Theory of War

First, war "is an extension of both policy and politics with the addition of military force." Thus, policy and politics set the goals of the war. However, there may be disagreements on ways and means between the government and the military.

There are two military approaches to imposing one's will on the enemy. The first way is called a **strategy of annihilation** (or a **strategy of incapacitation**), in which the enemy's military power is eliminated, either permanently or temporarily. The second way is called a **strategy of erosion**, in which the will of the enemy's leadership is eroded by convincing him that the cost of resistance is more than he is willing to pay.

As to means, the focus here is on military force; however, it is not isolated from the other elements of national power. The military force itself may "take any number of forms from the mere deployment of forces as a demonstration of resolve to the enforcement of a negotiated truce to general warfare with sophisticated weaponry."

The doctrine defines the **spectrum of conflict** to mean the entire set of military operations from deterring war to global conflict, with military operations other than war, small wars, and regional conflicts in between. The **levels of war** are the strategic, operational, and tactical levels.

The **strategic** level is divided into national strategy, which addresses the attainment of policy objectives by all means, and military strategy, which addresses securing the policy objectives through military force. The **tactical** level addresses "the art and science of winning engagements and battles." The **operational** level links the other two levels. The commanders at the operational level make the decisions on which battles to fight or avoid and when, where, and under what conditions to do one or the other.

Warfighting does not talk about "principles of war." The closest analog is a series of discussions about important concepts in warfare.

- **Initiative** and **Response**: "It is through the initiative that we seek to impose our will on the enemy." "The response generally has a negative aim, that of negating – blocking or counterattacking – the enemy's intentions." During a war, there is a continuous interaction of these two. The choice of initiative or response may not be a free choice; one side (or a part of it at some location and time) may be forced into response, even if it desires the initiative. The **offense** and **defense** are related concepts, although subtly different. For example, a defender setting an ambush has the initiative when the enemy falls for the trap. The doctrine says that the defense is more efficient than the offense, tending to expend less energy. This is "not the same as saying that the defense is inherently the stronger form of warfare." Further, the doctrine makes the point that both are necessary for both sides of a conflict. The offense is an integral part of the concept of defense (think of counterattacks), and the defense is an integral part of the offense (think of assuming the defense to allow for a concentrated offense elsewhere on the battlefield). Finally, the concept of a **culminating point** is introduced. This is the point at which an offensive has reached the limits of its strength and must revert to the defense, if only temporarily. Also, this is the "point that the defensive element of the offense is most vulnerable to the offensive element of the defense, the counterattack."
- **Styles of Warfare**: The two endpoints of the spectrum of styles of warfare are attrition warfare and maneuver warfare. **Attrition warfare** "pursues victory through the cumulative destruction of the enemy's material assets by superior firepower." **Maneuver warfare** is more than just moving assets on the battlefield. It is based on a systems view of the forces and looks to systematically incapacitate the enemy. Neither of these styles exists in a theoretically pure form; however, there are many examples that are close enough to the endpoints to provide good illustrations of the concepts.
- **Combat Power**: Combat power, as defined in this doctrine, is "the total destructive force we can bring to bear on our enemy at a given time." Combat power is "the situationally dependent and unique product of a variety of physical, moral, and mental factors." Note that this differs from the definition in the DuBois, Hughes, and Low theory.

- **Speed** and **Focus**: The time component of speed is **tempo** – "the consistent ability to operate quickly." The spatial component of speed is the ability to move rapidly. Both are important sources of combat power. Focus is "the generation of superior combat power at a particular time and place." "The combination of speed and focus adds 'punch' or 'shock effect' to our actions."
- **Surprise** and **Boldness**: Surprise and boldness are contributors to combat power. "By surprise we mean a state of disorientation resulting from an unexpected event that degrades the enemy's ability to resist." There are three ways to achieve surprise: deception, ambiguity, and stealth. "Boldness is the characteristic of unhesitatingly exploiting the natural uncertainty of war to pursue major results rather than marginal ones." Boldness is not the same as immediate aggressive action; it can consist of waiting for the enemy to commit before acting. It also must be "tempered with judgment lest it border on recklessness."
- **Centers of Gravity**: A center of gravity is a factor that is critical to the enemy, something he cannot do without, something that if eliminated will most quickly lead to his defeat. This is related to the concept of **critical vulnerability**. Vulnerabilities exist in time and space. A critical vulnerability is one "that, if exploited, will do the most significant damage to the enemy's ability to resist us." The two concepts are complementary. The center of gravity is a desired point of attack. A critical vulnerability is a pathway for that attack.
- **Opportunity**: Opportunities may result from own actions, enemy mistakes, or chance. Clearly, opportunities may be unexpected; however, they can also be created. When they appear, they should be exploited. "It is often the ability and willingness to ruthlessly exploit these opportunities that generate decisive results. The ability to take advantage of opportunity is a function of speed, flexibility, boldness, and initiative."

Warfighting discusses the theory of war as providing a foundation for conducting war, not as an academic exercise.

Preparing for War

Preparing for war includes the following:

- Force planning (planning for the creation and maintenance of the right military force)
- Organization (defining the structure needed for warfighting)
- Doctrine (the way to practice the military profession)
- Professionalism (defining what this means in the Marine Corps)
- Training ("to develop forces that can win in combat")
- Military education (responsibility lies in the Marine Corps schools, each commander, and the individual)
- Personnel management (assignments should be based on specific abilities and temperament, with stability as part of the goal; promotions "should reward the willingness to accept responsibility and exercise initiative")
- Equipping (getting the right equipment for the force)

The Conduct of War

The Marine Corps has chosen to conduct warfare using the maneuver warfare style, using the definition in *Warfighting*. "Maneuver warfare is a way of thinking in and about war that should shape our every action." The application of this doctrine includes assigning a mission without specifying how to do it. The mission has two parts, the task and the reason. The subordinate is expected to use judgment and initiative to adapt the execution of the mission to the situation as it unfolds.

Chinese Thinking

I don't know the current Chinese theory of conflict and do not have access to the classified US estimates of any such theory. However, we do have a book, published in China in 1999 and available in translation, which describes one Chinese theory of conflict. The translator asserts that "its release was endorsed by some elements of the [People's Liberation Army] PLA leadership (Qiao and Wang 1999)." Sean McFate says the Three Warfares strategy built on this theory and "was then officially recognized by China's Central Military Commission and Communist Party in 2003 (McFate 2019)." We also have a rough translation of the Chinese publication, *The Science of Military Strategy (2013)* (Chinese Academy of Military Science 2018). This book lays out the Chinese view of military strategy.

Unrestricted Warfare

Qiao and Wang's translator quotes one of the authors (Qiao) as saying in an interview "that 'the first rule of unrestricted warfare is that there are no rules, with nothing forbidden.' Elaborating on this idea, he asserted that strong countries would not use this same approach against weak countries because 'strong countries make the rules while rising ones break them and exploit loopholes' (Qiao and Wang 1999)."

Qiao and Wang define eight principles for what they call "beyond-limits combined war." These principles confirm the need for "modern conflict." Assuming that some adversary will employ these or similar principles, it is clear that future conflicts will diverge from any past notion of "conventional war."

- **Omnidirectionality**: "In terms of beyond-limits combined warfare, there is no longer any distinction between what is or is not the battlefield. Spaces in nature including the ground, the seas, the air, and outer space are battlefields, but social spaces such as the military, politics, economics, culture, and the psyche are also battlefields. And the technological space linking these two great spaces is even more so the battlefield over which all antagonists spare no effort in contending. Warfare can be military, or it can be quasi-military, or it can be non-military. It can use violence, or it can be nonviolent. It can be a confrontation between professional soldiers, or one between newly emerging forces consisting primarily of ordinary people or experts."

- **Synchrony**: Synchrony consists of "conducting actions in different spaces within the same period of time."
- **Limited Objectives**: "Do not pursue objectives that are unrestricted in time and space." "Limited objectives means limited in relation to measures used. Thus, the principle of setting limited objectives means that objectives must always be smaller than measures."
- **Unlimited Measures**: "The trend toward no limits is a trend toward continual enlargement of the range of selection and the methods of use of measures. It is not intemperate use of measures, and even less is it absolutist use of measures, or the use of absolute measures. Unlimited measures to accomplish limited objectives is the ultimate boundary."
- **Asymmetry**: "From force disposition and employment, selection of the main combat axis and the center of gravity for the attack, all the way to the allocation of weapons, in all these things give two-way consideration to the effect of asymmetrical factors, and use asymmetry as a measure to accomplish the objective."
- **Minimal Consumption**: "Use the least amount of combat resources sufficient to accomplish the objective."
- **Multidimensional Coordination**: "'Multidimensional coordination' refers to coordination and cooperation among different forces in different spheres to accomplish an objective." The authors affirm that this is not a novel principle; however, they say that the difference "and this is a great difference, [is] the introduction of non-military and non-war factors into the sphere of war directly rather than indirectly."
- **Adjustment and Control of the Entire Process**: This also is not a novel concept. However, because of the vast number of factors, the authors say that "any attempt to tie a war to a set of ideas within a predetermined plan is little short of absurdity or naïveté. Therefore, it is necessary to have feedback and revisions throughout the entire course of a war while it is actually happening, in order to keep the initiative within one's grasp." Further, information technology and high-technology measures can make the "entire course of a war extremely short, and incidentally make adjusting and controlling it much more difficult."

P. W. Singer and Emerson Brooking, in their book, *LikeWar: The Weaponization of Social Media*, reinforce the use of this theory: "Since 2003, the Chinese military has followed an information policy built on the 'three warfares': psychological warfare (manipulation of perception and beliefs), legal warfare (manipulation of treaties and international law), and public opinion warfare (manipulation of both Chinese and foreign populations) (Singer and Brooking 2018)."

Chinese Military Strategy

The authors of *The Science of Military Strategy (2013)* assert that military strategy is defined by political strategy (although in the heat of war, there will be feedback from the military situation to political strategy) (Chinese Academy of Military Science 2018). The authors lay out the basic features of contemporary war, all requiring information as a key component:

- Systemic attack has become the basic form of confrontation. Attacks are no longer attempts to eliminate the opposing army, but designed to destroy a key part of the opposing system so that it falls apart.
- The long-range precision strike is a decisive feature of combat.
- Networked information systems are an important support function for combat.
- Information-based warfare, involving destroying the opponent's firepower systems and information systems, have become the main means of combat.
- "Combat space expands from 3D to 5D." Space and cyber dimensions have been added to the traditional air, land, and sea dimensions of conflict.
- Battle rhythm has become faster and more important with information warfare.
- "No People, Invisible, Silent" war involves unmanned vehicles, stealth vehicles, silent running vehicles, and "highly automated intelligent weapons."

The authors also discuss "the war China may face in the future," couched in terms of defense, rather than offense:

- "Large-scale, high-strength defense war" has low probability, but high risk. This involves a conventional attack on China, from either an escalated crisis or a premeditated attack.
- "Large-scale, higher-strength anti-separatist war" is described in terms of reunification of Taiwan. It is judged to be both relatively high probability and risk. China must be prepared for a high-end local war with foreign intervention.
- "Small and medium-sized, low-intensity self-defense counterattacks" describes conflict concerning islands claimed by China and border-related conflicts. The risk and probability are rated as moderate, but with a rising trend.
- "Comprehensive judgment" concerns a large-scale offensive against China. This is not expected to include an invasion, but rather a high-end war on the sea, in the air, and in cyberspace. "The danger is rising; the threat of the eastern war is heavier than the threat of the western war, and the threat of war at sea more than war threats on land. Space and cyberspace war threats are increasingly becoming a reality. There is an increase in the probability of foreign military defenses and even limited combat operations; the most serious threat of war is a strong enemy. The large-scale strategic raid launched to destroy the potential of our country's war and force China into an effective war threat is a limited military conflict at sea, and it needs to focus on preparing for a larger-scale, higher-intensity local war in the direction of nuclear deterrence."

The Hundred-Year Marathon

Michael Pillsbury, a distinguished expert on China, asserts that we are at war with China and have been since 1949, the year Mao Tse-tung took power. In his book, *The Hundred-Year Marathon*, Pillsbury says the plan is to replace America as the global superpower by 2049 (Pillsbury 2015). The methodology employs the concept of *shi* (described in section "Sun Tzu"). Part of the plan has been to conceal the fact that the plan existed and to persuade the West to aid China's ambitions. Other parts have included pushing the boundaries of international law to obtain geopolitical positions in ways that fall just short of war.

Qualitative Analyses

There have been several qualitative analyses of various types of conflict, with counterinsurgency and counterterrorism being the most popular. However, in the twentieth century, analyses of combat – including qualitative analyses – were prevalent.

Qualitative Combat Analyses

In the late 1970s, an Army colonel and historian, Trevor Dupuy, introduced and promoted the Quantified Judgment Method of Analysis of Historical Combat Data (QJMA) and a model based on this, the Quantified Judgment Model (QJM) (Dupuy 1985). The underlying basis was a qualitative judgment of historical events (hence its place here in section "Qualitative Analyses"), which were then quantified. Colonel Dupuy was a better historian than mathematician, and his work never gained much approval. However, his extensive compilations of combat data did support later analyses (discussed in section "Quantitative Analyses"). Further, his support for the need to include human factors, such as training, morale, and leadership, while generally disregarded because of the overall QJM mathematical problems, was well-founded and found support in later efforts.

Another, more famous Army officer also did some qualitative combat analysis. While at the Army War College in 1932, Major George S. Patton, Jr., produced an analysis of the characteristics and tactics of the "next war" (Patton 1932). Patton examined wars from 2500 B.C. up to his present time, categorized them as being fought by mass armies or professional armies (or a mix), and drew conclusions about the results. For example, mass armies correlated (not his word) with simple and cheap equipment and simple tactical formations, while professional armies correlated with complicated and costly equipment and multifaceted tactical formations. Similarly, mass armies (generally) did not use or need mobility, while professional armies did. Professional armies cost more per man and perhaps more in total than mass armies; however, Patton concluded the value of achieving a decisive victory with a professional army would offset the cost. He evaluated his arguments saying, "It is believed that the evidence and arguments thus far presented are sufficient to warrant us in saying that there is a reasonable probability that the next war will be characterized by the use of smaller and better trained armies." He advocated the use of maneuver in the next war. The German use of the Blitzkrieg and his own generalship in World War II validated his prediction.

Sean McFate has some comments about analyses of counterinsurgency. "The biggest problem with COIN is that it ignores history. Insurgency scholarship generally disregards the huge number of insurgencies that fail (McFate 2019)." He lists three strategies that can succeed in beating an insurgency – "if you can stomach them." He calls the first, "drain the swamp." He says, "usually this means blasting the population until the insurgency is dead, collateral damage be damned." The second is "export and relocate," in which you forcibly spread the population supporting the insurgency across a large territory full of other people, making them a minority in each new location. He cites Stalin's response to the Chechen insurgency. "Of the 496,000 people who were deported, at least a quarter perished." The third strategy is "import and dilute." He cites the Chinese conquest of Tibet, in which China "imported millions of Han Chinese, making Tibetans a minority in their own homeland (McFate 2019)."

Qualitative Non-combat Analyses

The RAND Corporation performed a qualitative comparative analysis of 30 insurgencies from 1978 to 2008 and provided a useful bibliography (Paul et al. 2012). This work examined 20 distinct counterinsurgency approaches and their results. It used Charles Ragin's Qualitative Comparative Analysis (QCA) to compare the performance of the approaches. They found that the approaches could be separated into good practices and bad practices and when good practices outnumbered bad practices, there was a perfect discrimination between success and failure.

A paper by Kucik and Paté-Cornell took a different approach (Kucik and Paté-Cornell 2012). The authors built a model of the dynamic interaction between a government and an insurgent group at the level of a village. The moves and countermoves were simulated and analyzed as a game between the parties. The government utility (representing the stability of the region) was the objective function.

A third approach is contained in a novel assessment paradigm for counterinsurgencies that was developed for the Afghan conflict (Schroden et al. 2013). The authors advocated moving from a doctrinal approach that said little on "*how to actually conduct an assessment*" [emphasis in the document] to a two-tiered approach of strategic and campaign assessments. The strategic assessment consisted of a set of "questions in narrative, analytic form," and the campaign assessment used a set of standards with narratives to gauge campaign task accomplishments. The paper addressed the pros and cons of this assessment approach.

A fourth approach used quantitative analysis to define a qualitative classification of counterinsurgencies. The question was which historical counterinsurgencies were similar enough to a particular instance to be useful in deriving guidance. King, Hering, and Newman evaluated classification schemes to determine how counterinsurgencies should be classified (King et al. 2014).

A fifth approach used quantitative and qualitative methods to analyze the impact and success of the Turkish Army's military strategy/doctrine and military operations against the Kurdish insurgency (Űnal 2016). This analysis was complicated by shifts in the Turkish strategy and adaptation of the insurgent's actions in response and the nature of the conditions in neighboring Syria and Iraq.

Analyses of terrorism have also been popular, especially since the terror attacks of September 11, 2001. Just as the best classification of counterinsurgencies is important, so is the best classification of terrorists. Understanding the current nature of a particular terrorist organization yields estimates of its strategies and their potential impact. It can also yield an estimate of how that organization might evolve over time. Coles and Zhuang developed archetypes of terrorist organizations based on the attack vector, target type, and general features (Coles and Zhuang 2016). The attack vector includes such things as car bomb, suicide bomb, and radiological bomb. The target type includes such things as civilian personnel, military personnel, and civilian infrastructure. General features include such things as announcing targets before striking, claiming responsibility after strikes, and political or religious affiliation. Any given terrorist organization might be a combination of several archetypes.

Quantitative Analyses

A number of authors have worked with the historical combat reports, attempting to deduce valid theoretical constructs from the data. Frederick William Lanchester was one of the earliest (1916) to construct a quantitative theory of combat (Lanchester 1956). Robert Helmbold produced several seminal statistical analyses of combat data, creating the advantage parameter (Helmbold 1997), among other measures. Hartley built on this and other work, culminating in a book on the topic (Hartley 2001). This book describes predictive models for attrition, duration, surprise, movement, and overall victory in combat. It also includes the Oak Ridge Spreadsheet Battle Model (ORSBM), which includes these mathematical models. These models depend on standard data on force sizes and dispositions, but also include certain human factors, such as leadership, morale, and training. Kuikka extended the analysis by deriving a combat equation based on stochastic theory from the data (Kuikka 2015).

These analyses were analyses of aggregate data, which might be adequate for combat involving very large numbers of similar systems. However, some systems, such as aircraft and ships, are available in fewer numbers and demand finer detail in order to yield adequate results. For example, determining attrition of aircraft might require analysis of multiple segments of their flight paths under different conditions. Paciencia, Richmond, Schumacher, and Troy describe a framework for air attrition modeling (Paciencia et al. 2018).

The chief difference between modern conflict, as defined here, and traditional conflict is the inclusion of social concerns. Some of these social concerns currently require qualitative analysis, rather than quantitative analyses. However, some social concerns are amenable to quantitative analysis.

One aspect that can be quantified is social networks. Bernardoni, Deckro, and Robbins describe using social network analysis to inform stabilization efforts (Bernardoni et al. 2013). The authors use social network analysis (SNA) structural holes theory to identify holes in the network structure of failed and failing states. These holes represent impediments to mutually beneficial collective action. A focus on filling these holes should improve the situation.

Grand Strategy

John Lewis Gaddis in his book, *On Grand Strategy* (Gaddis 2018), uses Isaiah Berlin's concept of foxes and hedgehogs (Berlin 2013), Clausewitz's work on strategy (Clausewitz 1993), and the strategy in Tolstoy's *War and Peace* (Tolstoy 2007), among others, to explain Grand Strategy, defined as "the alignment of potentially unlimited aspirations with necessarily limited capabilities." These "alignments are necessary across time, space *and* scale."

Gaddis explains the characterization of people as hedgehogs or foxes, saying, "Hedgehogs, Berlin explained, 'relate everything to a single central vision' through which 'all that they say or do has significance.' Foxes, in contrast, 'pursue many ends, often unrelated and even contradictory, connected, if at all, only in some *de facto* way'." Gaddis uses this concept to compare and contrast military leaders and their choices of strategies. Xerxes was the Persian King of Kings in 480 B.C. and determined to invade and conquer Greece. He was fixated on the idea and proceeded with it until finally he lost his army and had to flee. He was a hedgehog. Octavian Caesar was not a battlefield commander, yet he used those who were to win battles. He negotiated political swamps, pitting opposing forces against each other. At 32 he was the most powerful man in his world. He died in *his* bed at the age of 76. He was a fox. The hedgehogs don't do too well in the comparison.

Gaddis names Clausewitz and Tolstoy "The Grandest Strategists," in a chapter with that name. He credits both with respecting both theory and practice. "It's as if, in their thinking, abstraction and specificity reinforce each other, but never in predetermined proportions. Each situation requires a *balancing* derived from judgment and arising from experience, skills acquired by learning from the past and training for the future."

In creating strategy to achieve a goal, he explains, "the configuration is triangular, although in two ways. For as you balance knowns, probabilities, and unknowns, you're also doing so across time, space, and scale." He compares this to an artist's sketch and to Clausewitz's *coups d'oeil*, which have the purpose of capturing the essentials without including too much detail – in other words, a model.

For Gaddis, theory is a set of rules that are applicable in some circumstances, but not necessarily in all. It is good for preparing – for learning, but should not be expected to serve as a checklist for action – the *practice* of war.

Sean McFate considers the old method of warfare, conventional war, to be obsolete. He sees "the emergence of a new global system that [he calls] 'durable disorder,' which contains rather than solves problems." "The defining feature of durable disorder is persistent armed conflict, but not as you know it." He believes the "West is losing because it suffers from strategic atrophy." It needs a new grand strategy (McFate 2019).

In his book, *The New Rules of War*, McFate describes how war as actually practiced differs from conventional war. For example, he says, "there is no such thing as war or peace – both coexist, always." By this he means that actors are working to push the limits to obtain their objectives, using methods that fall short of acts that provoke adversaries into declaring war. "Hearts and minds do not matter," as brute force is sufficient and is being used in various places now and will be used in the future. He claims, "technology will not save us," and "the best weapons do not fire bullets." Purchasing more sophisticated and expensive weapons will not win wars, but persuasion will (where brute force is not desirable). "In modern warfare, influence is more potent than bullets (McFate 2019)."

This sentiment echoes the words of General Valery Gerasimov, Chief of the General Staff of the Russian Federation, in a 2013 article, "The Value of Science in Prediction (Gerasimov 2013)." Gerasimov said, "The very 'rules of war' have changed. The role of nonmilitary means of achieving political and strategic goals has grown, and, in many cases, they have exceeded the power of force of weapons in their effectiveness." This is part of what has become known as the Gerasimov Doctrine. Non-military means are "supplemented by military means of a concealed character, including carrying out actions of informational conflict and the actions of special-operations forces. The open use of forces – often under the guise of peacekeeping and crisis regulation – is resorted to only at a certain stage, primarily for the achievement of final success in conflict." This foreshadows the annexation of Crimea in 2014.

McFate says the future lies in shadow wars, in which forces wage war, but narrative warfare is used to obscure the identity of the principal actor and even that a war is taking place. "Cyber is important, but not in the ways people think. It gives us new ways of doing old things: sabotage, theft, propaganda, deceit, and espionage. None of this is new. Cyberwar's real power in modern warfare is influence, not sabotage. Using the internet to change people's minds is more powerful than blowing up a server (McFate 2019)."

Singer and Brooking reinforce the concept of durable disorder and McFate's view of cyberwar. "The thread that runs through all these strange internet skirmishes is that they take place simultaneously, in the same space. Sometimes, the conflict is between feuding celebrities; other times, nations embroiled in a life-and-death struggle. Sometimes, these battles dominate social media chatter completely; other times, they pass with nary a mention." "There aren't two or ten of these conflicts, but many thousands, all unfolding at once and leaving no one and nothing untouched. By merely giving them our attention, we become a part of them. Like cyberwar, these LikeWars are also about hacking. But they're not targeting computer networks – they're targeting human minds." "There's one more aspect that makes them different from conflicts of the past. Anyone can fight in them, but all the combatants are equally powerless in one key, new way. For while these warriors of LikeWar each fight their own personal and global wars across the internet, they aren't the ones writing its rules (Singer and Brooking 2018)."

McFate reminds us that "the 'Westphalian Order' made states the sheriff, outlawing mercenaries and those who hired them." Modern wars are using mercenaries (we just call them contractors) and this trend will grow. The concept of nation-states as the sole actors capable and allowed (by international law) to wage war has become obsolete. Many of the approximately 194 nations recognized today are nations in name only. Many are narco-states, run by fabulously wealthy drug organizations, or states run by warlords as their personal adjuncts, or essentially lawless areas, denoted by names and boundaries for reference purposes and historical reasons, but not acting as nation-states serving a populace. "When the super-rich can rent militaries, they become a new kind of superpower, one capable of challenging states and their rules-based order. Big oil companies will have private armies, as will random billionaires (McFate 2019)."

McFate decries the purchase of such expensive weapon systems as the F-35 fighter aircraft and aircraft carriers as wasting money on a type of war that won't be fought. He cites the example of Russia's use of *maskirovka* in moving "little green men" into Crimea without Western opposition as an example of shadow warfare (McFate 2019). On the other hand, Russia is seen as pursuing its own expensive weapon systems, such as a nuclear-powered missile, thought to have been involved in a mysterious explosion in August of 2019 (Macias 2019). Similarly, China officially revealed its J-20 fifth-generation stealth fighter in 2016. There has been speculation that some technology from the US F-35 was stolen and used in developing the J-20 (Wikipedia 2019b). The appearance is that both Russia and China continue to develop expensive, hi-tech weapon systems, lending credence to arguments that the United States must continue doing the same.

2.4 Illustrations of Conflict

The approach of describing examples of conflict lies somewhere between the reduction approach in the definition section of this chapter and the synthesis approach in the theory section. This section describes some examples of conflict.

Attrition Warfare and the Cold War

For decades the devolution of the Cold War into a hot attrition war was the conceptual ideal of conventional warfare. Militaries designed, built, and deployed equipment and force structures to fight this war. They designed doctrines and trained troops to fight using the doctrines. Generations of warriors spent their entire careers immersed in it.

Fortunately (for our purposes and for the sake of those who would have lost their lives in it), this pure ideal is unsullied by historical events, as it was fought only in mental and computer models. I was involved (in the mid-1970s) with these computer models. I don't remember what was classified and what was not nor know what may have been declassified; however, the details are not important here, and there are plenty of sources for rough approximations that are adequate for this discussion.

In the early days of computerized combat modeling and simulation (say the 1970s), computational power and storage was limited, and the computer models were small by today's standards. In the United States during the Cold War, there were two major situations that were considered worthy of modeling, nuclear war and general-purpose forces war. The latter consisted of the updated versions of World War II-type forces, such as tanks, artillery, infantry, close air support aircraft, fighter aircraft, bombers, naval forces, etc. In the general-purpose forces war category, the principal scenario was the possible NATO-Warsaw Pact conflict. The models started with the forces in contact along the boundary between the Eastern Bloc countries and the Western countries.

Consider a very simple balance of forces calculation based on the number of troops on each side.

- NATO forces – Northern Army Group (NORTHAG), 4 army corps on the order of 160,000 troops; Central Army Group (CENTAG), 4 army corps on the order of 160,000 troops; 2 French army corps on the order of 80,000 troops. Total of ~400,000 troops
- Warsaw Pact forces – between 600,000 and 1.3 million troops, depending on source (Bitzinger 1989)

Visualize all of these troops standing on the border, facing each other. The border covers roughly 1600 miles, implying 250 troops/mile on the west and 375/mile to 800/mile on the east. That is roughly 21 feet between each on the west and 6–14 feet between each on the east. This is not quite "shoulder to shoulder," but close, especially when considering equipment sizes (tanks, artillery, armored personnel carriers, etc.) associated with troops. Naturally, the troops were not actually lined up in this fashion. Combat support and combat service support (logistics, maintenance, medical, etc.) and air support were deployed to the rear, not at the front. Further, there were reserves in rear areas. Still, the troops at the front had little room for maneuver; the reserves would have been the ones maneuvering to support where needed or to exploit advances when possible.

Figure 2.8 presents a notional picture of the line separating the forces, with NATO to the left and the Warsaw Pact to the right. The numbers of forces were so large that the units on each side were positioned virtually shoulder to shoulder, with little room to maneuver. Accordingly, the first models adopted a "piston" model of combat. Several notional lines of control are shown in Fig. 2.8 and used to define the pistons.

The models did not actually have maps behind them, and the pistons were stylized for computational efficiency, as shown on the left side of Fig. 2.9 (with vertical lines within each piston approximating the curved line of the real border). Combat modeling consisted of counting up the armaments on a side in a piston, multiplying the number of each type by a weighting factor that approximated its combat value,

Fig. 2.8 NATO versus Warsaw Pact

and summing the weighted counts. The side with the larger value within a piston would prevail. This was pure attrition warfare. Algorithms were used to determine the amount of advance for a day's fighting and the line of contact would move the appropriate amount and direction. Other algorithms would calculate the losses on each side, resulting in new comparative combat values to generate the next day's movement. The right side of Fig. 2.9 shows a notional picture after several days fighting, with the arrows showing the movement and the heavier black line showing a resulting border that fits the calculated positions within each piston.

As computer power improved, more intricate models became possible. Later models employed more sophistication in calculating the values of the weapons and in adjudicating the results, both attrition and movement. Some lateral movements (maneuver) could be modeled. Also, the definition of the pistons was refined; however, at a gross level, the picture was similar, the two sides "pushed" at each other and the stronger was more successful, resulting in changed force levels and changed position of the line of contact.

I have omitted some important strategic issues, concentrating on the tactical and operational point of view to make this illustration more like pure attrition warfare than it actually was. Also, it is important to realize that this war never took place.

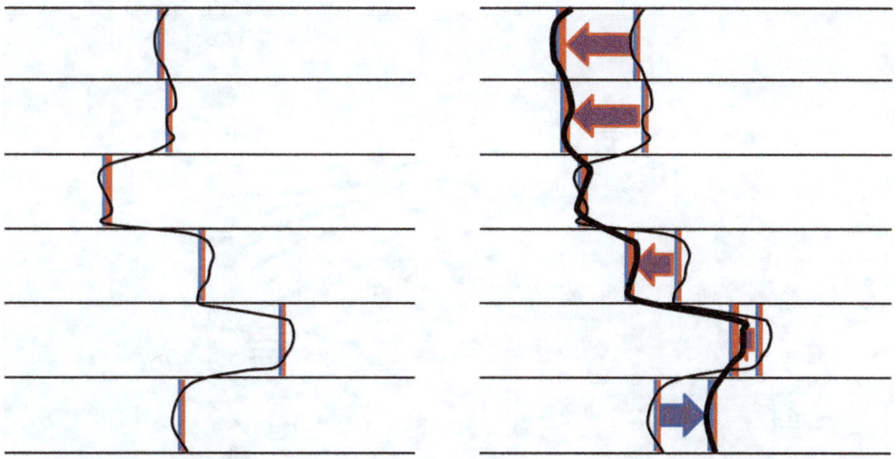

Fig. 2.9 Piston warfare

Regardless of the plans and analyses of those plans, it is almost a certainty that this modeled war would not have matched the real thing.

Maneuver Warfare and Desert Storm

Later, maneuver warfare was reinvented. Compare the 1997 USMC document (U.S. Marine Corps 1997) described above. It made maneuver warfare the official doctrine of the Marines. Figure 2.10 (public domain map of the maneuver in Desert Storm (Wikipedia n.d.)) illustrates the concept. The Marine forces were used as a threat to pin the Iraqi forces in place. The majority of the coalition forces then maneuvered from Saudi Arabia to the west of Kuwait to attack the Iraqi positions from an unexpected direction. The maneuver was basically unopposed until contact was made with the Iraqi positions.

The "shock and awe" description of the beginning of the second Gulf War supported the impression that that war would fall into the same category. Subsequent experience showed that the entire conflict did not do that.

Modeling maneuver warfare required different modeling concepts. The logistics required for so much maneuver was different in scale and kind from that required in the piston models. Similarly, the intelligence requirements were different. The final battles after contact could be modeled in the old way; however, that portion of the overall campaign constituted a much smaller portion of the whole campaign. This meant that just modeling the combat would have provided less insight than would be desired.

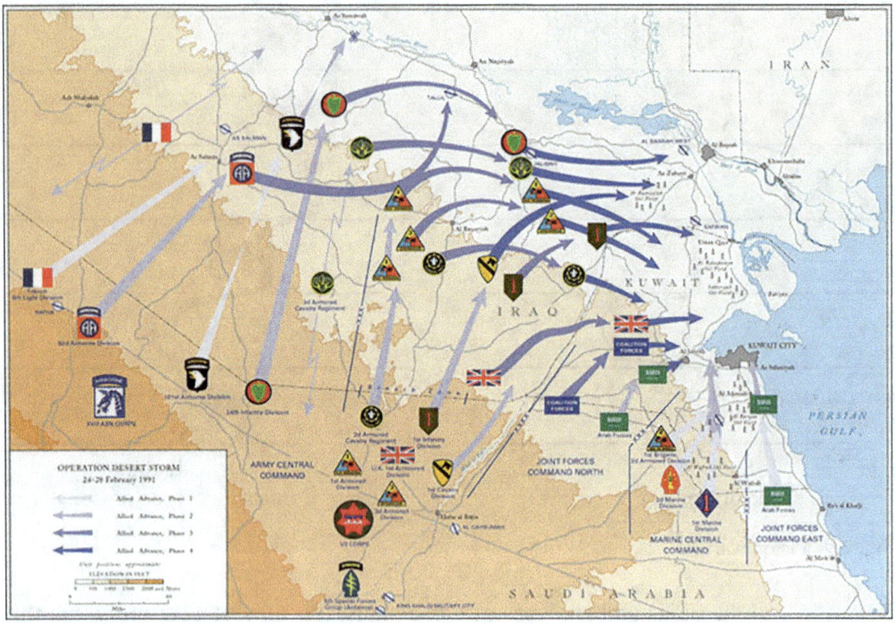

Fig. 2.10 Maneuver warfare

Castle Warfare

The war within South Vietnam did not have borders; however, at certain points it was waged under the attrition warfare style. This was especially evident during the time when "body counts" were used to indicate the state of the conflict. However, the Vietnam War provided an example of yet another type of combat. Figure 2.11 illustrates the concept of castle warfare in which strong points are scattered around the theater. Victory was assumed to be found either in taking the strong points or in holding them. Combat in the rest of the theater was designed to either prevent or enable attacks on the "castles."

Note that the concept of Forward Operating Bases in Afghanistan also partook of some elements of castle warfare.

The NATO bombing of Yugoslavia during the Kosovo war in 1999 also had some aspects of castle warfare (Wikipedia 2018a). The bulk of the action consisted of air strikes, resembling the siege of a castle with siege engines throwing rocks at the fortifications and the inhabitants. The resemblance was strengthened by the results: it appeared that it wasn't damage that caused the capitulation, but loss of will, a not uncommon event in classic sieges (Wikipedia 2018b).

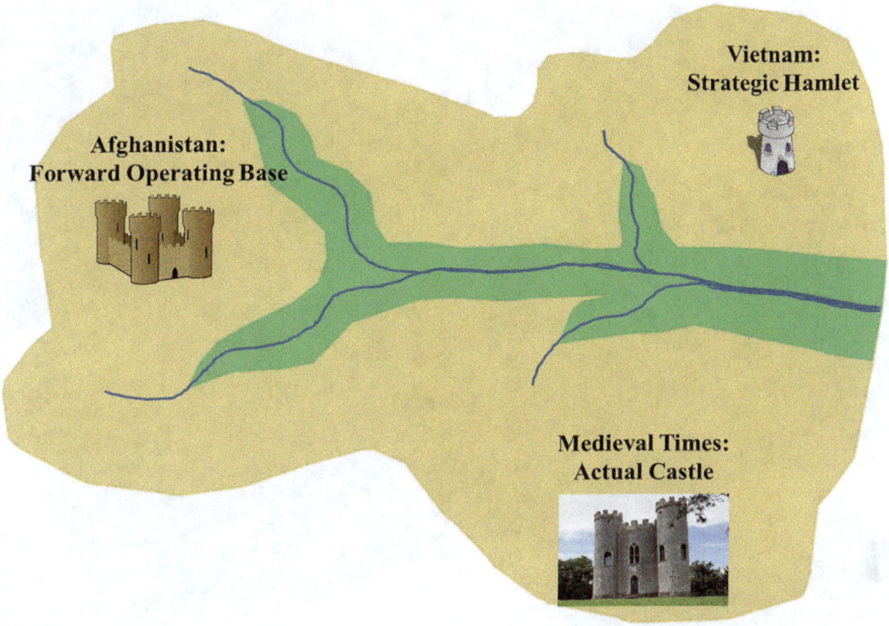

Fig. 2.11 Castle warfare

Naval Warfare

Naval warfare elicits the image of ships in the middle of the ocean. Certainly, ships sail the oceans; however, there is not much there – just water. From the crow's nest of a sailing ship, another ship has to be about 25 miles away before it is visible. That means that enemy ships within an area of about 2000 square miles can be visible (on a clear day). The area of the North Atlantic Ocean is about 16 million square miles, which means that a ship could observe around one hundredth of a percent of the ocean. Clearly, finding an enemy ship to fight at some random point in the ocean would be difficult. (In the era of radar and satellites, the problem is reduced; however, the enemy ship is probably going somewhere, so just sailing to where it was located would probably not result in coming up on the enemy ship.)

On the other hand, there are lots of interesting feature where the sea meets the land (the littoral), harbors, forts, towns, and so forth. Not only are you more likely to find enemy ships because the area of the littoral is so much smaller than the area of the entire ocean, but if you look in interesting places, the chances of finding the enemy become even greater. In his book, *Fleet Tactics and Coastal Combat*, Capt. Wayne Hughes reports, "One will find in any compendium of sea battles … that naval battles large and small have almost without exception been fought in proximity to land (Hughes 2000)."

Figure 2.12 illustrates one tactic of naval warfare. In this case, one fleet is successfully performing a "crossing the 'T'" maneuver. This fleet is sailing from the

Fig. 2.12 Naval warfare

upper right to the lower left. As each ship comes into position to fire at the enemy fleet, it only faces the bow armament of a single opposing ship, but can fire its own broadside armament. Thus, the successful fleet can defeat the other in detail. This engagement is the result of successful maneuver; however, other engagements, such as broadside-to-broadside engagements, are examples of attrition warfare.

The basic concept in crossing the "T" is that it is better to have many of your ships engaging a single enemy ship, with the prospect of destroying it with minor damage to each of your ships, than to line up one against one in a more "even" battle.

Hughes uses the classic Battle of the Nile to illustrate a number of points about naval warfare (Hughes 2000). In 1798, the British Rear Admiral Sir Horatio Nelson found and engaged the French Vice-Admiral François-Paul Brueys d'Aigalliers near the mouth of the Nile in Egypt. Nelson had the smaller fleet, but still decisively defeated the French. The French ships were anchored in a bay in support of the French army ashore; however, as Hughes points out, the French had not fortified the point of the bay and had not trained their seaman adequately. Nelson used superior seamanship to engage the French ships close to the ocean from both the shoreward and the seaward side, producing two to one odds, defeating the French in detail. Hughes discusses how this battle illustrates six lessons:

- "Men matter most." "Leadership, morale, training, physical and mental conditioning, willpower, and endurance are the most important elements in warfare." "At the Nile Nelson established tactical ascendency at the outset, but his force also had to win a very hard battle."
- "Doctrine is the glue of good tactics." By doctrine, Hughes does not mean dogma, but "a comprehensive and practiced plan of action. Sound doctrine will establish unit amidst chaos. So although in one sense the Battle of the Nile did not go as planned at all … in a deeper sense it is the epitome of a sound plan executed flawlessly in spirit."
- "To know tactics, you must know weapons." Hughes says that Nelson "adopted his tactics to a weapon system that in its essentials was centuries old, and with insight that has rarely if ever been equaled at sea."
- "The seat of purpose is on the land." "While it is proper to think of the destruction of the enemy's fleet as the fleet's foremost objective, beyond that immediate objective is always some higher goal." "The battle devastated Napoleon's ambitions."
- "A ship's a fool to fight a fort." The expression is credited to Nelson; however, he still proceeded with his attack. The point of the bay had not been adequately fortified and Nelson was not required to fight a fort. Hughes says, "ships do all they can to avoid a slugfest with fortifications ashore, but ships cannot operate in coastal waters without dealing with weapons directed toward the sea from the land."
- "Attack effectively first." "This means that the first objective in battle is to bring the enemy under concentrated firepower while forestalling his response." "Suffice it to say that, with the readiness of his force and his instant decision to bore in at once and exploit the unreadiness of the French, Nelson was adhering to this maxim."

The quoted material above is all from Hughes (2000).

Unconventional Conflict

The Syrian Civil War (2011 to present [2019]) shows that wars need not have just two sides these days (Wikipedia 2018a, b, c). The sides and forces on each side are represented in the following simplified list:

- Pro-Syrian government
  - Syrian government and forces
  - The Russian government forces
  - The Iranian forces
  - Hezbollah
  - Numerous armed groups on the Syrian side

- Syrian opposition
  - Numerous armed rebel groups
  - The US forces
  - The Turkish forces
  - The Israeli forces
- Kurdish opposition (several groups)
- The remnant Islamic State of Iraq and Syria (ISIS) forces
- The remnant Al Qaeda forces

The Bosnian War (1992–1995) was part of the breakup of Yugoslavia. During this time there were Muslim Bosniaks, Catholic Croats, and Orthodox Serbs fighting each other (Wikipedia 2019a, b, c).

Unconventional conflict is complex because of the number of relevant actors, the variety of actions that can be taken, the importance of the passive environment, and the number of relevant metrics. Figure 2.13 illustrates this complexity. A generic situation (or instance of unconventional conflict) will have a Host Nation in which the situation occurs, with its own government, police, and military. The intervening external coalition brings its forces and contractors. Frequently, there are other contractors and NGOs. There may be internal troublemakers (insurgents, terrorists, etc.). And there may be an unfriendly external nation. Each group has its own agenda, with goals and tasks aimed at accomplishing the goals. And all actions play out in the same time and space. One of the principal geopolitical areas for unconventional

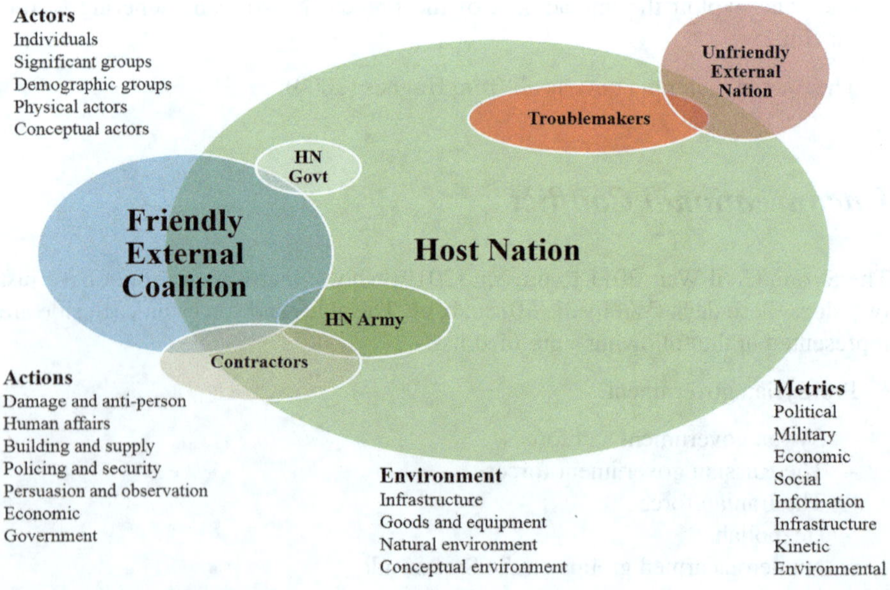

Fig. 2.13 The generic unconventional conflict situation

conflict is within a failed or failing state. Therefore, having the ability to identify such states and to predict which states might fail would be extremely valuable.

Asymmetric warfare is frequently associated with unconventional conflict. In asymmetric warfare, the sides use different means (weapons and styles) of conflict. Generally, this is caused by differences in the resources and capabilities of the sides. However, it may also involve different players having different ends in mind – different goals.

It is important to understand that while observable facts are important, the opinions or sentiment of the actors are also important. Lt. General Flynn (ex-Director of the Defense Intelligence Agency (DIA)) underlined the criticality of the opinions of the populace in obtaining intelligence and for winning in Iraq (Flynn and Ledeen 2016).

The situation can be likened to multiple players playing different board games at the same time on the same board, all interacting, as in Fig. 2.14. Two players are playing checkers, a game with simple rules and simple strategy. One player is

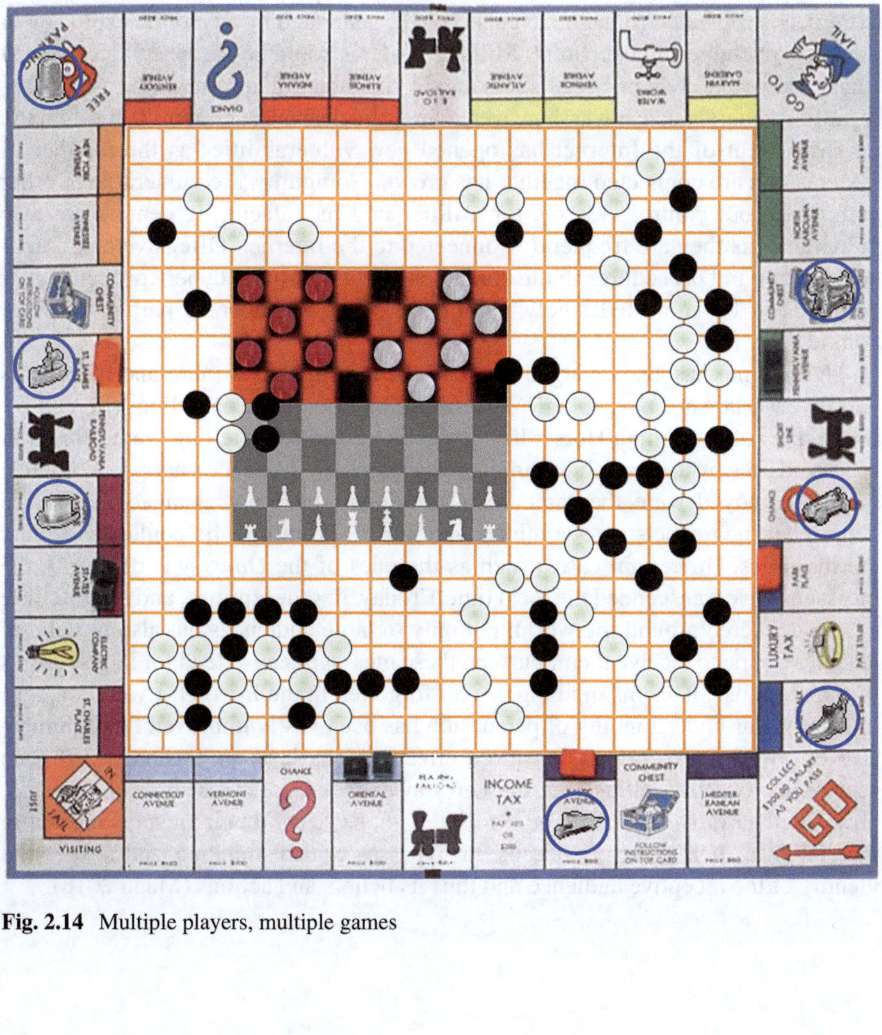

Fig. 2.14 Multiple players, multiple games

playing chess, a game with more complex rules and more complex strategies. Two players are playing Go, a game with simple rules and complex strategies. And – multiple players are playing Monopoly, a game with complex rules and complex strategies. The combined game has multiple players with unknown rules and very complex strategies.

Cyber and Information War

Deception and counters to deception have been parts of conflict at least since the Trojan Horse episode described in both *The Odyssey* (Homer 1946) and *The Aeneid* (Virgil 1969). However, the advent of electronic equipment has opened new avenues for mischief. Dependence on electronic equipment makes all vulnerabilities more important (not just those associated with deception and counter-deception). In the early days of nuclear war worries, the potentially catastrophic effects of the extremely strong electromagnetic pulse (EMP) generated by an atomic explosion in the atmosphere became apparent. Military systems could be hardened against such events; however, civilian systems remained vulnerable. In this ontology, we are disregarding general nuclear war; however, a single nuclear event remains in the domain.

The advent of the Internet has opened new vulnerabilities as the number of systems that are connected together has grown. Computers are vulnerable to other computers, but control systems for utilities and manufacturing centers are also vulnerable, as they are frequently connected to the Internet. Offensive and defensive actions and procedures in the computer world constitute cyber conflict. Cyber conflict can be contested solely within the cyber world or as part of a larger conflict.

The Internet has also enlarged the reach of information warfare and its cousins narrative warfare and psychological warfare or psychological operations (PSYOPS). In the days of the Trojan Horse, lies, deceptions, and story-telling were limited in space and time by the available communications media. The necessary psychological means employed were the same because humans have not changed their basic nature, but the impacts on operations were limited to those who could see or hear the messages. (In certain cases, such as the tales of the *Odyssey* and *Aeneid*, the messages have transcended the local time.) Today, lies, deceptions, and story-telling can be delivered around the world, not only to large audiences but also to tailored audiences, ripe to receive them. Further, these messages can extend for large periods of time because, it seems, nothing is ever forgotten in the Internet. Even more significantly, our understanding of persuasion has been vastly improved through massive efforts to make persuasion more effective (Hartley and Jobson 2020). The importance of story-telling (see the *Odyssey* and *Aeneid*) has been re-discovered, now called narrative warfare. The story need not be "true," but it must resonate with the audience. It is effective because it bypasses critical thinking and shapes the identity of the receptive audience and thus its beliefs and actions (Maan 2018).

By now, game theory is well known as a way of describing simple conflicts and for calculating optimal strategies. Less well known is the concept of hypergames in which perception, misperception, and deception are important parts. In a hypergame, players may be playing different games and imperfectly perceiving the other players' games. This is an essential part of actual unconventional conflicts, illustrated in Fig. 2.14 (Hartley 2017, 2018). Hypergame analysis is similar to game theory in supporting analyses of possible results; however, it is much more complicated. Mateski, Mazzuchi, and Sarkani describe a diagrammatic approach to modeling the perception in hypergames. "Perception, misperception, and deception remain critical elements of success – and failure – in conflict (Mateski et al. 2010)."

The extension of classical spying and deception and counters to them into the realm of computers has opened new avenues for conflict. Not only are the military forces subject to attack, but also the homeland infrastructure of the military forces and the homeland populace are subject to attack. The hardware systems are vulnerable as are the opinions, policies, and politics of the supporters of each side.

Diplomatic Conflict

At its most peaceful, diplomatic conflict consists of negotiations and agreements. Diplomacy can result in changes in the actions of one or more nations. A "stiff diplomatic note" indicates dissatisfaction with a situation and indicates the possibility of stronger actions. Offering to provide or withhold favors among countries can also result in changes of actions.

Economic Conflict

Nearly immediate effects can be seen from such actions as freezing bank accounts. The United States froze the assets of a Russian businessmen with close ties to Russian President Putin in retaliation for Russia's annexation of Crimea in 2014 (Trindle et al. 2014). Long-term strategies may involve the stronger economic power causing the weaker power to spend itself into defeat (Perlo-Freeman 2019). Particular sectors of the economy can be significant in their own right. For example, the need to import oil and fuels (energy) costs the importing nation directly and may have less direct costs in the form of obligations and possible coercion. Energy independence allows a nation to avoid these problems and becoming a net exporter adds to the nation's influence (Olson 2018; Wall Street Journal Editorial Staff 2018). A trade war is yet another form of economic conflict. The United States and China are currently engaged in a trade war using tariffs on commercial products as the weapon that is, as yet, unresolved (Wikipedia 2019a, b, c).

2.5 Characterization of Modern Conflict

This section recapitulates the important points of the chapter and discusses a unifying characterization. We first note that modern conflict has both combat aspects and unconventional conflict aspects. The Grand Strategy of Gaddis (Gaddis 2018) applies to both. For Gaddis, theory is a set of rules that are applicable in some circumstances, but not necessarily in all. It is good for preparing – for learning, but should not be expected to serve as a checklist for action – the *practice* of conflict. Hence it is wise to consider both the theories and the actual practice of conflict.

Combat Aspects of Modern Conflict

We have described the types of combat operations that may be applied in modern conflict. We described the theories of combat, both the theoretical and empirical aspects. We also described some examples of combat types.

Unconventional Conflict Aspects of Modern Conflict

Similarly, we have described the types of unconventional operations that may be applied in modern conflict. We described the theories of unconventional conflict, both the theoretical and empirical aspects. We also described some examples of unconventional conflict types.

Organizing Principles for Modern Conflict

Currently, the most successful organizing principles for both traditional combat and unconventional conflict consist of the DIME and PMESII concepts, which were described in Chap. 1.

An Ontological View

In earlier books, we have described a new organizing principle for unconventional conflict, the ontology. In the first book, we described how an ontology is useful in organizing a model of unconventional conflict and in understanding such conflicts (Hartley 2017). In the second book, we concentrated on the description of the ontology (Hartley 2018). In this book, we extend the ontology to the whole of modern conflict. This ontology is useful both for modeling modern conflict and in understanding it. Further, we claim that the ontological view is superior to the DIME/PMESII view. As we note in the discussion in later chapters, the DIME/PMESII paradigm is useful for a grand conceptual view, but fails at the detail level because too many assumptions are required to connect the details to decompositions of the DIME/PMESII categories. On the other hand, the ontology is specifically designed to handle the details. The ontology, however, does not discard the DIME/PMESII paradigm; rather it subsumes it.

GTO as the Heart of Conflict

The elements of the ontology, Actor, Object, and Action classes, are its infrastructure; however, the heart of the ontology as a description of conflict lies in the Goal-Task-Owner (GTO) sets of the ontology, described in Chap. 12. Conflicts arise between two or more Actors – but there are reasons for the conflicts. Perhaps the agendas of the parties have areas that are mutually incompatible or perhaps the Actors don't like each other and develop agendas to do something about the mutual dislike. The GTO sets represent these agendas.

Conflicts are not purely military. Business conflicts, conflicts among religious groups, and government economic conflicts all are based on conflicting agendas. And, of course, government geopolitical conflicts, either direct or indirect (such as one government "needing" the advantage of position against a third party and so taking action against an "innocent" government), are based on conflicting agendas.

Agendas aren't constant; they may change over time or as the situation changes. Agendas may have hidden goals and open goals (e.g., consider the "Jewish problem" in Nazi Germany that may have started hidden and then become open or may have evolved from having a useful internal enemy to a policy of extermination).

Subordinate Actors may have the "appropriate" sub-agenda or may have additional goals (such as take over the superior's position) or may have almost completely contradictory agendas.

Agendas need not be self-consistent. Tasks may not actually lead to achieving goals, even if they aren't opposed by others. One goal may be deleterious to the accomplishment of other goals. (For example, the Nazi extermination program removed considerable capacities from warfighting.)

Describing the MCO

The MCO is the ontology that characterizes modern conflict. Its construction has been informed by the contents of this chapter. It includes both the ontological infrastructure and the connections among the infrastructure, such as GTO sets. In Chap. 3, we will review some general ontological features and present an overview of the MCO.

Chapter 3
General Ontology Overview

Before we discuss the ontology for modern conflict, we need an understanding of ontologies in general. Chapter 1 included an introduction to ontologies and mentioned foundational ontologies as the upper level or most primitive of ontologies. A foundational ontology is domain-neutral. It defines a set of categories and relationships that is general in nature and is thought to represent reality at its most basic level. The *thing*, *class*, and *relationship* trio is a primitive foundational ontology.

Domain ontologies (such as the one depicted in this book) describe a particular domain of knowledge. A domain ontology may import a foundational ontology to aid in interoperability with other domain ontologies and to avoid the problem of defining the contents of the foundational ontologies within the domain ontology.

Here we begin by discussing the choice of the best foundational ontology for the creation of an ontology for modern conflict. We continue with an overview of the structure of the Modern Conflict Ontology (MCO). And we compare the MCO with a related ontology. The MCO is derived from the Unconventional Conflict Ontology (UCO) (Hartley D. S., An Ontology for Unconventional Conflict, 2018), so a discussion of the changes from the UCO to the MCO is in order.

3.1 Choosing a Foundational Ontology

The main rationale for a foundational ontology is to separate the domain-neutral parts of an ontology from the domain specific parts to allow commonality among different ontologies. In considering using a foundational ontology to support the MCO, I envisioned two benefits. First, using a domain-neutral foundational ontology would validate the top level of the MCO and connect the MCO to other ontologies. Second, the continuation of the definitional and structural disciplines from the foundational ontology to the rest of the MCO would improve it systemically and make it acceptable to the academic ontology community.

D. S. Hartley III, *An Ontology of Modern Conflict*, Understanding Complex Systems, https://doi.org/10.1007/978-3-030-53214-7_3

Since foundational ontologies represent the most basic aspects of reality, independent of a particular domain of knowledge, one might think they would all have to be identical. Unfortunately, that is not the case. The similarities and differences of the three foundational ontologies briefly described below aid in understanding foundational ontologies. There are other ontology frameworks (not discussed here), such as the Descriptive Ontology for Linguistic and Cognitive Engineering (DOLCE) and the Standard Upper Merged Ontology (SUMO), each with its own advantages and disadvantages.

Close reading of the three foundational ontologies described below will show similarities among them. However, it is clear that each is based on a different view of the world. Some of these differences are literally philosophical differences. For example, there is disagreement about whether the referents in an ontology are to be concepts or things. The difference appears to be one of emphasis or engagement with reality. The concept orientation admits that our perceptions are imperfect and that the implication is that we should focus on defining our concept of reality. An example of the benefit of this view is that it enables the determination of equivalence (or not) of concepts. The realist orientation, on the other hand, sets the referents of an ontology to be actual things. A realist ontology may have errors due to poor concepts (e.g., all swans are white); however, these can be corrected when discovered (Arp et al. 2015). There are also structural differences, for instance, whether the basic subclass relationship, **is_a**, allows for multiple parents.

The desiderata for the choice of a foundational ontology for the MCO are as follows:

- Minimize the modifications to the UCO required to bring it into compliance with the chosen foundational ontology.
- Maximize the benefits of choosing the particular foundational ontology (e.g., maximize the number of classes and relationships that can be defined in the foundational ontology and imported into the MCO).
- Maximize the value of naming and structural conventions that will be enforced throughout the MCO, based on the conventions of the chosen foundational ontology.
- Allow connections to other ontologies using the chosen foundational ontology.

General Formal Ontology (GFO)

GFO is a foundational ontology, developed by Onto-Med for medicine and the life sciences (Onto-Med 2010). Its usage is widespread, justifying its consideration.

The GFO is an extensive foundational ontology, as evidenced by the number of classes and relationships that are defined for it. Figure 3.1 shows the top level of the ontology. The definitions clearly read as if they are taken from a philosophy course. (Do not worry if some seem incomprehensible. Just read them to get the flavor of the GFO.) All of the definitions are quoted from the "gfo-basic.owl" file for GFO, available from the Onto-Med website (Onto-Med 2010).

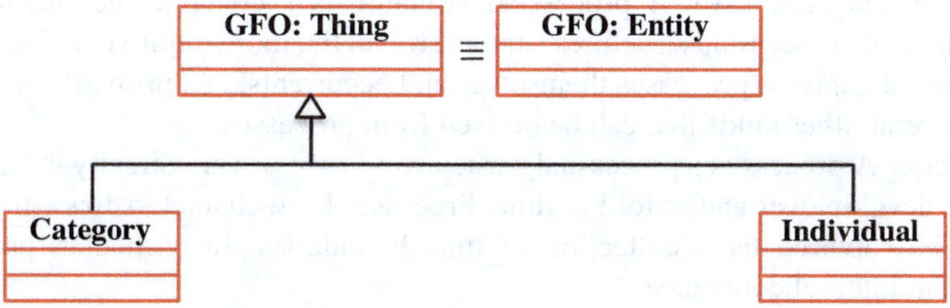

Fig. 3.1 GFO foundational ontology (top portion)

GFO Classes

- **Thing**: is an undefined term.
- **Entity**: is equivalent to thing. "We use the term entity for everything that exists where existence is understood in the broadest sense."
- **Category**: A category is a thing. "Categories satisfy the following conditions: (1) categories can be instantiated; (2) categories can be predicated of other entities."
- **Concept**: "Concepts are categories that are expressed by linguistic signs and are present in someone's mind."
- **Symbol_structure**: A symbol_structure is a category. "Symbolic structures are signs or texts that may be instantiated by tokens."
- **Universal**: "Universals are immanent categories. They exist *in re*." That is, they exist everywhere; however, they only exist in things. For example, "redness" exists everywhere; however, there is not a separate "redness," only the "redness" of a particular thing.
- **Individual**: "Individuals are entities that are not instantiable."
- **Abstract**: "Abstract individuals are independent from time and space (they are not in time and space)." Examples: the number "2" or pi.
- **Concrete**: "Concrete individuals have a relation to time or space (they are in time and space)."
- **Perpetuant**: "A perpetuant is a concrete individual which persists through time and exhibits presentials."
- **Presential**: A presential is a concrete individual. "A presential exists wholly at exactly one time boundary."
- **Material_object**: "A material structure is an individual which satisfies the following conditions: it is a presential, it occupies space, it is a bearer of qualities, but other entities cannot have it as quality, and it consists of an amount of substrate, and it instantiates a persistant ('has identity')."
- **Amount_of_substrate**: "An amount of substrate is a presential, namely, the matter of some material object. Amounts of substrate follow different identity criteria than material objects, i.e., they instantiate different persistants."
- **Material_boundary**: "A material entity which depends on a material object and occupies a spatial boundary."

- **Processual_structure**: A processual_structure is a concrete individual. "The category of processual structures centers around the more intuitive notion of processes. It captures processes themselves and occurrents, i.e., primarily structures of several other kinds that can be derived from processes."
- **Process**: A process is a processual_structure. "Processes are directly in time, and they develop over and unfold in time. Processes have characteristics which cannot be captured by a collection of time boundaries. In particular, processes exhibit internal coherence."
- **Continuous_process**: "A process is a continuous process if it can be partitioned such that the partition contains only states or processes within a continuous change. Those states and changes must be based on the same reference category."
- **Discrete_process**: A discrete_process is a process. "Discrete processes have a partitioning into an alteration of discrete changes and states."
- **State**: "A process is a state with respect to a category iff [if and only if] every [one] of its process boundaries instantiates that category."
- **Processual_role**: A processual_role is a process. It is also a role. "A processual role corresponds to the manner in which a single participant behaves in some process." "Processual roles are roles with a process as context, and they are dependent processes."
- **Change**: A change is a process. "A change refers to a structure centered around two presentials, which are boundaries of one and the same process. If they exist at coinciding time boundaries, the change comes close to notions in the literature like 'punctual' or 'instantaneous event' as well as 'moment' (in a temporal reading). Alternatively, the presentials may be boundaries at the opposite ends of a process of arbitrary extension. Either notion of change is relative to contradictory conditions between which a transition takes place. These conditions refer to some collection of pairwise disjoint subcategories of one category."
- **Continuous_change**: "A continuous change is a change such that (1) its process boundaries exist at non-coincident time boundaries, (2) any two non-coincident process boundaries of its process are distinguished with respect to the reference category, and (3) any two coincident process boundaries exhibit no such distinction, i.e., no discrete changes of the same reference category."
- **Discrete_change**: "A discrete change is a change at coincident time boundaries, for which a recognizable difference exists. That means there is a category with two disjoint subcategories such that each of these is instantiated by exactly one of the process boundaries in the change."
- **Event**: An event is a process. "Events are processual structures comprising a process, and one of its extremal process boundaries. The latter must further satisfy a condition which differs from the condition applicable to all other boundaries of the process, i.e., the extremal boundary instantiates a subcategory of the event's reference category which is disjoint with that category instantiated by the remaining process boundaries."

- **History**: A history is a process. "Histories are processual structures which consist of a process and a number of presentials which are constituents of the boundaries of that process."
- **Property**: A property is an individual. "A property is a dependent entity which another entit[y] has, i.e., that entity exhibits its property. Other common terms for property in natural language are characteristic, feature, quality, etc."
- **Relational_role**: A relational_role is a property. It is also a role. "A relational role corresponds to the way in which an argument participates in some relation."
- **Relator**: A relator is an individual. "A relator mediates between or connects entities. It is a dependent entity which depends on those entities connected."
- **Role**: A role is an individual. "A role corresponds to the involvement of some entity (the player of the role) within another entity (the context of the role). Examples of role terms: student, patient, runner, reader, and factor."
- **Relational_role**: A relational_role is a role. It is also a property. "A relational role corresponds to the way in which an argument participates in some relation."
- **Social_role**: A social_role is a role. "A social role corresponds to the involvement of a social object within some society."
- **Space_time**: A space_time entity is an individual. "A space-time-entity is something in which concrete entities can be located."
- **Space**: A space entity is a space_time entity. "Spatial entities in GFO are analyzed according to the ideas of Franz Brentano. Starting from connected three-dimensional entities (topoids), related spatial entities can be distinguished."
- **Spatial_region**: A spatial_region is a space entity. "Space regions are mereological sums [fusions] of topoids."
- **Topoid**: A topoid is a spatial_region. "Topoids are connected compact regions of space. They have spatial boundaries."
- **Spatial_boundary**: A spatial_boundary is a space entity. "Boundaries of regions are surfaces, boundaries of surfaces are lines, and boundaries of lines are points. As in the case of time boundaries, spatial boundaries have no independent existence, i.e., they depend on the spatial entity of which they are boundaries."
- **Surface**: A surface is a spatial_boundary. "Topoids are bounded by surfaces."
- **Line**: A line is a spatial_boundary. "Surfaces are bound[ed] by lines."
- **Point**: A point is a spatial_boundary. "Lines are bound[ed] by points."
- **Time**: A time entity is a space_time entity. "The time model of GFO is based on Brentano and the glass continuum of Allen and Hayes."
- **Temporal_region**: A temporal_region is a time entity. "Time regions are defined as the mereological sum of chronoids, i.e., time regions may consist of non-connected intervals of time."
- **Chronoid**: A chronoid is a temporal_region. "Chronoids are entities *sui generis*. Every chronoid has exactly two extremal and infinitely many inner time boundaries which are equivalently called time points."

- **Time_boundary**: A time_boundary is a time entity. "Time boundaries depend on chronoids and can coincide. Left time boundaries, if viewed from the perspective of bounding a specific chronoid, are those which are earlier than any inner or right time boundary of that chronoid. On the other hand, within a pair of coincident time boundaries, a left time boundary is later than the right time boundary in that pair."
- **Left_time_boundary**: A left_time_boundary is a time_boundary; see time boundary definition.
- **Right_time_boundary**: A right_time_boundary is a time_boundary; see time boundary definition.

GFO Relationships

In addition to things, the GFO also defines a number of relationships. The **is-a** relationship, which defines subclasses **allows** multiple parents:

- **is_a**: Each instance of *A* must be an instance of *B*.
- **abstract_part_of**: "The abstract part-of relation is denoted by p(x,y); the argument types of this relation are not specified, i.e., we allow arbitrary entities to be arguments. We assume that p(x,y) satisfies the condition of a partial ordering, i.e., the following axioms: reflexivity, antisymmetry, and transitivity."
- **part_of**: Part_of is a sub-relation of abstract_has_part. "The relation between parts and wholes. The union of several domain-specific part-of relationships not contained explicitly in gfo-basic, like spatial part-of or part-of among material structures."
- **proper_part_of**: Proper_part_of is a sub-relation of part_of. "The irreflexive variant of part-of."
- **temporal_part_of**: Temporal_part_of is a sub-relation of part_of. "A part-of relationship between two time entities. Time-boundaries cannot have parts."
- **role_of**: Role_of is a sub-relation of abstract_part_of. "The relationship between a role and its context. Typically, the nature of the context determines that of the roles, which are in some sense a part of that context, e.g., processes form the context for certain roles (processual roles), such that the latter are recognizable as processes."
- **abstract_has_part**: "Abstract has-part is the inverse of the abstract part-of relationship."
- **has_part**: Has_part is a sub-relation of abstract_has_part. "The inverse of part-of."
- **has_proper_part**: Has_proper_part is a sub-relation of has_part. "The inverse of proper part-of."
- **has_temporal_part**: Has_temporal_part is a sub-relation of has_part. "The inverse of temporal part-of."
- **has_role**: Has_role is a sub-relation of abstract_has_part. "The inverse of role of."

- **coincides_with**: "Coincidence is a relationship between space boundaries or time boundaries, respectively. Intuitively, two such boundaries are coincident if and only if they occupy 'the same' space, or point in time, but they are still different entities."
- **depends_on**: "This relation captures the notion of existential dependence."
- **boundary_of**: Boundary_of is a sub-relation of depends_on. "'Boundary of' connects boundaries to the entities which they bind."
- **spatial_boundary_of**: Spatial_boundary_of is a sub-relation of boundary_of. "Spatial boundaries may bound spatial entities."
- **time_boundary_of**: Time_boundary_of is a sub-relation of boundary_of. "Time boundaries bound temporal regions."
- **left_boundary_of**: Left_boundary_of is a sub-relation of time_boundary_of. "Left boundary of a temporal region."
- **right_boundary_of**: Right_boundary_of is a sub-relation of time_boundary_of. "Right boundary of a temporal region."
- **exhibits**: "In the GFO-account of persistence, perpetuants exhibit presentials, i.e., the former 'exist through' the latter at the time where the presential exists. With respect to persistants, presentials instantiate persistants."
- **necessary_for**: "An entity is necessary for another one if the latter is required for the former to exist."
- **has_boundary**: Has_boundary is a sub-relation of necessary_for. "The has-boundary relation connects entities with their boundaries, e.g., chronoids to time boundaries, topoids to surfaces, etc."
- **has_spatial_boundary**: Has_spatial_boundary is a sub-relation of has_boundary. "Space entities may have spatial boundaries."
- **has_time_boundary**: Has_time_boundary is a sub-relation of has_boundary. "Temporal regions have temporal boundaries."
- **has_left_time_boundary**: Has_left_time_boundary is a sub-relation of has_time_boundary. "Temporal regions have exactly one extremal left time boundary."
- **has_right_time_boundary**: Has_right_time_boundary is a sub-relation of has_time_boundary. "Temporal regions have exactly one extremal right time boundary."
- **exists_at**: "Presentials exist at a single time boundary."
- **framed_by**: "A presential is framed by a spatial region if the location which the presential occupies is a spatial part of that region."
- **occupies**: Occupies is a sub-relation of framed_by. "Presentials exist in space, and the space entity occupied by a presential is uniquely determined (where a fixed granularity is assumed)."
- **frames**: "A spatial region frames a presential if the location which the presential occupies is a spatial part of that region."
- **occupied_by**: Occupied by is a sub-relation of frames. "Presentials exist in space, and the space entity occupied by a presential is uniquely determined (where a fixed granularity is assumed)."
- **has_participant**: "Temporally extended entities have participants."

- **participates_in**: "Intuitively, objects participate in processes, for instance. In GFO, participation accommodates the GFO approach to persistence, i.e., at least presentials can participate in processes. Moreover, it is useful to extend the notion of participation also to other temporally extended entities."
- **has_property**: "Entities can have properties. In GFO, properties are individualized, and 'has property' links an entity to its particular property."
- **property_of**: "Links properties to their bearers."
- **instance_of**: "The instantiation relation holds between an entity and a category. Put differently, the category is predicated of that entity, the entity is the instance of that category. Entities of all kinds can be instances, which results in categories which have individuals as instances or categories which may have categories as instances, such as 'species'."
- **token_of**: Token_of is a sub-relation of instance_of. "Inverse of the has_token relation."
- **instantiated_by**: "Inverse of instance of."
- **has_token**: Has_token is a sub-relation of instantiated_by. "The specific relation from a symbol structure (a category in GFO) to an entity seen as an occurrence of that symbol structure, as a token of it."
- **plays_role**: "Entities can play roles with respect to some other entity which provides a context for that role. The plays-role relationship links an entity with its role."
- **played_by**: "The inverse of plays-role."
- **projects_to**: "Links an entity to its temporal extension. Entities which are in time are related to the corresponding temporal regions by projects to. Moreover, entities related to others which are in time may likewise project to temporal regions."
- **projection_of**: "If, for instance, a process happens during a certain time, i.e., some temporal region, that region is the projection of that process (to time)."

DoDAF Meta-Model (DM2)

The IDEAS foundational ontology was developed by the IDEAS Group for military Enterprise Architectures (IDEAS Group). Its purpose was to facilitate interoperability of Enterprise Architecture (EA) models. (Note that the MCO is not an EA model.) Several countries were involved in the effort, with each country developing its own architectural framework (domain model). The US domain model was named the Department of Defense (DoD) Architecture Framework (DoDAF) (DoD Chief Information Officer 2010). The DM2 is one of the few DoD ontologies that are available for consideration.

These references do not make a clear distinction between what might be considered the foundational ontology and what should be included in the domain ontology.

Probably there are many more things and relationships than are described below that should be considered foundational. Further, it appears that an entity can be a class with instances and an instance of another class at the same time, something that is not allowed in many other ontologies. For example, the following statement appears (DoD Chief Information Officer 2010). "The IDEAS Foundation is a higher-order ontology, so types may have members that are also types." Figure 3.2 shows the top level of the DM2 foundational ontology.

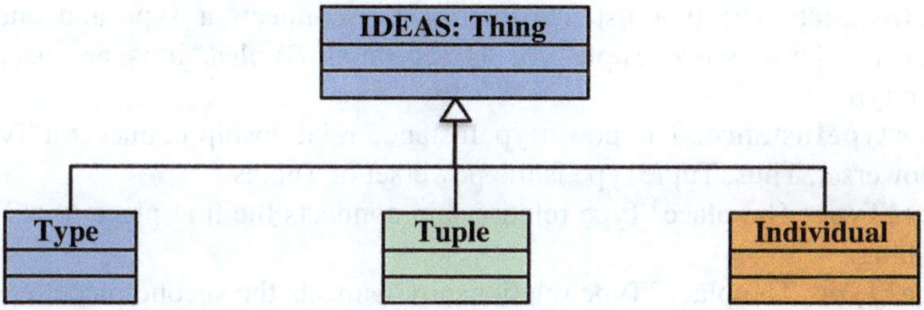

Fig. 3.2 DM2 foundational ontology (top portion)

Based on numbers of classes and relationships, the DM2 foundational ontology is simpler than the GFO. However, many of the DM2 concepts are difficult to grasp, and the implied ontology seems more complicated. All of the definitions are inferred from the web page for IDEAS Group (IDEAS Group) and from the DoD Chief Information Officer reference (2010).

DM2 Classes

- **Thing**: is an undefined term.
- **Individual**: An Individual is a Thing with spatiotemporal extent that exists as an indivisible whole or as a single member of a category.
- **Tuple**: A Tuple is a Thing. It is a relationship between objects, an ordered set having two or more ordered "places."
- **Couple**: A Couple is a Tuple with two places.
- **SuperSubtype**: A SuperSubtype is a Couple that defines the superSubtype relationship (similar to the *is-a* in other ontologies; however, the "sub" here is defined by "subset"). The first "place" identifies the parent thing and the second "place" identifies the child thing.
- **Type**: A Type is a Thing that is a set of Individuals or classes of other sets or classes. For example, the Nimitz-Class aircraft carrier is a Type.
- **IndividualType**: An IndividualType is a Type whose members are all Individuals.
- **TupleType**: A TupleType is a Type that is a set whose members are all of the Tuples.
- **Powertype**: A Powertype is a Type that is the set of all subsets of a given Type.

DM2 Relationships

In addition to things, the DM2 also implicitly defines a number of relationships. The following items are actually classes that embody relationships. The **superSubtype** relationship (the **is_a** relationship in this ontology) **appears** to allow inheritance from multiple parents.

- **superSubtype**: The superSubtype relationship connects a child Thing to a parent Thing.
- **typeInstance**: The typeInstance relationship connects a Type and one of its instances. Thus, since TupleType is set of all Tuples, it is an instance of Powertype.
- **powertypeInstance**: The powertypeInstance relationship connects a Type and its powerset. Thus, TupleType is the powerset of Tuples.
- **place1Type**: The place1Type relationship connects the first place of a Tuple to its Thing.
- **place2Type**: The place2Type relationship connects the second place of a Tuple to its Thing.
- **placeType**: The placeType relationship connects the places of a Tuple to their Things.

Basic Formal Ontology (BFO)

BFO is a foundational ontology, developed by the Institute for Formal Ontology and Medical Information Science (IFOMIS) (2017). Like the GFO, the BFO is used by many ontologies. In particular, it is used by a DoD domain ontology that is related to the domain of the MCO. A later section will compare the two ontologies. Figure 3.3 shows the top level of the BFO.

The number of classes in the BFO is comparable to the GFOs; however, the number of relationships is distinctly smaller. All of the definitions are quoted from *Building Ontologies with Basic Formal Ontology* (Arp et al. 2015) or, if the definition there was entangled in the text, from the *Basic Formal Ontology 2.0:*

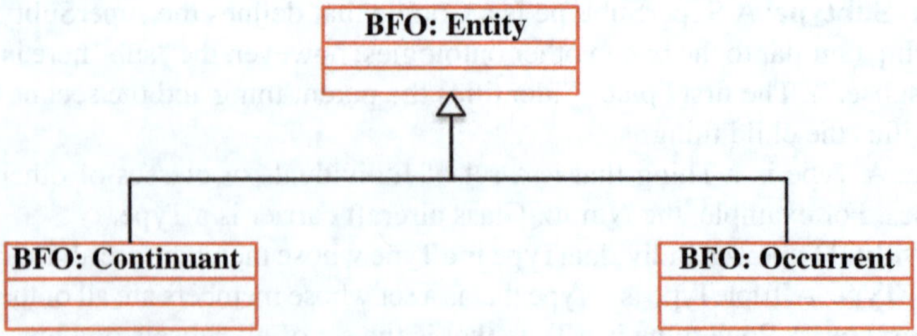

Fig. 3.3 BFO foundational ontology (top portion)

Specification and User's Guide (Smith et al. n.d.). The definitions in the first reference are written for a general audience, while those in the second reference are often written in mathematical format for precision. Despite these differences in style, the definitions are equivalent.

BFO Classes

- **Entity**: is an undefined term.
- **Continuant**: "A continuant is an entity that persists, endures, or continues to exist through time while maintaining its identity."
- **Independent continuant**: An independent continuant "is a continuant entity that is the bearer of qualities." And no entity can bear it as a quality.
- **Material entity**: "A material entity is an independent continuant that has some portion of matter as proper or improper continuant part."
- **Object**: "An object is a material entity that is (1) spatially extended in three dimensions; (2) causally unified, meaning its parts are tied together by relations of connection in such a way that if one part of the object is moved in space then its other parts will likely be moved also (the parts share in this sense what we can think of as a common fate); and (3) maximally self-connected (which means intuitively that the different parts of the object are tied together in a certain way and that anything that is tied to these parts in the same way is itself part of the object)."
- **Object aggregate**: "An object aggregate is a material entity that is made up of a collection of objects and whose parts are exactly exhausted by the objects that form this connection. In addition the objects forming and object aggregate are separate from each other in the sense that they share no parts in common."
- **Fiat object part**: "A fiat object part is a material entity that is a proper part of some larger object but is not demarcated from the remainder of this object by any physical discontinuities (thus it is not itself an object)."
- **Immaterial entity**: An immaterial entity is "an independent continuant that contains no material entities as parts."
- **Continuant fiat boundary**: "A continuant fiat boundary is an immaterial entity that is of zero, one, or two dimensions and does not include a spatial region as part."
- **Two-dimensional continuant fiat boundary**: "A two-dimensional continuant fiat boundary (surface) is a [continuant fiat boundary that is a] self-connected fiat surface whose location is defined in relation to some material entity."
- **One-dimensional continuant fiat boundary**: "A one-dimensional continuant fiat boundary is a [continuant fiat boundary that is a] continuous fiat line whose location is defined in relation to some material entity".
- **Zero-dimensional continuant fiat boundary**: "A zero-dimensional continuant fiat boundary is a [continuant fiat boundary that is a] fiat point whose location is defined in relation to some material entity."

- **Site**: A site is "a three-dimensional immaterial entity that either (1) is (partially or wholly) bounded by a material entity or (2) is a three-dimensional immaterial part of an entity satisfying (1)."
- **Spatial Region**: "A special region is a[n immaterial] continuant entity that is a continuant_part_of space$_R$ as defined relative to some frame R."
- **Zero-dimensional spatial region**: A "zero-dimensional region is a spatial region with no dimensions, also called a spatial point."
- **One-dimensional spatial region**: A "one-dimensional region is a spatial region with one dimension, also called a spatial line."
- **Two-dimensional spatial region**: A "two-dimensional region is a spatial region with two dimensions, also called a spatial surface."
- **Three-dimensional spatial region**: A "three-dimensional region is a spatial region with three dimensions, also called a spatial volume."
- **Specifically dependent continuant**: "A specifically dependent continuant is a continuant entity that depends on one or more specific independent continuants for its existence."
- **Quality**: A "quality is a specifically dependent continuant that, in contrast to roles and dispositions, does not require any further process in order to be realized." An example is the color (a quality) of this portion of blood (the independent continuant).
- **Relational quality**: A relational quality is a quality that has "a plurality of independent continuants as [its] bearers." An example is being a parent of (relational quality) of these children (the independent continuants).
- **Realizable entity**: "A realizable entity is … a specifically dependent continuant that has at least one independent continuant entity as its bearer and whose instances can be realized (manifested, actualized, executed) in associated processes of specific correlated types in which the bearer participates." "Examples of realizable entity include the role of being a doctor, the functions of the reproductive organs, the disposition of a portion of blood to coagulate …."
- **Role**: "A role is an externally grounded realizable entity, that is, it is a realizable entity that is possessed by its bearer because of some external circumstances (e.g., the bearer has been assigned the role by some other persons, who have roles of their own which grant them a certain authority). A role is thus always optional; the bearer does not have to be in the given external circumstances."
- **Disposition**: "A disposition is a realizable entity in virtue of which – for example, through appropriate triggers – a process of a certain kind occurs (or can occur or is likely to occur) in the independent continuant in which the disposition inheres."
- **Function**: "A function is a disposition that exists in virtue of the bearer's physical makeup, and this physical makeup is something the bearer possesses because it came into being either through evolution (in the case of natural biological entities) or through intentional design (in the case of artifacts), in order to realize processes of a certain sort."
- **Generically dependent continuant**: A generically dependent continuant is "a continuant that is dependent on one or [more] other independent continuants that

can serve as its bearer." A pdf file is an example: it is dependent on some physical storage device; however, it can be moved from one device to another and copied to a third device.

- **Occurrent**: "An occurrent is an entity that unfolds itself in time or it is the instantaneous boundary of such an entity (e.g., a beginning or an ending) or it is a temporal or spatiotemporal region which such an entity occupies_temporal_region or occupies_spatiotemporal_region."
- **Process**: A "process is an occurrent entity that exists in time by occurring or happening, has temporal parts, and always depends on some (at least one) material entity." This is analogous to the relation between a specifically dependent continuant and its independent continuant bearer. Here the dependent thing involves time.
- **History**: A history is a process that is "the sum of all processes taking place in the spatiotemporal region occupied by the material entity or site in question."
- **Process profile**: The definition is missing in the first reference and mathematically convoluted in the second reference. It appears that a process profile is a process that consists of measurements of some part of another profile.
- **Process boundary**: A "process boundary is an occurrent entity that is the instantaneous temporal boundary of a process."
- **Spatiotemporal region**: "A spatiotemporal region is an occurrent entity at or in which occurrent entities can be located." It is a part of space-time defined relative to some frame of reference.
- **Temporal region**: "A temporal region is an occurrent entity that is a part of time (of the whole of time)."
- **Zero-dimensional temporal region**: "A zero-dimensional temporal region – also called a temporal instant – is a temporal region that is without extent."
- **One-dimensional temporal region**: "A one-dimensional temporal region – also called a temporal interval – is a temporal region that is extended in time."

BFO Relationships

In addition to entities, the BFO also defines a number of relationships. The BFO distinguishes among relationships connecting the entities, the relationships connecting instances of the entities with entities, and the relationships connecting instances. It also distinguishes between relationships connecting continuants from those connecting occurrents. In the latter distinction, if a particular relationship, such as *is_a*, is philosophically identical in both cases, the same term is used in both cases. However, where there is a philosophical difference, such as *part_of*, different terms are used: *continuant_part_of* and *occurrent_part_of*.

BFO lists eight relationships as core relationships. The pattern *A* related_to *B*, where *A* and *B* are entities, will be used in each definition below. The **is_a** relationship enforces a taxonomy of subclasses – that is **no subclass** has multiple parents. The three foundational relationships are the following:

- **is_a**: Each instance of *A* must be an instance of *B*.
- **continuant_part_of**: For each instance *a* of continuant *A* at time *t*, there is an instance *b* of continuant *B* at time *t* such that *a* is an instance level continuant_part_of *b* at time *t*. The instance level relation uses the common definition of "part of," modified by the requirement that the relation exists at time *t*.
- **occurrent_part_of**: For each instance *a* of occurrent *A*, there is an instance *b* of occurrent *B* such that *a* is an instance level occurent_part_of *b*. The instance level relation uses the common definition of "part of."

The two spatial relationships are the following:

- **located_in**: For each instance *a* and every time *t*, if *a* is an instance of continuant *A* at *t*, there is an instance *b* of continuant *B* at *t* such that *a* is instance-level located_in *b* at *t*. The instance level relation uses the common definition of an object being located in a spatial region to define a being located in *b* to mean there is a spatial region in which *b* is located and that *a* is located in a spatial region that is part_of *b*'s spatial region.
- **adjacent_to**: For each instance *a* and every time *t*, if *a* is an instance of continuant *A* at *t*, there is an instance *b* of continuant *B* at *t* such that *a* is instance-level adjacent_to *b* at *t*. Presumably the instance-level definition uses a similarly complicated method of specifying the relation as found in the located_in definition.

The two temporal relationships are the following:

- **derives_from**: This relation is expressed in biological terms and refers to the starting (*A*) and end (*B*) continuant entities of a process, such as "plasma cell derives_from B lymphocyte."
- **preceded_by**: This relation expresses the concept that all of the time instants of one occurrent (*A*) are earlier than all of the time instants of the second occurrent (*B*).

The single participation relationship is the following:

- **has_participant**: This relation is between a process *A* and a continuant entity *B* and asserts that *B* participates in or is involved in the process *A*.

The Choice

The choice of an ontology structure must be driven by the content of the domain of knowledge. If a structure, however elegant, impedes the description of the domain of knowledge, it is unsatisfactory.

The DM2 has the advantage of connections to DoD ontologies; however, none of the classes or relationships are useful in the MCO. The GFO has no particular connection advantages, and too many of the classes and relationships are irrelevant to the MCO and would serve as useless baggage. The BFO, like the GFO, has no particular connection advantages. The BFO has more useful classes and relationships than the GFO, but still has too many irrelevant ones.

The BFO makes a strong argument for not allowing multiple inheritance through the class-subclass relationship. That is, no class should have multiple parents in its **is_a** relationship. This argument includes a practical (computability-type) component as well as a philosophical component, which essentially declares that the apparent need for multiple parents is actually the result of muddy thinking (insufficient definitions) (Arp et al. 2015).

The variety and differences among the foundational ontologies disabused me of the notion of some "blessed" foundational ontology, which would automatically confer validity on the MCO. Second, the number of choices also militates against a major benefit from connectivity with other ontologies. Third, the actual contents of the three foundational ontologies showed minimal benefits from class and relationship definitions. However, the naming and structural conventions of the BFO promised systemic improvements that should be practically useful and academically acceptable.

Thus, rather than making a choice of an existing foundational ontology to use with the MCO, I chose one as a pattern for developing the MCO. The ontology framework I chose as a pattern for the MCO is the BFO, described in Arp et al. (2015). The BFO foundational ontology contains parts that are largely irrelevant for the modern conflict ontology (such as processes) and lacks a major component that the MCO needs. However, for the purposes of the MCO, the important concepts are the principles and rules that are expressed in justifying the choices made in constructing the BFO, rather than the actual BFO. For example, use a basic *is-a* taxonomy structure where possible, meaning that no multiple parents are allowed, and standardize the definition of elements: state the parent class name, and then say how the thing is specialized from the parent class and what differentiates it from others with the same parent. Thus, it would be proper to say that the design of the MCO was guided by the BFO, rather than saying that the MCO uses the BFO.

3.2 Modern Conflict Ontology Structure

Figure 3.4 illustrates the basic context diagram, from which the rest of the MCO is derived. The operational environment is the world of the MCO – its domain.

The classes of the MCO represent entities, not concepts of entities. For example, the *roadInfrastructure* class represents a portion of road infrastructure in the real world. The Interstate network in the United States would be an instance of this class, as would be the road outside the place where you live.

The active elements of the domain, the elements that do things, are called Actor classes. These classes include individual humans, groups of humans, and nonhuman elements that act. The Actor Ontology (a sub-ontology of the MCO) performs the function of identifying the differences and commonalities among the Actor classes.

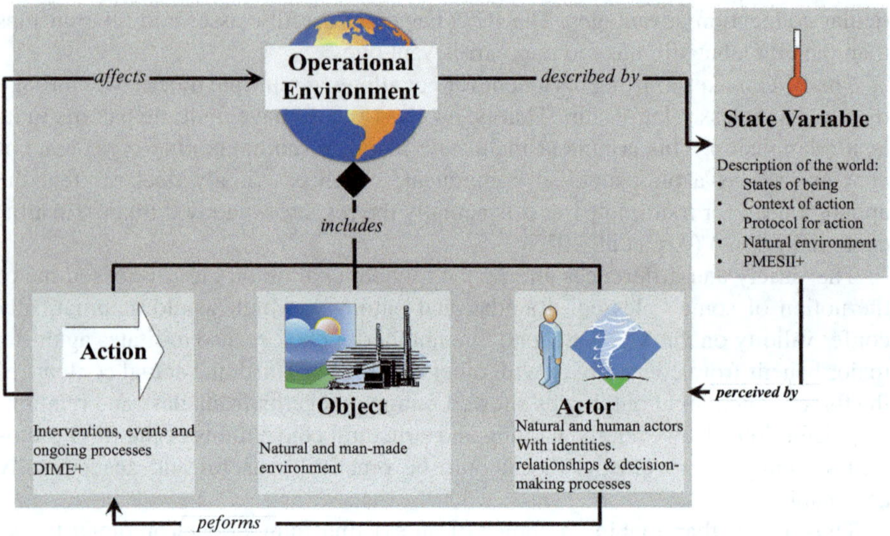

Fig. 3.4 Basic context diagram

Actors ***perform*** Actions, as shown in the figure. Actions are the interventions, events, and ongoing processes. Actions directly cause changes. As with the Actor Ontology, the Action Ontology performs the function of identifying the differences and commonalities among the Action classes.

Actions cause changes to or ***affect*** the entire operational environment, which consists of Actors, Actions, and Objects, although the effects of Actions on other Actions is indirect. An Action may change the environment in such a way as to change the choice of subsequent Actions or to change the effects of subsequent Actions. An Action doesn't change any other Actions directly.

The Objects consist of natural environment elements, man-made environment elements, and conceptual elements. As with the Actor and Action Ontologies, the Object Ontology identifies the differences and commonalities among the Object classes.

All of the elements of the operational environment, the Actors, Actions, and Objects, are ***described by*** state variables (also called metrics). The effects of Actions are captured as changes to the appropriate state variables that are linked to corresponding elements.

Actors ***perceive*** the state variables. This is the only way that Actors know anything, including about themselves. Objects perceive nothing; they just exist. This closes the cycle of the basic context diagram.

Figure 3.5 illustrates the full context diagram of the MCO. Note the PMESII block that is connected to the state variable block. This indicates that the metrics have a second organization that is based on the PMESII categories, which is a paradigm for understanding the state of the situation.

Also note the DIME block that is connected to the Action block. This indicates that the Actions also have a second organization, in this case based on the DIME

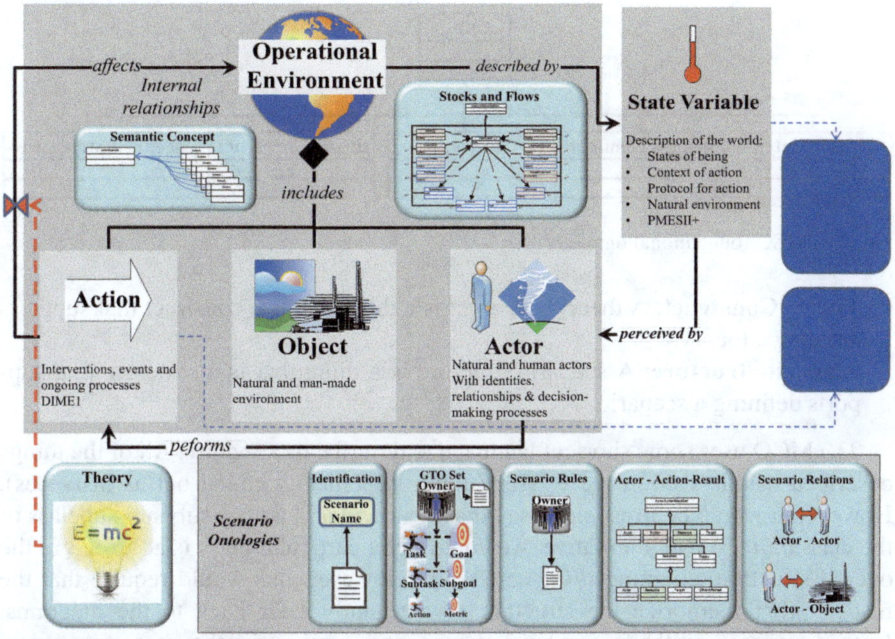

Fig. 3.5 Full context diagram

categories. The DIME paradigm is useful for understanding the levers of power, the Actions that can affect changes to the situation.

The five blue-green blocks contained in the gray block on the bottom right are the Scenario Ontologies (also sub-ontologies of the MCO). These are ontologies that are situation-dependent. Their structures are situation-independent, but their full instantiations and thus meanings depend on the actual situation.

The last change from the basic context diagram to the full context diagram is the addition of the theory block on the lower left. The red arrow from this block to the ***affects*** arrow indicates that theories influence the effects of Actions. In a modeling situation, the influence consists of defining the code for implementing the effects. In a real-world situation, the influence consists of explaining the effects that are observed and perhaps for providing a rationale for why a particular Action was selected.

Figure 3.6 illustrates the MCO foundational ontology. The classes of the foundational ontology are defined as follows:

- **Thing**: is an undefined term.
- **GenericMetric**: A genericMetric is a thing that is a state variable for an element.
- **GenericElement**: A genericElement is a thing that is an actor, action, or object part of the operational environment.
- **Concept**: A concept is a thing that defines interrelationships among elements.

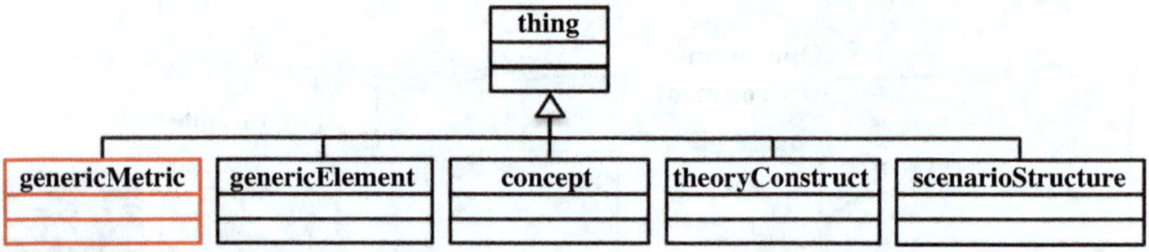

Fig. 3.6 MCO foundational ontology

- **TheoryConstruct**: A theoryConstruct is a thing that is a construct that supports defining a theory.
- **ScenarioStructure**: A scenarioStructure is a thing that is a structure that supports defining a scenario.

The MCO uses some shortcut language to describe its structure. All of the things described in this section are classes (unless specifically called out as instances). However, it would be tiresome to say that a particular class is a sub-sub-subclass of the class *thing*. As an alternative, we could say a particular class is at level 3 in the overall structure, counting *thing* as level 0. However, this would require that the reader either memorize the structure or constantly refer back to the diagrams. Instead we focus on the lowest level classes, because they are the ones of interest – they are the classes that have instantiations as things in the real world. The levels above these instantiable classes are often referred to as **subcategories** and **categories**, which generally do not have instantiations. For example, an *applePie* class would have an instantiation in a particular apple pie that is for sale; however, its superclass, the subcategory *pie*, would not have an instantiation in some generic pie. The particular apple pie would be an instantiation of *pie*, but only indirectly, not directly. Thus, at the subcategory, category, and higher levels, there are no instantiations of the classes.

MCO Metric Ontologies

Before we describe the Metric Ontologies, some notes about the nature of metrics will be valuable to understanding them.

There is an Action class that consists of basically doing nothing. It has metrics for consuming money and supplies. This reminds us that troops get paid and eat even when they are not fighting.

Some metrics are easy to understand (even if they may be difficult to quantify). The instantaneous support for or against the government is a simple metric. Each human actor (identified by identification (ID) metric) at a given time (identified by time metric) has an opinion about this (identified by the metric in question). The value for the metric may be null (or some similar indicator); however, the presence of the metric ensures the opportunity for the human actor to have an opinion.

Because each actor has a location metric, the value for this metric can be displayed by location and time.

Other metrics may be simple to understand, but difficult to use, for example, consumption of supplies. Consumption is generally not an instantaneous action. Often what is desired from a consumption metric is the sum over time. Given a cumulative record, instantaneous rates of consumption can be approximated, which satisfy the "state vector" concept. The problem comes in accumulating the consumption correctly. If records are kept at regular intervals, say daily or weekly, then all consumption can be captured. However, if records are only made when "something happens," the consumption engendered by the event is recorded, but consumption during the intervals between events is not captured. Events are important and regular reporting is important. The immediate solution is to do both; however, care is needed to prevent double counting.

Figure 3.7 illustrates the two Metric Ontologies in the MCO. The metrics or state variables (at the bottom of the figure) describe the state of being of the elements at any given time. (Generally, each metric will be a vector, with values expressing various significant aspects of the element.) The State Variable (MType) Ontology identifies the differences and commonalities among the metrics by the type of information contained in the metric. The PMESII Ontology identifies the differences and commonalities among the metrics using the PMESII paradigm, which divides the world into activity domains. At the lowest level, we find a particular metric class.

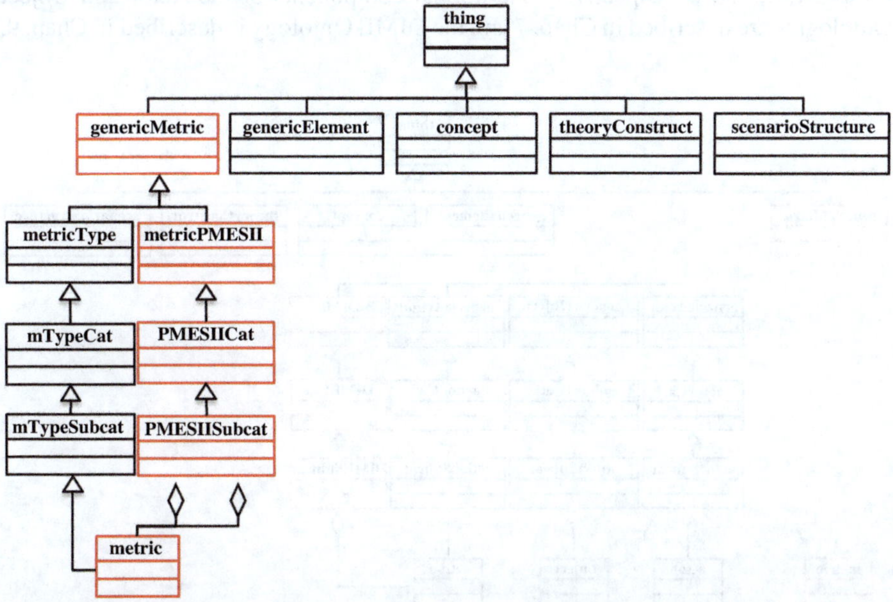

Fig. 3.7 MCO metric ontologies

The figure shows that this class has only one connection to the MTypeSubcat, indicating a single parent and a taxonomic structure. On the other hand, the metric class has two connections of a different type to the PMESIISubcat class. These are aggregation relationships or *partOf* relationships, indicating that the metric class can be *partOf* more than one PMESIISubcat class. We see here that the MCO cannot be a taxonomy and must be a more general ontology as there are classes with multiple connections and there are multiple types of relationships. The MType Metric Ontology is described in full detail in Chap. 8, and the PMESII Metric Ontology is described in Chap. 9.

MCO Element Ontologies

Figure 3.8 illustrates the four element ontologies of the MCO. The Actor Ontology describes the differences and commonalities among the active elements, the Actors, of the operational environment. The Object Ontology describes the differences and commonalities among the objects of the operational environment. The Action Ontology describes the differences and commonalities among the Actions of the operational environment. The DIME Ontology offers an alternative relationship structure for the Actions, based on the DIME paradigm.

Each of the three kinds of element is connected to a set of metrics (indicated by the multiple *hasMetric* relationships), which describe its state value vector at a particular time. The Actor, Action, and Object Ontologies are described in detail in Chaps. 4, 5, and 6, respectively. Additional components of the Actor and Object Ontologies are described in Chap. 7 and the DIME Ontology is described in Chap. 9.

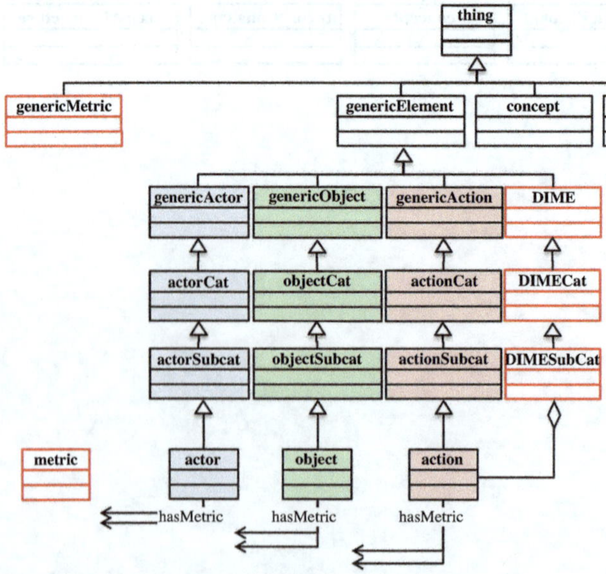

Fig. 3.8 MCO element ontologies

MCO Concept Ontologies

Figure 3.9 illustrates the two concept ontologies of the MCO. The Semantic Concept Ontology connects elements through semantic similarities to a *conceptName*. This is somewhat like a thesaurus. The Stocks-and-Flows Ontology takes its name from the idea of flows filling and emptying a container (the stock). Many of the elements can be cast into these roles. With the addition of several other possible roles, more

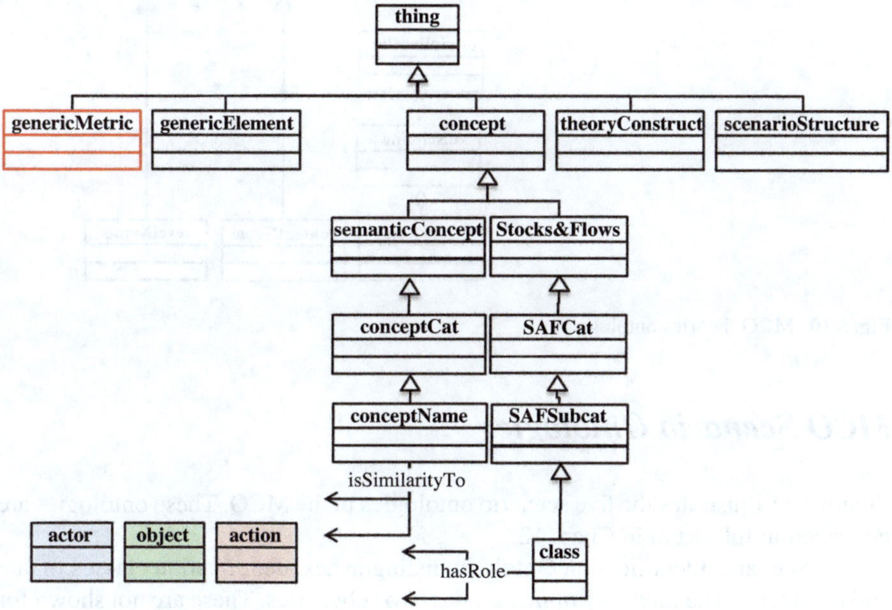

Fig. 3.9 MCO concept ontologies

elements can be involved in a Stocks-and-Flows (SaF) relationship.

The figure shows that each *conceptName* class is connected to multiple elements through *isSimilarTo* relationships. And each SaF class is connected to multiple elements through SAF Role-type relationships. These ontologies are described in full detail in Chap. 10.

MCO Theories Ontology

Figure 3.10 illustrates the Theories Ontology of the MCO. Theories have validity levels within a context. The connections among these three classes are not shown in the figure. The theories may be connected to instantiations of the MCO classes to explain the impact of actions on the operational environment. This ontology is described in full detail in Chap. 11.

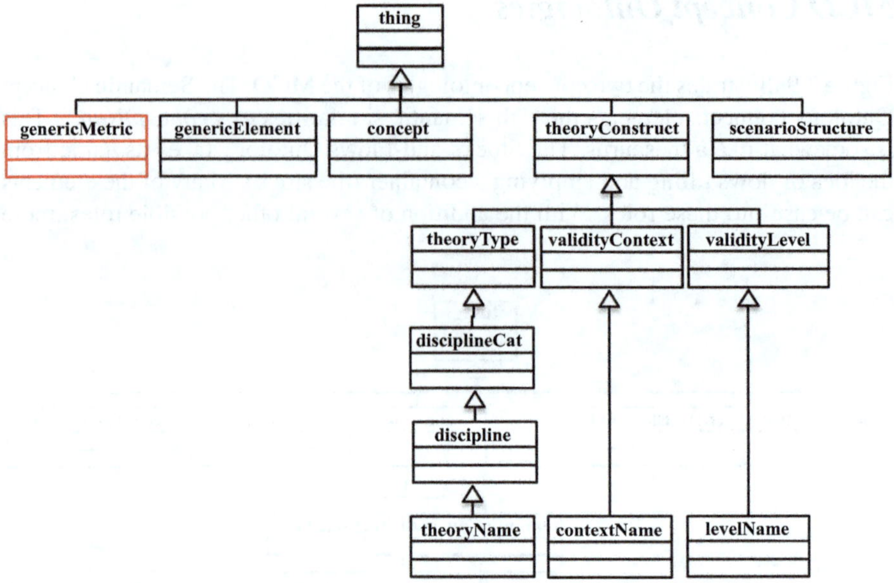

Fig. 3.10 MCO theories ontology

MCO Scenario Ontologies

Figure 3.11 illustrates the five scenario ontologies in the MCO. These ontologies are described in full detail in Chap. 12.

The Scenario Identification Ontology distinguishes identification classes (name and metadata). The *identificationClass* has two subclasses. These are not shown for clarity; however, the actual identification data are instantiations of these two classes.

The Scenario GTO Ontology defines the structures in the Goal-Task-Owner (GTO) sets that describe the agendas of the parties to the conflict. The *GTOClass* is the result of collapsing several classes for clarity. The actual GTO set data are instantiations of these classes.

The Scenario Rules Ontology defines rule classes (rule set and rule). The *ruleClass* is the result of collapsing these two classes for clarity. The actual rule data are instantiations of these two classes.

The Scenario AAR Ontology defines the structures in the Actor-Action-Result (AAR) sets that describe the results of actions. There are no subclasses to the *scenarioAAR* class. The actual AAR set data are instantiations of this class.

The Scenario Relationships Ontology distinguishes classes of relationship types between two elements (for example, between and actor and another actor). The

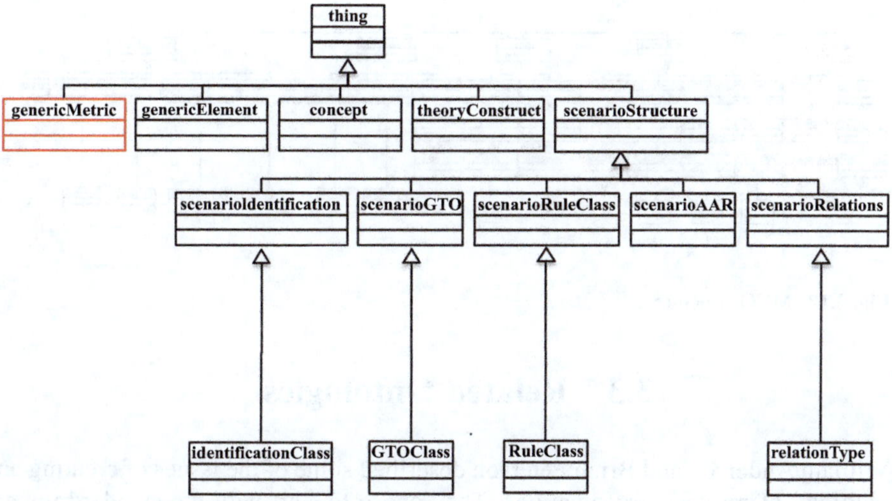

Fig. 3.11 MCO scenario ontologies

scenarioRelationships class is the result of collapsing several classes for clarity. The actual relationships are instantiations of these classes.

MCO Relationships

In this overview we have seen several types of MCO relationships: *is-a*, *partOf*, *hasMetric*, *isSimilarTo*, and SAF Role-type relationships. There are many more that will be described in the subsequent chapters that describe each of the parts of the MCO. Chapter 13 collects these relationships and defines them.

MCO Structure Overview

Figure 3.12 illustrates an overview of the MCO structure, combining the figures illustrating each of the parts. You are not meant to be able to read the details of the figure but rather to appreciate that the parts do fit together to form a whole. This does illustrate the need for computer implementations of real, as opposed to instructional, ontologies. All of the parts of an ontology are required to mesh, regardless of the difficulties in visualizing the meshing.

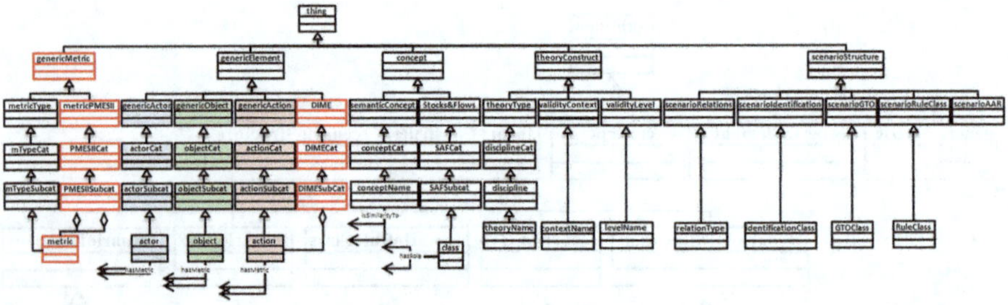

Fig. 3.12 MCO structure

3.3 Related Ontologies

William Anderson and Brian Peterson described some of the issues of creating an ontology of military organizations. They chose to start with the standard upper ontology (SUO) [different from SUMO, which was mentioned in the section on choosing a foundational ontology, above] and the *ModernMilitaryOrganization* and *ModernMilitaryUnit* classes as the core of the ontology. They created *NationalMilitaryOrganization* as a subclass of the first class. This subclass has instantiations, such as the German Wehrmacht in World War II. To accommodate branches of service, e.g., the US Army and the US Navy, *NationalMilitaryOrganizationBranchOfService* is a subclass of *ModernMilitaryOrganization*. They also created classes to describe echelons, such as division, brigade, and battalion. They created relations to describe types of subordination of units to other units, for example, assignment, attachment, and operational control (Anderson and Peterson 2001). The details of this ontology differ from those of the MCO; however, the underlying content is expressible in the MCO.

Oltramari et al. discussed an ontology of cyber security, using the DOLCE upper ontology mentioned in the section on choosing a foundational ontology above. The DOLCE upper ontology is organized differently from the BFO, discussed above, but has similarities, such as distinguishing between continuants and occurrents. Within continuants, it distinguishes among agents, objects, and substances. They used a security core ontology (SECCO) as a middle layer ontology. This ontology provides for stakeholders, assets, and threats, among other things, as roles. Their domain level ontology requires extensive detail because the goal was to describe the operations of cyber security. For example, they needed a task, *RETRIEVE-FILE-SECURELY-TASK*, and needed it "divided into simpler temporally structured and logically connected subtasks (Oltramari et al. 2014)." The MCO has some cyber elements; however, its granularity is much lower than that of this cyber security ontology.

The TRADOC Operational Environment (OE) Ontology (US Army TRADOC 2017), version 3.2, 1/31/2017, "aims to capture specific domains related to the

Table 3.1 Comparing ontology counts

	OE Ontology	MCO
Axiom	17,880	387,576
Class count	2444	7543
Object property count	264	233
Data property count	30	2
Annotation property count	67	41
Individual count	263	43,434

Operational Environment." It appears to be aimed at a tactical level (given the inclusion of such things as eye color, scars, and tattoos).

Table 3.1 compares illustrative item counts between the OE Ontology and the MCO, version 4.26. The items are couched in the Protégé terminology. Axioms are assertions, such as asserting that one class is a subclass of another. Classes are the classes described in this book. Object properties are relationships between classes or instances. Data properties are properties about a data type that an object may have, such as *hasArmyTaskName* in the OE Ontology or *hasValidityValue* in the MCO. Annotation properties are types of comments. Individuals are instances of classes. In the OE Ontology, '*Acre Measurement Unit*' and '*Square Meter Measurement Unit*' are instances of the class '*Measurement Unit of Area.*' In the MCO, *aCrimeNetwork* and *DominicFamily* are instances of the class *networkRelations*.

As Table 3.1 shows, the MCO is a much larger ontology than the OE Ontology. The MCO has three times the number of classes, 21 times the number of assertions (axioms), and 165 times the number of individuals. However, this last comparison is somewhat misleading, as most of the MCO individuals are included to demonstrate the concepts within the ontology, rather than being required to implement the ontology.

The discussions of the detailed comparisons, presented below, include references to the structure of the MCO. These references are not fully explained, as they are the subject of the previous section, which describes the MCO structure, and the following chapters, which discuss the details. The reader should glean from these references some general concepts about the MCO, but not worry too much about the precise definitions.

Figure 3.13 shows the upper-level class structures of the two ontologies. The OE Ontology is based on the BFO foundational level ontology, as shown by its *continuant* and *occurrent* division. The Target class is an addition to the BFO structure. The division between *entity* and *Obsolete Class* is also an addition to allow the removal of classes that will no longer be used from the general structure, while preserving their existence for potential reevaluation. The MCO, while using some of the philosophy of the BFO, proceeds directly to the structure of its contents. Some of its contents can be seen in the lowest level of the OE Ontology that is shown in this figure.

Figure 3.14 shows a part of the lower-level class structure for each ontology. At these levels, there is more commonality evident in the individual classes; however,

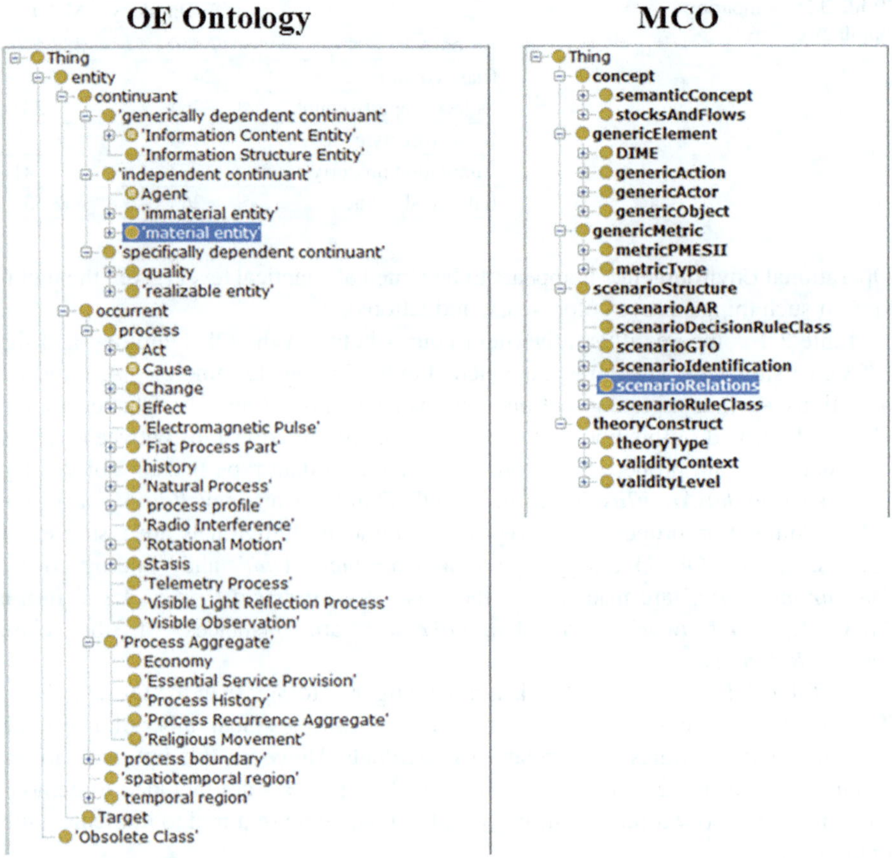

Fig. 3.13 Comparing upper-level class structures

the structural differences imply differences in the semantic definitions of the seemingly related classes or the philosophical differences in representation.

The semantic differences are conjectural. The types of Actor in the MCO include individuals, significant groups, and demographic groups. The OE Ontology includes these divisions; however, they appear at different levels in the ontology. The *Person* agent in the OE Ontology only has four individual roles and a role as a member of an organization. The MCO, on the other hand, defines numerous roles for an *individualActor* (the corresponding class to *Person*) as a *keyLeader* (shown in the figure) and numerous roles as an *otherIndividual* (not expanded in the figure). Such examples may indicate differing semantics.

However, the differences in philosophy are evident. The OE Ontology role as an *Organization Member* includes only predefined religions for the organizations (expansion not shown in the figure). The religious organizations of the MCO are left to the instances (data), rather than being defined in the ontology. Similarly, the *Armed Force* class in the OE Ontology has limited subclasses and the "Military Personnel Force," when expanded (not shown in the figure), includes only army

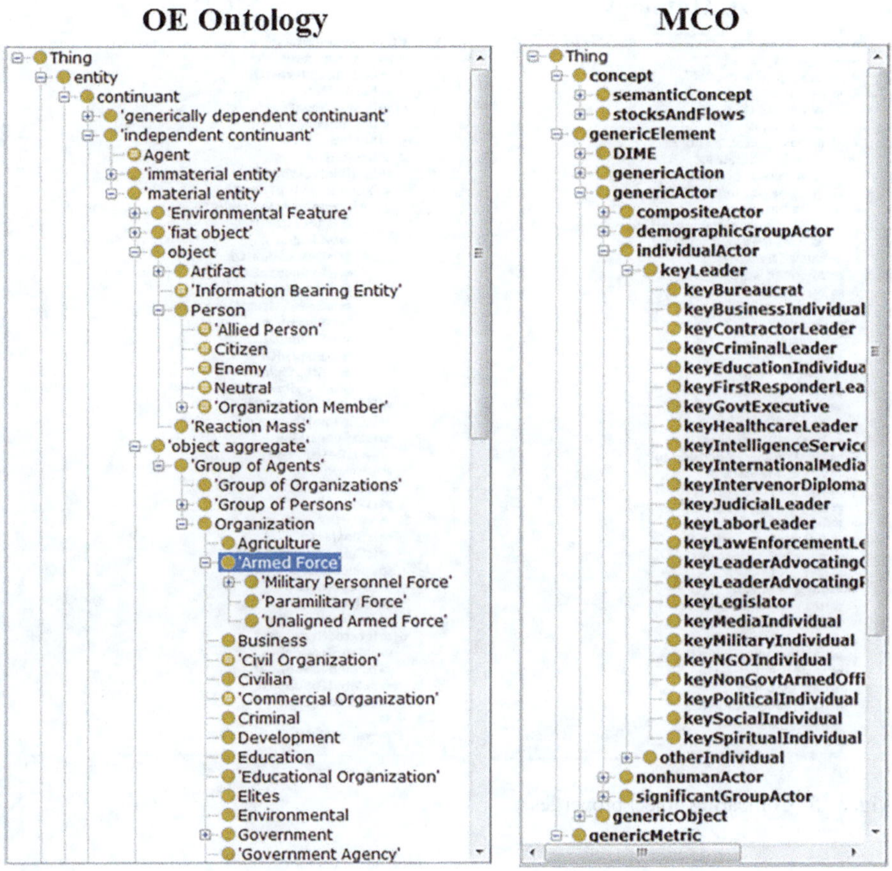

Fig. 3.14 Comparing lower-level structures

units (except for *Carrier Air Wing*). Further, these army units are predefined (e.g., *Army Battalion*, *Army Brigade*, *Army Fire Platoon*, and *Army Squad*). The MCO includes a *compositeActor* class, which has a *hierarchicalActor* class (not shown in the figure) that is used to construct various hierarchical organizations (including all types of military units) as instances. Many of these instances are included in the MCO (Chap. 7) to illustrate the process of creating hierarchical units.

Figure 3.15 shows a partial list of object properties (the relationships between classes) for the two ontologies. The intent is clearly the same; relationships between classes that express concepts found in the real-world model for each case are defined as object properties. However, there are differences at the detail level. For example, when the *has familial relationship to* object property in the OE Ontology is expanded (not shown in the figure), the relationships include both *has maternal uncle* and *has paternal uncle*. The similar object property, *isConnectedByFamilyTo* (expanded in the figure), does not even include an "uncle" type relationship. The count for object properties in Table 3.1 reflects this finer grained set of relationships with a larger

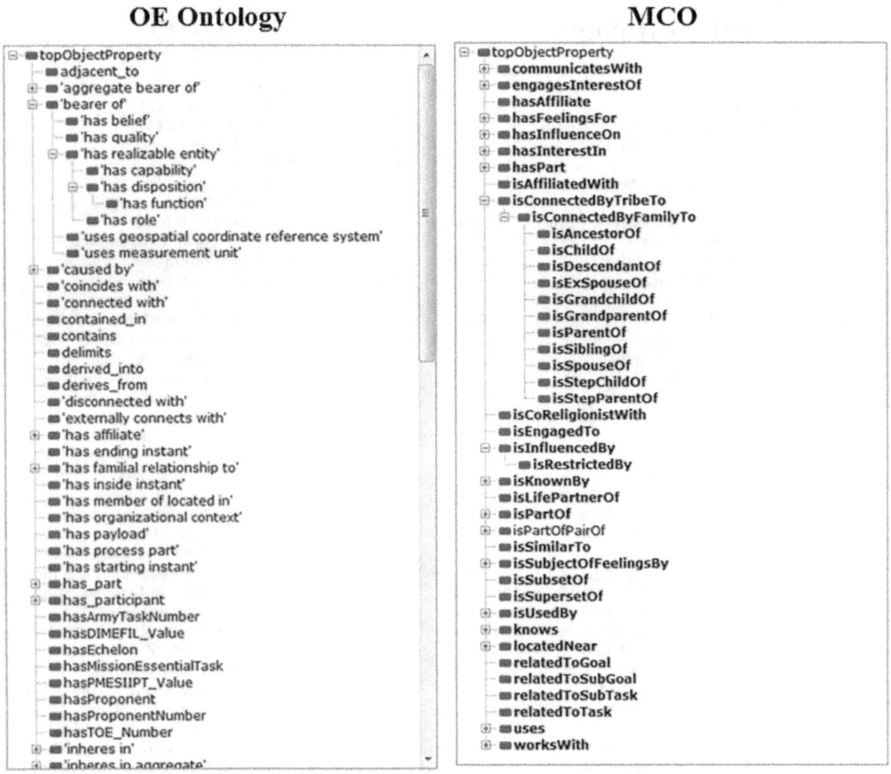

Fig. 3.15 Comparing object properties

object properties count for the OE Ontology. On the other hand, the MCO includes structural relations that are used for defining agendas (e.g., *relatedToGoal* and *relatedToTask*), which are either not present in the OE Ontology or are defined in another manner.

Figure 3.16 shows the complete list of data properties for the two ontologies. The main point is that some particular classes in each ontology require specific data properties. Examples are *hasArmyTaskName* in the OE Ontology and *hasGTOLevel* in the MCO. A secondary point is that the MCO is deficient in addressing the particular data needs of its metrics. In the discussions concerning metrics in subsequent sections and chapters, the types of data required are described, but the details, such as whether the data should be integers, decimals, latitude/longitude, and so forth, are omitted. The OE Ontology includes these data needs. The data units are included in the OE Ontology individuals list.

Figure 3.17 shows a partial list of the annotation properties (general comments) for the two ontologies. These comments support human understanding of the elements of the ontologies. Some are attached to the ontology as a whole and others

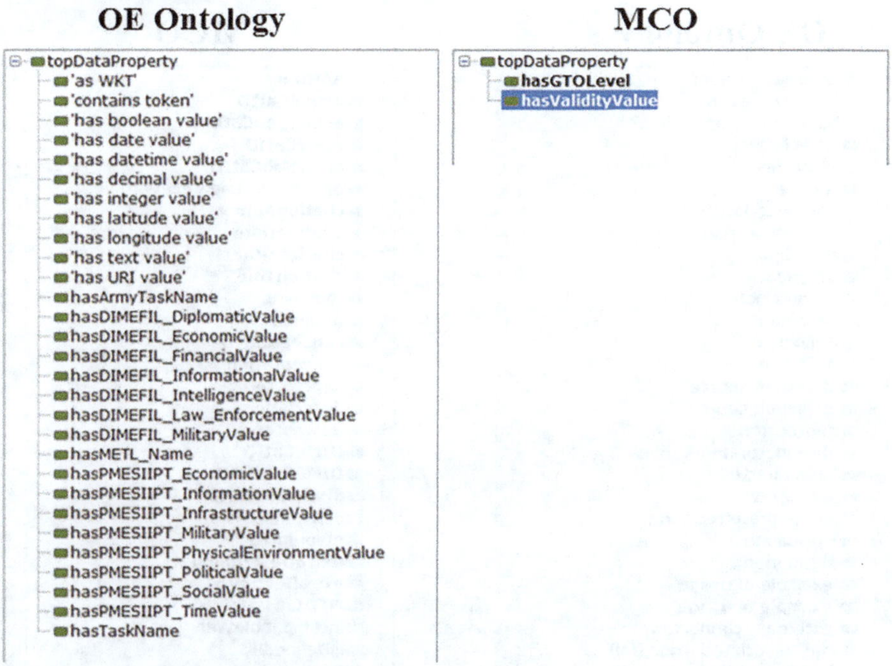

Fig. 3.16 Comparing data properties

are attached to the classes, individuals, data properties, or object properties. The choice of what to annotate and how is a stylistic decision. For example, a large number of MCO annotation properties end in "ID." These are created to support sorting the various entries within Protégé into a sequence other than alphabetical order.

Figure 3.18 shows a partial list of the individuals in each ontology. In the OE Ontology, they are measurement units, such as national currency units, physical measurement units, and time units. In the MCO, they are instances of classes that are used to illustrate the various MCO structures, such as military and civilian facilities and military and civilian hierarchical organizations.

Figure 3.19 shows a sample "model" (a Protégé term for a picture of the classes and connections) in the OE Ontology. This model shows the two levels above the "Armed Force" class and the two levels below that class. The black triangles in some of the ovals indicate that there are additional levels below (and in one case above) the class named in the oval that are not shown.

Figure 3.20 shows a sample model from the MCO. This model shows the two levels above the *scenarioRelations* structure and the five *scenarioRelations* subclasses. For example, the *networkRelations* structure can be used to create instances of networks containing multiple types of relationships between Actors.

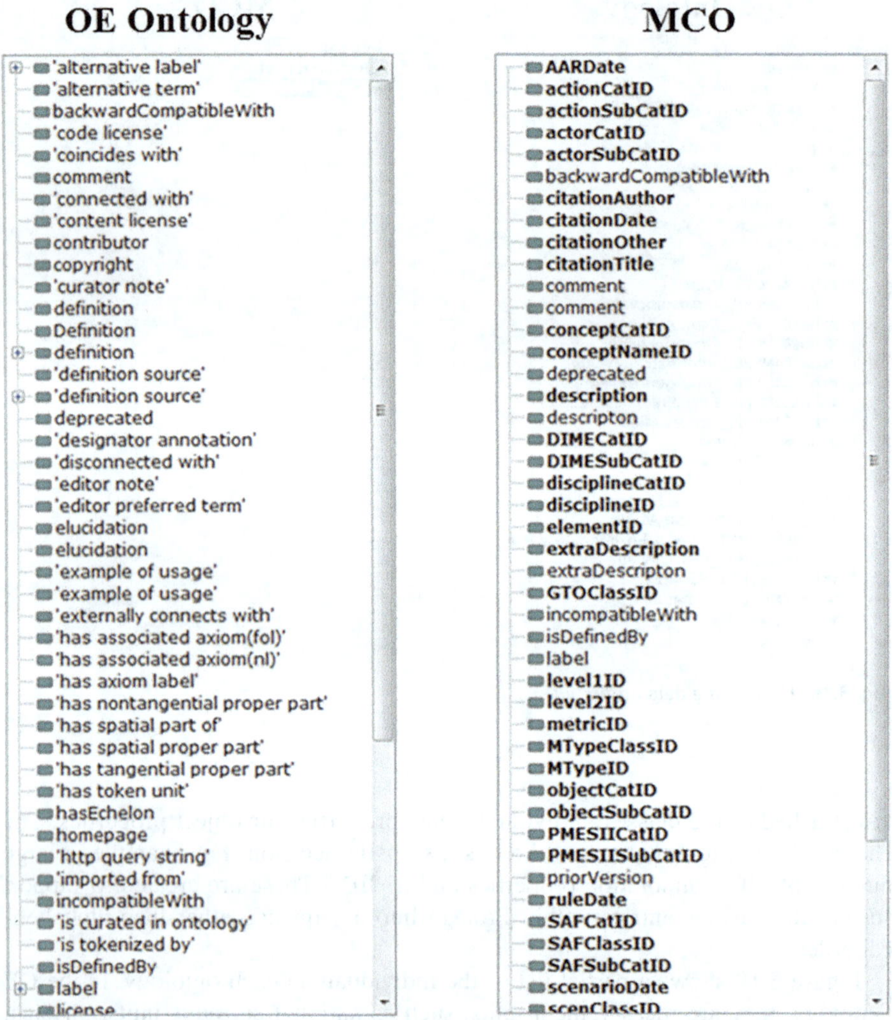

Fig. 3.17 OE Ontology annotation properties

Fig. 3.18 OE Ontology individuals

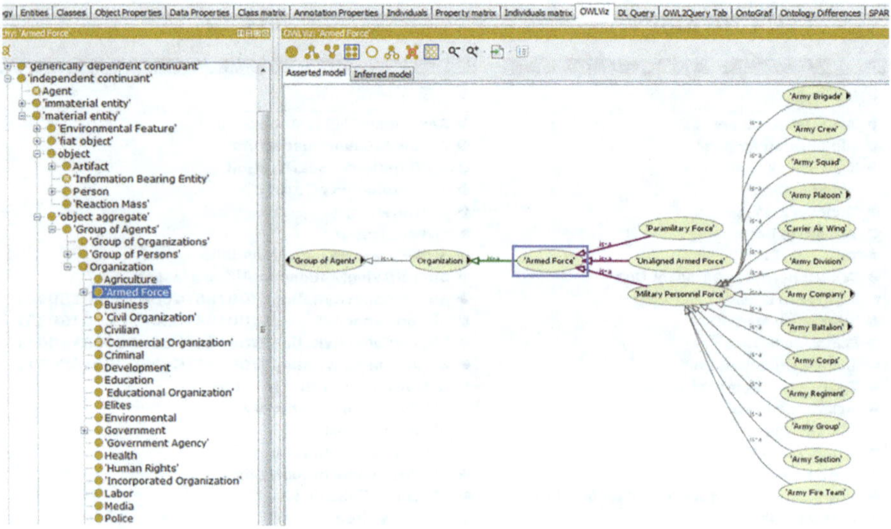

Fig. 3.19 OE Ontology sample model

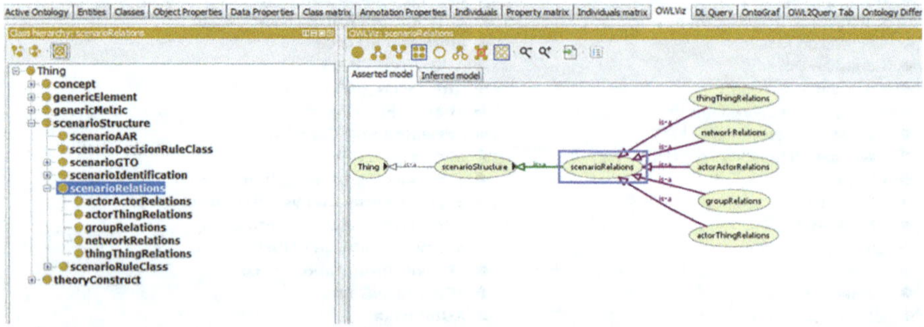

Fig. 3.20 MCO sample model

3.4 Changes from the Unconventional Conflict Ontology

The MCO is very similar to the UCO; however, there have been several major and many minor changes.

Table 3.2 displays a comparison of the basic counts between the MCO and its predecessor ontologies. The elements are divided into Actors, Actions, and Objects (using the newer terminology). The metrics are listed separately. In the Irregular War Ontology (IWO) (Hartley D. S., Unconventional Conflict: A Modeling Perspective, 2017) and the Unconventional Conflict Ontology (UCO) (Hartley D. S., An Ontology for Unconventional Conflict, 2018), the sums of the three types of elements are larger than the displayed totals because several elements are repeated

Table 3.2 Comparison of the MCO with predecessors

		IWO	UCO	MCO
Elements	Actors	102	101	114
	Actions	389	392	474
	Objects	209	179	191
	Totals	682	653	779
Metrics		754	791	5930

in two or three of the types. These elements have been separated in the MCO, so the sum equals the displayed total.

The counts for the IWO and the UCO are similar, reflecting only minor changes: some elements were combined in the UCO, and additional metrics were created. The counts for the MCO are dramatically different; the changes are explained below.

Types of Changes from the UCO

A number of changes resulted from the BFO influence:

- Assign globally unique names to all classes, not just unique within their structure (e.g., Actor versus Object).
- Standardize the names of elements: start with small letter, noun phrase using the singular construction.
- Change the terminology from passive environment or passive element to object.
- Standardize the definition of elements: state the parent class name, and then say how the thing is specialized from the parent class and what differentiates it from others with the same parent.
- Use a basic *is-a* taxonomy structure where possible, meaning that no multiple parents are allowed.
- Replace the *is-a* relationships with other relationships where connections to multiple classes are required.
- Split multi-type elements into the part that is an Actor, the part that is an Object, and the part that is an Action to avoid dual parents.
- Rethink the PMESII connections for metrics to avoid dual parents.
- Decide not to use processes.

A number of changes resulted because of adding combat to the ontology:

- Add combat elements.
- Add composite elements to support the needs of combat concepts and improve the expression of non-combat concepts.
- Add relationships for military organizations, such *hasTenant* and *hasMilSystemsGroup*.

- Add instances to help explain the concepts, leading to refinements of the structures.

Structural changes were made to regularize the overall configuration and to add new representations.

- Add the DIME Ontology.
- Refine the concept ontologies.
- Create scenario ontology.
- Integrate the Theories Ontology.

Rethinking the state variables (metrics) led to several major changes:

- Realize the implications of the fact that the content of a metric is at the instance level of the metric and at the instance level of the element to which it is attached.
- Get rid of most common metrics. Only situation, identity, time, and location are common. For example, each element has an *identity* metric. The value of the identity applies to the instance of the element and is found in an instance of the *identity* metric. All metrics could be implemented as common metrics except for the desire to connect to the PMESII paradigm. The particular PMESIISubcat class to which a particular metric should be attached depends, in part, on the element to which the metric provides state information. The four common metrics are not relevant to the PMESII paradigm; however, the rest must be unique for each element. This, however, does create an explosion in the number of metrics.
- Reform the metric structure. Make a metric *partOf* the PMESII class structure, not connected by *is-a*. This allows multiple connections to the PMESII class structure for each (non-common) metric and is a better representation for the relationship than defined by the *is-a* relationship.
- Refine the metric structure of the UCO so that the set of metrics for a given entity form a metric vector containing one or more (or none) from each of the following (lettered entries):

1. Metric vector key consists of
   (a) 100 situation
   (b) 110 identity
   (c) 120 time
2. Physical parts of vector
   (a) 210 location
   (b) 230 quantity
   (c) 231 members
   (d) 240 disasterIndicators
   (e) 250 movable
   (f) 270 weaponry
   (g) 280 damage

3. Flow parts of vector
   (a) 290 capacityFlowrate
   (b) 291 munitionConsumption
   (c) 292 supplyConsumption
   (d) 293 powerConsumption
   (e) 294 moneyConsumption
4. Relationship parts of vector
   (a) 310 affiliation
   (b) 320 hierarchy
   (c) 330 owner
   (d) 331 initiator
   (e) 332 object
   (f) 340 activity
   (g) 350 availability
5. Human, Social, Cultural Behavior (HSCB) parts of vector
   (a) 410 decisionMaking
   (b) 420 influence
   (c) 430 fairnessCorruption
   (d) 440 effectiveness
   (e) 450 efficiency
   (f) 460 operatingHealth
   (g) 461 levelRating
   (h) 462 progress
   (i) 470 professionalism
   (j) 480 transparency

While the UCO certainly had implicit relationships between ontology elements and structures, they were not explicitly described. The MCO explicitly names and defines its relationships and describes a structure for them.

Why There Are No Process Elements

A "process" is a thing that has a temporal starting point and a different temporal ending point. Whether a process should be included in an ontology depends in part on the granularity of the ontology. Meiosis and mitosis are cell division processes. Each takes a certain amount of time, requiring different temporal starting and ending points. Meiosis in human males takes about 74 hours and mitosis takes about 2 hours. If the time granularity of a particular ontology is measured in months, not hours, then it would generally not make sense to include these processes in that ontology.

Further, a "process" refers to a standard thing. That is, there is a standard starting state and a standard ending state. There are also generally standard sub-processes that occur in sequence. Variants are allowed, but must be specifiable. The random actions of a group of people at a beach, starting in the morning and ending in the evening, might be of interest for some ontology; however, they would not be identified as a process.

There are processes that occur in modern conflict: loading ammunition into a particular weapon system is a process and performing a pre-flight check of an aircraft is a process. However, the time granularity of the MCO militates against declaring these to be processes in the ontology. If they were significant enough to be included (a different granularity issue), they would be included as Actions, with essentially zero temporal duration.

There are real-world processes that might be candidates for "processes" in the MCO; however, they have a different problem. For example, a battle has a starting point and a different ending point and even has somewhat standardized starting and ending states. The starting state might be described as two separated forces, each with its own force strength. The ending state might be described as one of several discrete possible states: one side is victorious and the other side is defeated; both sides are rendered unfit for further combat, with neither truly victorious; the battle is inconclusive and the two sides separate; etc. For an ontology with very large granularity, defining this as a process might be useful. For the Modern Conflict Ontology, such a definition is not useful. The problem lies in the sub-processes: they are indeterminate. Here it is more useful to use Actions and admit that Theories are required to determine or explain the results, which are then followed by additional Actions.

Plans are documents of actors, actions, places, and timing. After-the-fact process descriptions are just records of what happened, where, and when. Documents are just objects in the Object Ontology. Processes in the sense of industrial processes don't really exist in conflict, because the exceptions overwhelm the "standard" processes.

3.5 Recapitulation

The point of an ontology is to create a structure to contain the knowledge of a domain and then to fill it with that knowledge. In the case of modern conflict as the domain of knowledge, things such as human leaders, air bases, etc. and relationships such as agendas, networks, etc. need to be represented. This domain covers conflicts anywhere in the world from theater-level to global in area. Because modern conflict includes both conventional and unconventional conflict, the ontology can represent any historical or projected future conflict. The major caveat is granularity: most individual actions, motives, processes, etc. are of too fine a granularity to be represented. This chapter has extended the introduction to ontologies of the Chap. 1 by discussing the construction of the MCO to represent the domain of modern conflict. Subsequent chapters will discuss the parts of the MCO in detail.

Chapter 4
MCO Actor Ontology

Actors are the motivating elements in the Modern Conflict Ontology (MCO): they cause things to happen. Generally, Actors are people, but not always. Figure 3.8 showed how the Actor Ontology fits into the overall MCO structure (with Fig. 4.1 illustrating its icon as used in the context diagram, Fig. 3.5).

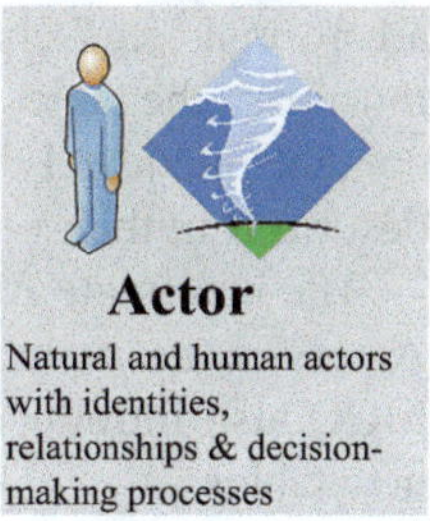

Fig. 4.1 Actor icon

4.1 Overview of Actors

There are five categories of Actors: individual Actors, significant group Actors, demographic group Actors, nonhuman Actors, and composite Actors. Figure 4.2 shows the top level of the Actor taxonomy. The picture of the Romeo and Juliet actors is a key to the general concept of Actors: Actors are things that are dynamic, perform deeds, and can also be acted upon.

D. S. Hartley III, *An Ontology of Modern Conflict*, Understanding Complex Systems, https://doi.org/10.1007/978-3-030-53214-7_4

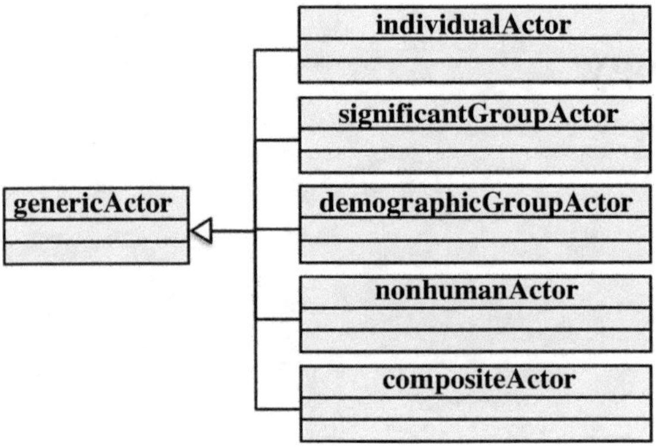

Fig. 4.2 Actor taxonomy overview

Changes to Actors from the UCO

In Chap. 3, we discussed the general changes from the Unconventional Conflict Ontology (UCO) to the Modern Conflict Ontology (MCO). Most of these changes are formal, e.g., changing spelling conventions and definition standards. For the most part, the tables of the Actors contained in the book about the UCO describe the content adequately (Hartley 2018). There are a few exceptions, which will be described in their proper place below.

The inanimate Actors category of the UCO has been replaced by the nonhuman Actors category in the MCO, and the appropriate tables are included here. The MCO also introduces an entirely new category, composite Actors, which requires new tables.

An Insight into Actors

The category of individual Actors is the most easily described category because we are each an individual actor. As explained below, there are 24 different classes of key leader Actors, differentiating among groups that may need a leader. This may seem like too many, because a class can be tailored, so why prescribe these 24 roles? The answer reflects on the purpose of an ontology, which is to describe what is known about a domain. Part of the method of doing that is to provide reminders of potentially useful distinctions. In this case, the existences of a *keyLeaderAdvocatingConflictAndDissension* and another *keyLeader-AdvocatingPeaceAndStability* are important in assessing the conflict.

The nursery rhyme "Rub-A-Dub-Dub" mentions the butcher, the baker, and the candlestick maker as examples of respectable persons. These people were identified by their roles in an eighteenth century town in precisely the same manner as our key leaders. Because these roles (and others) are not included, 24 Actor classes may seem to be too few. However, this is addressed by the way an ontology works. Each class is unique; but the instantiations of a class need not be singular. There is only one *keyBusinessIndividual* Actor class; but there could easily be three instantiations in a town, consisting of Henry, the butcher, Frank, the baker, and Elizabeth, the candlestick maker.

4.2 Individual Actors

An *individualActor* is an Actor who is a particular living being. This category of Actor is comprised of two subcategories: key leaders and other individuals. Figure 4.3 shows the individual Actor part of the Actor taxonomy. The "actor" class

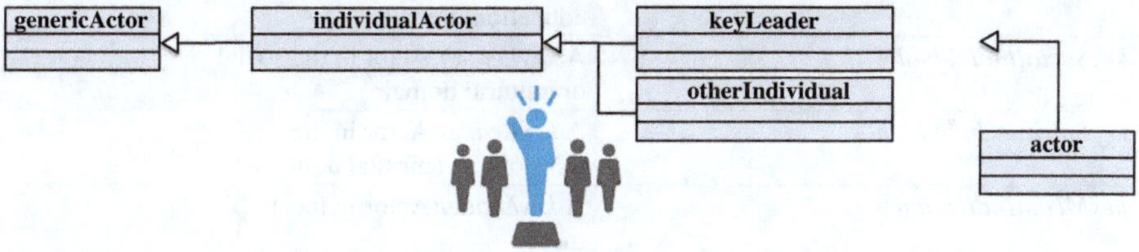

Fig. 4.3 Individual Actor taxonomy

box represents all of the Actor classes that are children of the subcategories. The "leader" in the figure stands for one of the types of individual Actor.

Key Leader Actors

A *keyLeader* is an *individualActor* who is an important, influential personage. There are 24 different *keyLeader* Actor classes, representing different roles in which such a person might need to be represented. For example, there is a *keyInternationalMediaLeader* class, and there is a *keyMediaIndividual* class, representing an important international media person and an important local media person, respectively (Table 4.1).

Table 4.1 Key leader Actors

Actor class	Definition	Examples
keyLeaderAdvocatingPeaceAndStability	A *keyLeader* Actor in the pro-peace political domain	
keyLeaderAdvocatingConflictAndDissension	A *keyLeader* Actor in the anti-peace political domain	
keyPoliticalIndividual	A *keyLeader* Actor in the political domain	
keyIntervenorDiplomaticPerson	A *keyLeader* Actor in the diplomatic domain	
keyMilitaryIndividual	A *keyLeader* Actor in an armed force	
keyNonGovtArmedOfficial	A *keyLeader* Actor in a nongovernment armed group	
keyCriminalLeader	A *keyLeader* Actor in crime	
keyLaborLeader	A *keyLeader* Actor in labor	
keyBusinessIndividual	A *keyLeader* Actor in a business	
keyNGOIndividual	A *keyLeader* Actor in a NGO, IO, or IGO organization	
keyEducationIndividual	A *keyLeader* Actor in the education domain	
keySocialIndividual	A *keyLeader* Actor in the social or cultural domain	
keySpiritualIndividual	A *keyLeader* Actor in the religious or spiritual domain	
keyMediaIndividual	A *keyLeader* Actor in local media	
keyContractorLeader	A *keyLeader* Actor of a contractor business, internal or external	
keyInternationalMediaLeader	A *keyLeader* Actor in international media	
keyLawEnforcementLeader	A *keyLeader* Actor in law enforcement	
keyBureaucrat	A *keyLeader* Actor in a government or quasi-government bureaucratic organization	
keyJudicialLeader	A *keyLeader* Actor in the judicial domain	
keyLegislator	A *keyLeader* Actor in a government or quasi-government legislative branch	
keyGovtExecutive	A *keyLeader* Actor in a government or quasi-government executive branch	
keyHealthcareLeader	A *keyLeader* Actor in the healthcare domain	

(continued)

Table 4.1 (continued)

Actor class	Definition	Examples
keyFirstResponderLeader	A *keyLeader* Actor in a first responder organization	Fire, medical, local police, etc.
keyIntelligenceServiceLeader	A *keyLeader* Actor in an intelligence service	

Other Individual Actors

An *otherIndividual* is an *individualActor* who has only temporary importance. There are 15 different *otherIndividual* Actor classes, which is less than the number of *keyLeader* Actor classes, as some of the different roles can be conveniently collapsed. For example, there is a single *localOrInternationalMediaPerson* class, which allows for both local and international media personnel (Table 4.2).

Table 4.2 Other individual Actors

Actor class	Definition	Examples
worker	An *otherIndividual* Actor who is a worker	
lawEnforcementPerson	An *otherIndividual* Actor who is in law enforcement	
intelligenceServicePerson	An *otherIndividual* Actor who works in an intelligence service	
govtPerson	An *otherIndividual* Actor who works in the government or a quasi-government	
contractorPerson	An *otherIndividual* Actor who is a contractor	
intervenorSupportPerson	An *otherIndividual* Actor who is an advisor to the government or proto-government or to intervenor organizations	Economists, computer experts, agronomists, etc.
localOrInternationalMediaPerson	An *otherIndividual* Actor who is a member of the local or international media	
NGOWorker	An *otherIndividual* Actor who is an NGO, IO, or IGO worker	
educator	An *otherIndividual* Actor who is in the education field	
healthcarePerson	An *otherIndividual* Actor who works in the healthcare field	
firstResponderPerson	An *otherIndividual* Actor who is a first responder	Fire, medical, local police, etc.

(continued)

Table 4.2 (continued)

Actor class	Definition	Examples
nonGovtArmedIndividual	An *otherIndividual* Actor who is a terrorist, insurgent, private security force-, or regime-sponsored non-military armed person	
interventionForcePerson	An *otherIndividual* Actor who is in the intervention force-military affiliation	
govtMilitaryPerson	An *otherIndividual* Actor who is a government military person	
intervenorDiplomaticPerson	An *otherIndividual* Actor who is an intervenor diplomatic person	

4.3 Significant Group Actors

A *significantGroupActor* is an Actor that is an organization composed of people and small enough that it does not require a density distribution. This category of Actor is comprised of five subcategories: social organization Actors, economic organization Actors, armed force Actors, unarmed political organization Actors, and armed political organization Actors.

Significant groups are chosen with consideration of the situation. For example, there will be many families in any country; however, one or more of them may exercise great control over the events in the country, hence the inclusion of *family* as a *significantGroupActor* subclass of the *socialOrganization* class. Figure 4.4 shows the significant group Actor part of the Actor taxonomy. The "actor" class box represents all of the Actor classes that are children of the subcategories. The committee in the figure stands for one of the possible significant group Actors.

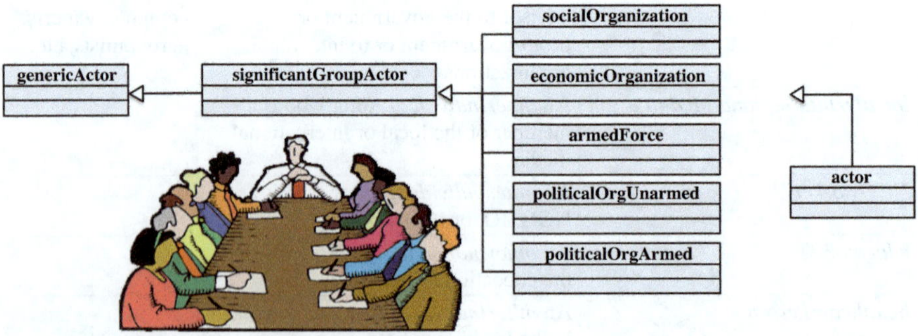

Fig. 4.4 Significant group Actor taxonomy

Social Organization Actors

A *socialOrganization* is a *significantGroupActor* that is organized for a social purpose. There are six *socialOrganization* Actor classes, representing six different types of social organizations. The social organization types include two that can include official organizations, health and education; however, the instantiations might consist of private organizations. There are also two "factions," religious and social. These are distinct from the religious and cultural "populations" in the *demographicGroupActor* category. The factions may be part of the populations but are organized more cohesively into smaller subsets and are characterized by more intense and directed activity, often holding strong fanatical positions. The *family* Actor class mentioned above is in the *socialOrganization* subcategory, as well as is the *NGOOrganization* Actor class. This latter class may be instantiated as a non-governmental organization (NGO), an international organization (IO), or an intergovernmental organization (IGO). [Including all three acronyms in the name proved to be too clumsy] (Table 4.3).

Table 4.3 Social organization Actors

Actor class	Definition	Examples
NGOOrganization	A *socialOrganization* Actor that is an NGO, IO, or IGO	NGO, non-governmental organization; IO, international organization; or IGO, intergovernmental organization
socialFaction	A *socialOrganization* Actor in the social issues domain holding strong to fanatical positions	
religiousFaction	A *socialOrganization* Actor in the religious or spiritual domain holding strong to fanatical positions	Religious faction, sect, or organization
family	A *socialOrganization* Actor that consists of an important family	
healthcareOrganization	A *socialOrganization* Actor in the healthcare field	
educationOrganization	A *socialOrganization* Actor in the education field	Public or private school, tech school, religious school, college, university, correspondence school

Economic Organization Actors

An *economicOrganization* is a *significantGroupActor* that is organized for an economic purpose. There are 15 *economicOrganization* Actor classes, representing the 15 different types of economic organizations in the ontology. As with all of the Actor classes, the instantiations depend on the situation. For example, the *energyBusiness*

Actor class might be instantiated by Exxon or it might be instantiated by a small, local seller of propane gas – or both. Interestingly, the *criminalOrganization* Actor class is included here because of the economic impact of such organizations and because of the largely economic orientation of most criminal organizations (Table 4.4).

Table 4.4 Economic organization Actors

Actor class	Definition	Examples
agricultureBusiness	An *economicOrganization* Actor in the agricultural field	
criminalOrganization	An *economicOrganization* Actor in the crime business	
energyBusiness	An *economicOrganization* Actor in the energy business	
financialServicesIndustryBusiness	An *economicOrganization* Actor in the financial business	
workerOrganization	An *economicOrganization* Actor that is a workers organization	Labor union, labor pool, etc.
mediaBusiness	An *economicOrganization* Actor in the media or information business	
manufacturingBusiness	An *economicOrganization* Actor in the manufacturing business	
serviceBusiness	An *economicOrganization* Actor in a service, retail, or wholesale business	Shops in a town or chain such as McDonald's, lawyer, architect, etc.
transportationBusiness	An *economicOrganization* Actor in the transportation or warehousing business	
tourismIndustryBusiness	An *economicOrganization* Actor in the tourism, leisure, or hospitality business	
miningBusiness	An *economicOrganization* Actor in the mining or associated business	
fishingBusiness	An *economicOrganization* Actor in the fishing or other aquaculture business	
timberBusiness	An *economicOrganization* Actor in the timber or associated business	
culturalBusiness	An *economicOrganization* Actor in the cultural business	Theater, museum, or sports
contractorBusiness	An *economicOrganization* Actor in the contractor business	Internal to the Host Nation or externally based

Armed Force Actors

An *armedForce* is a *significantGroupActor* that is organized and armed. There are seven *armedForce* Actor classes, representing the different types of armed forces in the ontology. The Host Nation government military forces and the intervention military forces are represented here. A terrorist group that is important enough to be noticed is obviously armed and organized and hence an *armedForce*. The distinctions among some of the classes, such as *paramilitaryForceOrganization*, *private SecurityForceOrganization*, and *regimeSponsoredNonMilitaryArmedForceOrganization*, depend on the situation. In some cases, the distinctions will be important, but not in others (Table 4.5).

Table 4.5 Armed force Actors

Actor class	Definition	Examples
interventionForceOrganization	An *armedForce* Actor that is a military intervention force, does not include *milSystems*	
govtMilitaryForceOrganization	An *armedForce* Actor that is a Host Nation military force, does not include *milSystems*	
regimeSponsoredNonMilitaryArmedForceOrganization	An *armedForce* Actor that is a regime-sponsored, non-military armed force, does not include *milSystems*	
paramilitaryForceOrganization	An *armedForce* Actor that is privately supported, does not include *milSystems*	
privateSecurityForceOrganization	An *armedForce* Actor that is a private company providing armed combat or security forces, does not include *milSystems*	
insurgentOrganization	An *armedForce* Actor that forcibly opposes the legitimate authority, does not include *milSystems*	
terroristOrganization	An *armedForce* Actor that uses violence and intimidation, especially against civilians, for political purposes, does not include *milSystems*	

Unarmed Political Organization Actors

A *politicalOrgUnarmed* is a *significantGroupActor* that is organized to serve government or political needs and is not armed. There are seven classes in this subcategory, representing the government branches and organizations with missions that do not require large percentages of the personnel to be armed. Thus, the fact that armed guards are part of the organization is not a disqualifying factor (Table 4.6).

Table 4.6 Unarmed political organization Actors

Actor class	Definition	Examples
govtDecisionAuthority	A *politicalOrgUnarmed* that is the Actor having power –depends on government type	Autocratic, democratic, theocratic, monarchy, warlords, etc.
firstResponderOrganization	A *politicalOrgUnarmed* Actor that responds to emergencies	Firefighter, emergency medical, or police responder organization, etc.
socialServicesOrganization	A *politicalOrgUnarmed* Actor that is an organization that provides services to improve the well-being of individuals, families, and communities	Child services, elderly care, etc.
judicialBranch	A *politicalOrgUnarmed* Actor that is a judicial organization	Both government and alternative (shadow)
executiveBranch	A *politicalOrgUnarmed* Actor that is an executive branch of government	At any level of government or alternative (shadow)
legislativeBranch	A *politicalOrgUnarmed* Actor that is a legislative branch of government	At any level of government or part – senate vs lower house – or alternative (shadow)
govtBureaucracyOrganization	A *politicalOrgUnarmed* Actor that is a government bureaucracy	Any department

Armed Political Organization Actors

A *politicalOrgArmed* is a *significantGroupActor* that is organized to serve government or political needs and is armed. There are five classes in this subcategory. Two of them represent the government organizations that may require large percentages of armed personnel, namely, law enforcement and intelligence services. One is the *interventionOrganization* Actor class. This class is distinct from the *interventionForceOrganization* class that is an *armedForce*. The *interventionOrganization* is the diplomatic arm of the intervention and will generally be the superior organization. It is not regarded as unarmed because its instantiation can call on the *interventionForceOrganization* instantiation for support at any time. The other two classes are basically political factions, one internal and the other external to the Host Nation. Neither is a *socialOrganization* because their aims are political and they are very likely to be or become armed (Table 4.7).

Table 4.7 Armed political organization Actors

Actor class	Definition	Examples
externalForceOrganizationAdvocatingConflict	A *politicalOrgArmed* Actor that comes from outside the Host Nation and is a force advocating conflict, does not include *milSystems*	Countries, groups, etc.
politicalFaction	A *politicalOrgArmed* Actor in the political domain holding strong to fanatical positions, does not include *milSystems*	
lawEnforcementOrganization	A *politicalOrgArmed* Actor that discovers, deters, and apprehends people who violate the laws, does not include *milSystems*	All types
interventionOrganization	A *politicalOrgArmed* Actor that intervenes in an unconventional conflict situation, does not include *milSystems*	Diplomats, advisors, support personnel
intelligenceServiceOrganization	A *politicalOrgArmed* Actor that is a non-intervention intelligence service, does not include *milSystems*	

Two Actor classes have been moved from this subcategory (in the UCO) into new subcategories (in the MCO), described below. The *SideInConflict* class has been renamed *sideInConflict* and is included in the *simpleCompositeActor* Table 4.14. The *GovTypeOrganization* class has been renamed *govTypeUnit* and is included in the *hierarchicalActor* Table 4.15.

4.4 Demographic Group Actors

A *demographicGroupActor* is an Actor that is composed of enough people that it is best described by a density distribution over a geographical area. This category of Actor is comprised of two subcategories: static population Actors and mobile population Actors. Figure 4.5 shows the demographic group Actor part of the Actor taxonomy. The "actor" class box represents all of the Actor classes that are children of the subcategories.

The map of England in the figure represents the possible location of a static population, a type of demographic group Actor, (without displaying the distribution of the static population within that location).

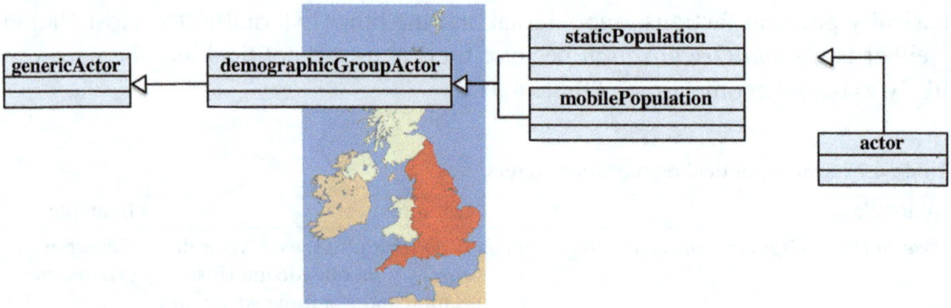

Fig. 4.5 Demographic group Actor taxonomy

Static Population Actors

A *staticPopulation* is a *demographicGroupActor* whose movement is best characterized by shifts in its density distribution over a constant geographic area. For example, the two major Muslim sects in Iraq, the Sunni and the Shi'a, are each found in most parts of the country; however, each dominates a different part of the country. If one manages to cause the other to abandon former strongholds, the effects are not likely to change the overall coverage by either but rather the density of each in the various regions of the country. This categorization may not be entirely valid for some of the smaller religious populations. One Actor in this subcategory has been renamed from *GeographicSubdivision* (in the UCO) to *geographicPopulation* (in the MCO), as this is a better name to refer to a population defined by its geographical location. There are seven classes in this subcategory (Table 4.8).

Table 4.8 Static population Actors

Actor class	Definition	Examples
politicalPopulation	A *staticPopulation* Actor portion that is politically active	
criminalPopulation	A *staticPopulation* Actor portion that is defined by its criminal status	
generalPopulation	A *staticPopulation* Actor portion of the general population	
culturalPopulation	A *staticPopulation* Actor portion that is defined by its cultural heritage	
religiousPopulation	A *staticPopulation* Actor portion that is defined by its religious beliefs	
geographicalPopulation	A *staticPopulation* Actor portion that is defined by its geographical extent, has population for members	
workerPopulation	A *staticPopulation* Actor portion that is defined by its employment status	All workers, workers at a particular trade, etc.

Mobile Population Actors

A *mobilePopulation* is a *demographicGroupActor* whose movement is best characterized by shifts in its density distribution over a changing geographic area. For example, the *internallyDisplacedPopulation* Actor class is designed to be instantiated by a large group of people who have already been displaced from their homes. They may be forced into regions or may voluntarily move into regions at some distance from their homes and they may move again, changing the area covered by the population and the density distribution of the population within that area. There are five classes in this subcategory (Table 4.9).

Table 4.9 Mobile population Actors

Actor class	Definition	Examples
demobilizedArmedForce	A *mobilePopulation* that was an *armedForce* Actor that has been demobilized	
internallyDisplacedPopulation	A *mobilePopulation* Actor portion that consists of internally displaced persons (IDPs)	
migrantPopulation	A *mobilePopulation* Actor portion that consists of migrants from outside the country – due to pull factors	
refugeePopulation	A *mobilePopulation* Actor portion that consists of migrants from outside the country – due to push factors	
expatriatePopulation	A *mobilePopulation* Actor portion that consists of people living outside the Host Nation	Expatriates, migrants, and refugees from the Host Nation

4.5 Nonhuman Actors

A *nonhumanActor* is an Actor that is not a human being. This category of Actor, renamed from *InanimateActor* in the UCO, is comprised of four subcategories: civilian vehicle Actors, military system Actors, environmental Actors, and animal Actors. Figure 4.6 shows the nonhuman Actor part of the Actor taxonomy. The "actor" class box represents all of the Actor classes that are children of the subcategories. The military aircraft in the figure represents a possible military system, which is a nonhuman Actor.

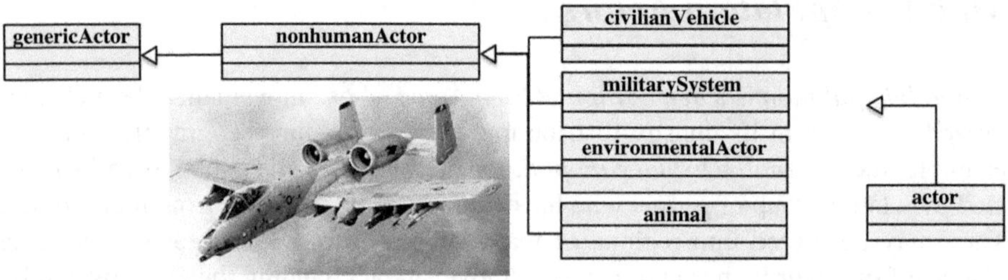

Fig. 4.6 Nonhuman Actor taxonomy

Civilian Vehicle Actors

A *civilianVehicle* is a *nonhumanActor* that is a civilian vehicle. There are four classes in this subcategory. These replace the *VehicleNonMilitary* class in the UCO (Table 4.10).

Table 4.10 Civilian vehicle Actors

Actor class	Definition	Examples
aircraft	A *civilianVehicle* that operates in the air	Light airplane, commercial airplane, helicopter
landVehicle	A *civilianVehicle* that operates on the land	Personal vehicle, farm vehicle, construction vehicle, service vehicle, transport vehicle
spacecraft	A *civilianVehicle* that operates in space	Satellite, space shuttle, launch vehicle
watercraft	A *civilianVehicle* that operates on the water	Boat, ship

Military System Actors

A *militarySystem* is a *nonhumanActor* that is a military system. These are the military analogs of the civilian vehicle classes. There are four classes in this subcategory. These replace the *MilitaryVehicle* class in the UCO (Table 4.11).

Table 4.11 Military system Actors

Actor class	Definition	Examples
airSystem	A *militarySystem* that operates in the air	Drone, helicopter, light aircraft, transport aircraft, attack aircraft, bomber
landSystem	A *militarySystem* that operates on the land, includes vehicles and fixed emplacements	Light, medium, and heavy vehicles, wheeled, tracked, and rail vehicles, unarmored and armored vehicles, artillery, reloadable missile launchers, fixed gun emplacements
seaSystem	A *militarySystem* that operates in the water	Boat, hovercraft, transport ship, combat ship, aircraft carrier, submarine
spaceSystem	A *militarySystem* that operates in space or near space	ICBM, satellite, space shuttle, space station, missile – from single-use launcher

Environmental Actors

An *environmentalActor* is a *nonhumanActor* that embodies an environmental force. There are six classes in this subcategory. The actions of these environmental forces are represented as Actions in the Action Ontology, and the inactive portions (e.g., when their existence is just backdrop information) are represented as Objects in the Object Ontology. The *ExperienceHealthEmergency* class in the UCO is replaced by the *healthTrauma* class here. The *AirMovementOrStorm* class is replaced by the *storm* class.

The distinction between Actor, Action, and Object representations can be illustrated by considering a hurricane. During hurricane season, the television weather will frequently display tropical storms over the Atlantic that might become hurricanes. In such a display, each storm is an Object. However, if there is a hurricane approaching land, it is described as an Actor that will move and engender damaging storm surge and winds. When the news is describing the destruction being caused, the emphasis is on Actions. In this way, the needs of the situation determine how best to represent the storm. The MCO provides the classes to support any of these representations, as needed (Table 4.12).

Table 4.12 Environmental Actors

Actor class	Definition	Examples
earthMovement	An *environmentalActor* that consists of moving earth	Earthquake; landslide/mudslide/avalanche, volcanic eruptions
fireOrWildfire	An *environmentalActor* that is a fire, whether urban or rural	Building or countryside fire
healthTrauma	An *environmentalActor* that inflicts damage on the health of people	Famine, epidemic, etc.
manmadeDisaster	An *environmentalActor* that results from the failure of human systems	Nuclear power plant/hazardous materials/chemical emergency; other man-made or technological disaster
storm	An *environmentalActor* that consists of moving air	Blizzard/heavy snowfall; hurricane/cyclone/tropical storm; thunderstorm/lightning/wind/hail; tornado
waterMovement	An *environmentalActor* that consists of moving water	Flood/dam failure, tsunami, seiche, storm surge

Animal Actors

An *animal* is a *nonhumanActor* that is a nonhuman animal. This is a new subcategory in the MCO, introduced because of the significant use of animals in recent conflicts. These include dogs for explosives detection and in rescue work and horses as transportation. The group class supports the instantiation of animals used in agriculture. There are two classes in this subcategory, one to represent an individual animal of any sort and the other to represent a group of *animals* (Table 4.13).

Table 4.13 Animal Actors

Actor class	Definition	Examples
animalGroup	A group of *animals*	Herd, pack, flock, etc.
genericAnimal	An *animal*, wild or tame	Horse, dog, camel, elephant, lion, wolf

4.6 Composite Actors

A *compositeActor* is an Actor that is composed of other Actors. This is a new category in the MCO. This category of Actor is comprised of two subcategories: simple composite Actors and hierarchical Actors. A separate chapter (Chap. 7) is provided to clarify the meaning of these new classes by showing instantiations. Figure 4.7 shows the composite Actor part of the Actor taxonomy. The "actor" class box represents all of the Actor classes that are children of the subcategories. The symbolic representation of an infantry battalion in the figure represents a possible hierarchical Actor.

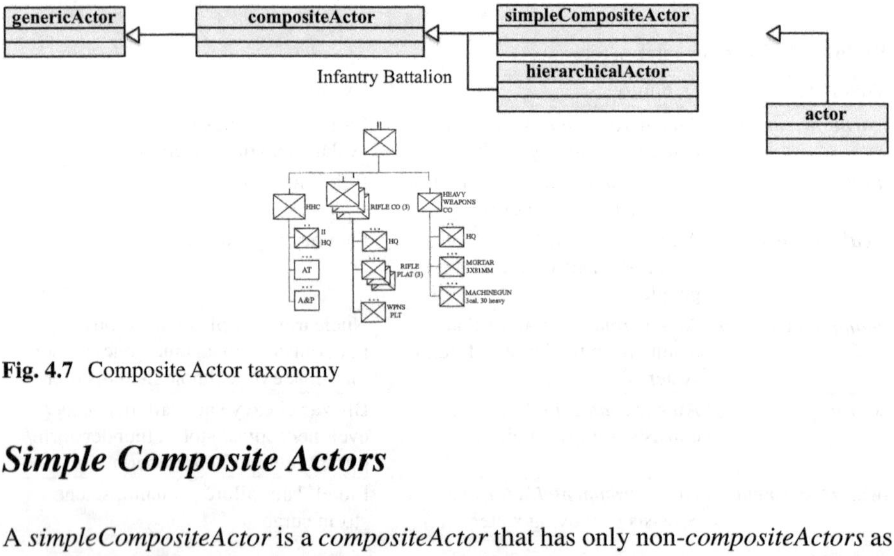

Fig. 4.7 Composite Actor taxonomy

Simple Composite Actors

A *simpleCompositeActor* is a *compositeActor* that has only non-*compositeActors* as components. The *milSystemsGroup* class supports the description of the military systems contained in armed forces. The *sideInConflict* class supports the description of coalitions and other groups of Actors in a conflict. There are two classes in this subcategory (Table 4.14).

Table 4.14 Simple composite Actors

Actor class	Definition	Examples
milSystemsGroup	A *simpleCompositeActor* that has *militarySystem* classes as components	
sideInConflict	A *simpleCompositeActor* provides structure for a side in conflict, and is used to group actors, such as government, military, law enforcement	Host Nation, factions, intervenors, etc.

Hierarchical Actors

A *hierarchicalActor* is a *compositeActor* that may have a *compositeActor* as a component along with non-*compositeActor* components. These classes support the description of hierarchical organizations. Each of the classes in this subcategory uses the postfix "Unit" in its name to differentiate it from other Actors. The point of this type of Actor is to allow the instantiation of hierarchies. Each level of the hierarchy, except the very lowest, requires at least one instantiation of the same class. Thus, a military division with four brigade subunits, each with four battalion subunits, would have a "unit" class instantiation, made up of four different "unit" class instantiations, each of which is made up of four different instantiations of either the same "unit" class or some other type of class. (See Chap. 7 for examples.) The definitions help to explain how these classes work. There are five classes in this subcategory (Table 4.15).

Table 4.15 Hierarchical Actors

Actor class	Definition	Examples
armedForceUnit	A *hierarchicalActor* that has a *keyLeader* class, an *armedForce* class, and a *milSystemsGroup* class and may have *armedForceUnit* classes as components	
govtTypeUnit	A *hierarchicalActor* that acts as a government organization and is composed of government branch classes and their key leaders, plus a *govtTypeUnit* class	Host Nation government, alternative (shadow) government, or intervention (as Host Nation government substitute) organization, including judiciary, legislature, executive, and bureaucratic organizations
organizationUnit	A *hierarchicalActor* composed of a *significantGroup* class, its *keyLeader*, and a possible *organizationUnit* class	
populationUnit	A *hierarchicalActor* composed of a *demographicGroupActor* class, its *keySocialLeader*, and a possible *populationUnit* class	Tribe, religion, political party
wholeOfGovt	A *hierarchicalActor* composed of a *govtDecisionAuthority* class and a *govtTypeUnit* class	

4.7 Actor Relationships

The only relationships used in this chapter are the *is-a* and *hasMetric* relationships, explained as follows:

is-a: Each instance of A is an instance of B and each property of B is a property of A. Its inverse relationship is *superClassOf*.
hasMetric: A has Metric B (also shown as *described by*). Its inverse relationship is *metricOf*.

4.8 Review of Actors

We often refer to "Actors" as shorthand for "Actor classes." Each Actor class is a thing that defines the instantiations of the class to be a particular subset of all of the things that cause actions. The Actor categories and subcategories are an aid to this definition process: each narrows the subset in a meaningful way.

The nouns of the modern conflict domain that represent active elements are Actors. (The nouns for inactive elements are represented by the Objects of Chap. 6.) Some Actors represent individuals; some represent groups; some represent demographic populations; and some represent inanimate things. However, all are active, capable of initiating Actions, which are discussed in Chap. 5.

The descriptions of composite Objects and composite Actors will be amplified by examples of instantiations in Chap. 7.

Chapter 5
MCO Action Ontology

Actions are the interventions, events and ongoing processes (by convention, addressed as occurring in zero time) in the Modern Conflict Ontology (MCO). They are initiated by Actors and are the direct causes of changes. Figure 3.8 showed how the Action Ontology fits into the overall MCO structure (with Fig. 5.1 illustrating its icon as used in the context diagram, Fig. 3.5).

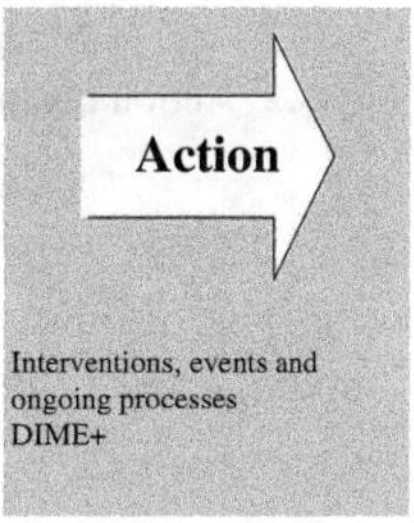

Fig. 5.1 Action icon

5.1 Overview of Actions

There are nine categories of Actions: strike Actions, command and control (C^2) Actions, information Actions, conflict Actions, conflict organization or personnel Actions, human affairs Actions, economic Actions, policing or criminal Actions, and civil government Actions. Figure 5.2 shows the top level of the Action taxon-

D. S. Hartley III, *An Ontology of Modern Conflict*, Understanding Complex Systems, https://doi.org/10.1007/978-3-030-53214-7_5

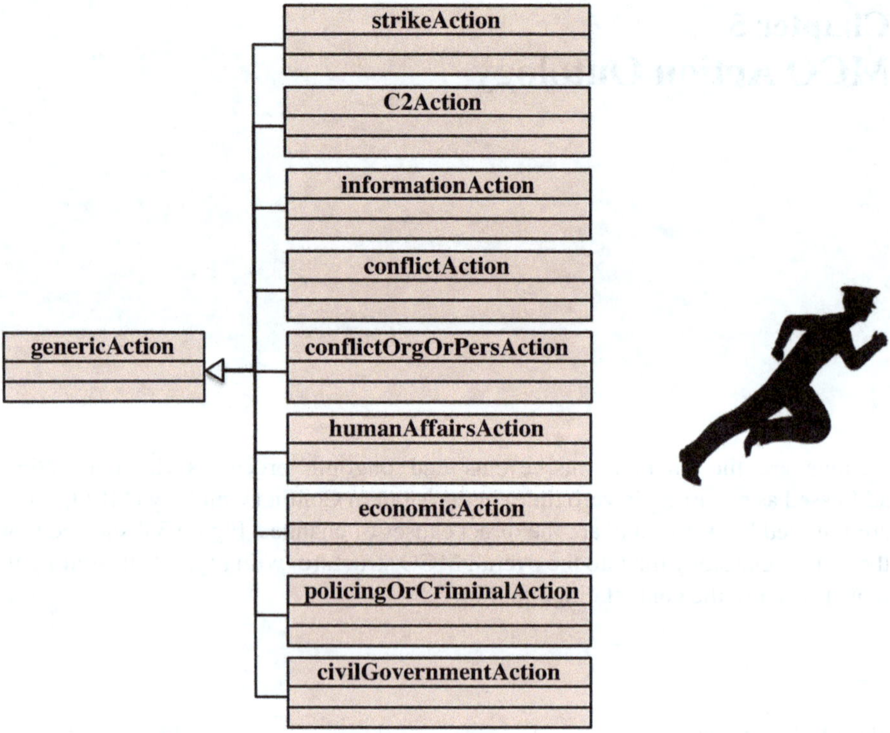

Fig. 5.2 Action taxonomy overview

omy. The running shape in the figure represents the action of running, which is an instantiation of the *moveSelf* Action.

Changes to Actions from the UCO

In Chap. 3, we discussed the general changes from the Unconventional Conflict Ontology (UCO) to the MCO. Most of these changes are formal, e.g., changing spelling conventions and definition standards. For the most part, the tables of the Actions contained in the book about the UCO describe the content adequately (Hartley 2018). However, the categories and subcategories have been changed, some Action classes have been revised, some new Action classes have been defined, and individual Action classes now appear in only one subcategory, requiring a new listing of all of the Action classes here.

An Insight into Actions

The strike Actions are the simplest to describe and thus provide a basis for understanding the other Action classes. In the ontology, there are 39 classes of targets that can be attacked; hence there are 39 Action classes in the *strikeTargetAction* subcategory. Similarly, there are 19 Action classes in the *strikeMethodAction* subcategory and seven Action classes in the *envirnStrikeAction* subcategory (standing for environmental strike action). The first subcategory is used where the target class is the most important element (sensors, elections, combat personnel, etc.). The second subcategory is used where the method (chemicals, cyber, etc.) is most important. The third subcategory is used where the "strike" is caused by environmental conditions (water movement, health condition, etc.). Each class is unique, but the instantiations of a class need not be singular. Thus, there might be multiple instantiations of the *airBlows* class, representing different storms. Further, there are possible cross-usages. For example, the *strikeCombatPersonnel* class might be instantiated in one case using high yield explosives and, in another case, using an improvised explosive device (IED). Alternatively, the *placeMinesIEDs* class might be instantiated to injure or kill combat personnel at one location and civilians at a different location. This flexibility allows the ontology to be used in modeling where different aspects of actions might be needed for different situations. A final note is important; the Action class indicates a particular type of action, but does not specify the result of the action. The result may be a "success" or a "failure" of the intent, which will be recorded in the state variables (metrics) classes that are connected to the instantiations.

5.2 Strike Actions

A *strikeAction* is an Action that by its nature can cause damage and injury. This category of Action is comprised of three subcategories: strike target Actions, strike method Actions, and environmental strike Actions. Figure 5.3 shows the strike

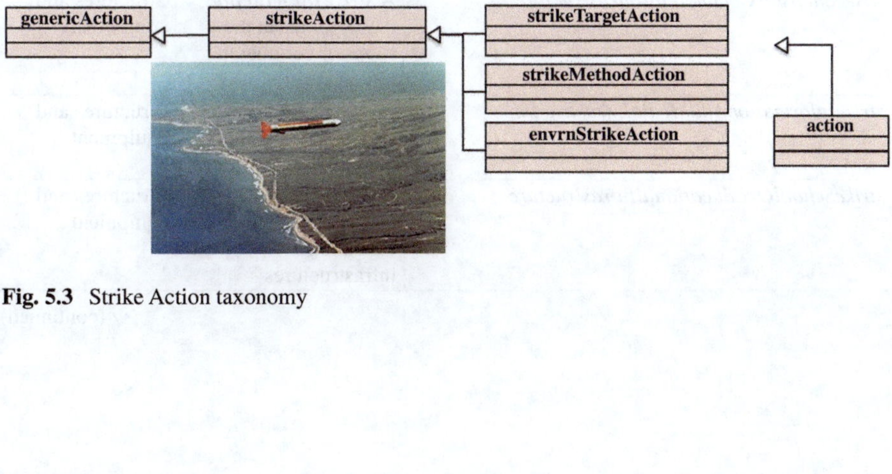

Fig. 5.3 Strike Action taxonomy

Action part of the Action taxonomy. The "action" class box represents all of the Action classes that are children of the subcategories. The cruise missile in the figure represents the strike method Action.

Strike Target Actions

A *strikeTargetAction* is a *strikeAction* that has a particular type of target. These targets include both military and civilian targets. They include both "legitimate" targets, such as military infrastructure and supply lines, and conventionally "illegitimate" targets, such as schools and religious facilities. This avoids any presuppositions about target choices. The user of the ontology is responsible for determining who might strike any type of target or, retrospectively, who did strike the target. There are 39 classes in this subcategory (Table 5.1).

Table 5.1 Strike target Actions

Action class	Definition	Examples
strikeFinancialInfrastructure	A *strikeTargetAction* that is aimed at the financial infrastructure	
strikeCivilianHousing	A *strikeTargetAction* that is aimed at civilian housing	
strikeReligiousFacility	A *strikeTargetAction* that is aimed at religious facilities	
strikeManufacturingInfrastructure	A *strikeTargetAction* that is aimed at the manufacturing infrastructure	Structures and equipment
strikeAgricultureInfrastructureLivestock	A *strikeTargetAction* that is aimed at the agricultural infrastructure	Structures, equipment, and livestock
strikeShopsOrCommercialInfrastructure	A *strikeTargetAction* that is aimed at the shops or commercial infrastructure	Structures and equipment
strikeInformationAndMediaInfrastructure	A *strikeTargetAction* that is aimed at the media infrastructure	Structures and equipment
strikeSchoolOrEducationalInfrastructure	A *strikeTargetAction* that is aimed at the educational infrastructures	Structures and equipment

(continued)

Table 5.1 (continued)

Action class	Definition	Examples
strikeHealthcareInfrastructure	A *strikeTargetAction* that is aimed at the health infrastructure	Structures and equipment
strikeElectricityProductionPlant	A *strikeTargetAction* that is aimed at the electricity production plant	Structures and equipment
strikeElectricityDistributionInfrastructure	A *strikeTargetAction* that is aimed at the electricity distribution system	Structures and equipment
strikeExtractiveEnergyProductionInfrastructure	A *strikeTargetAction* that is aimed at the extractive energy production infrastructure	Structures and equipment
strikeExtractiveEnergyTransportationInfrastructure	A *strikeTargetAction* that is aimed at the extractive energy transportation infrastructure	Structures and equipment
strikeGovtInfrastructure	A *strikeTargetAction* that is aimed at the government infrastructure	Government – including police – structures and equipment
strikeMilitaryInfrastructure	A *strikeTargetAction* that is aimed at the military infrastructure	Structures and equipment
strikeMilitaryVehicle	A *strikeTargetAction* that is aimed at a military system	Includes military system actors and military vehicles considered as part of the environment
strikeRoadInfrastructure	A *strikeTargetAction* that is aimed at the road infrastructure	
strikeRailroadInfrastructure	A *strikeTargetAction* that is aimed at the railroad infrastructure	Structures and equipment
strikeBridgeOrTunnelInfrastructure	A *strikeTargetAction* that is aimed at the bridge or tunnel infrastructure	Structures and equipment
strikeSeaportInfrastructure	A *strikeTargetAction* that is aimed at the seaport infrastructure	Structures and equipment

(continued)

Table 5.1 (continued)

Action class	Definition	Examples
strikeAirportInfrastructure	A *strikeTargetAction* that is aimed at the airport infrastructure	Structures and equipment
strikeVehicle	A *strikeTargetAction* that is aimed at a non-military vehicle	Includes vehicle actors and those considered as part of the environment
strikeWaterwaysInfrastructure	A *strikeTargetAction* that is aimed at the waterways infrastructure	Locks, canals, structures, and equipment
strikeWaterDistributionInfrastructure	A *strikeTargetAction* that is aimed at the water distribution infrastructure	Structures and equipment
strikeWaterOrSewageTreatmentFacilities	A *strikeTargetAction* that is aimed at the water or sewage treatment facilities	Structures and equipment
strikeDamInfrastructure	A *strikeTargetAction* that is aimed at the dam infrastructure	Structures and equipment
strikeCivilians	A *strikeTargetAction* that is aimed at civilians	
strikeSensorProcess	A *strikeTargetAction* that is aimed at military sensor equipment or systems or intelligence	
strikeCommunicationsProcess	A *strikeTargetAction* that is aimed at military communications equipment or systems	
strikeCommandAndControlProcess	A *strikeTargetAction* that is aimed at the military command and control processes or computers	
strikePrisonInfrastructure	A *strikeTargetAction* that is aimed at the prison infrastructure	

(continued)

Table 5.1 (continued)

Action class	Definition	Examples
strikeMiningInfrastructure	A *strikeTargetAction* that is aimed at the mining infrastructure	
strikeCulturalInfrastructure	A *strikeTargetAction* that is aimed at the cultural infrastructure	Theater, museum, or sports
strikeRefugeeCampOrTemporaryShelter	A *strikeTargetAction* that is aimed at the temporary shelter/ housing/refugee camps	In or near Host Nation
strikeMIS	A *strikeTargetAction* that is aimed at the MIS, etc., physically or with cyber attacks	
disruptElection	A *strikeTargetAction* that is aimed at disrupting an election or other civic function	Election, civic speech, etc.
strikeGovtInstitutionOrKeyOfficial	A *strikeTargetAction* that is aimed at a government official or institution	Civil government official, law enforcement personnel, etc.; police station, recruiting center, office, etc.
strikeCombatPersonnel	A *strikeTargetAction* that is aimed at the personnel of a combat force	Government military, intervenor force, terrorist, etc.
strikeSuppliesOrSupplyLine	A *strikeTargetAction* that is aimed at supplies or a supply line	Any type of supply

Strike Method Actions

A *strikeMethodAction* is a *strikeAction* that uses a particular method. The first nine classes define location choices. Clearly, a party with no space resources would have trouble using a strike originating from space. However, with the advent of cheap drones, almost any party can originate a strike from the air. The other classes define methods that range from "legitimate," such as high-yield explosives, to "illegitimate," such as chemical, biological, and radiological weapons. There are 19 classes in this subcategory (Table 5.2).

Table 5.2 Strike method Actions

Action class	Definition	Examples
strikeFromAirSpaceToSubsurfaceSea	A *strikeMethodAction* from a generic air/space system to a subsurface sea area	Includes area target
strikeFromAirSpaceToGroundSeaSurface	A *strikeMethodAction* from a generic air/space system to a ground/sea surface area	Includes area target
strikeFromAirSpaceToAirSpace	A *strikeMethodAction* from a generic air/space system to an air/space area	Includes area target
strikeFromGroundSeaSurfaceToSubsurfaceSea	A *strikeMethodAction* from a generic ground/sea surface system to a subsurface sea area	Includes area target
strikeFromGroundSeaSurfaceToGroundSeaSurface	A *strikeMethodAction* from a generic ground/sea surface system to a ground/sea surface area	Includes area target
strikeFromGroundSeaSurfaceToAirSpace	A *strikeMethodAction* from a generic ground/sea surface system to an air/space area	Includes area target
strikeFromSubsurfaceSeaToSubsurfaceSea	A *strikeMethodAction* from a generic subsurface sea system to a subsurface sea area	Includes area target
strikeFromSubsurfaceSeaToGroundSeaSurface	A *strikeMethodAction* from a generic subsurface sea system to a ground/sea surface area	Includes area target
strikeFromSubsurfaceSeaToAirSpace	A *strikeMethodAction* from a generic subsurface sea system to an air/space area	Includes area target
strikeWithHighYieldExplosives	A *strikeMethodAction* using high-yield explosives	Includes area target
strikeWithChemicals	A *strikeMethodAction* using a chemical agent	Includes area target
strikeWithBiologicals	A *strikeMethodAction* using a biological agent	Includes area target
strikeWithRadiologicals	A *strikeMethodAction* using a radiological agent	Includes area target
placeMinesIEDs	A *strikeMethodAction* that places mines – or other similar explosives	Includes area target
commitWarCrime	A *strikeMethodAction* that commits a war crime	Rape, pillage, torture, etc.
conductCyberAttack	A *strikeMethodAction* that conducts an attack using computer	

(continued)

Table 5.2 (continued)

Action class	Definition	Examples
strikeWithNuclearWeapon	A *strikeMethodAction* using a nuclear explosion	
conductElectronicWarAttack	A *strikeMethodAction* that consists of an electronic attack	
performSuicideBombingInsiderStrikeOrSabotage	A *strikeMethodAction* that uses a seeming civilian or previously trusted insider to perform the Action	

Environmental Strike Actions

An *envirnStrikeAction* is a *strikeAction* that has an environmental cause. In the MCO, the Action representations of these concepts have been separated from the Actor and Object representations. Some of these classes will only have natural environment Actors as initiators; however, some may have either a natural environment Actor or a human Actor as the initiator, such as "strikes" involving fire or obscuration of vision. There are seven classes in this subcategory (Table 5.3).

Table 5.3 Environmental strike Actions

Action class	Definition	Examples
earthShifts	An *envirnStrikeAction* that is caused by an earth shift	Earthquake; landslide/mudslide/avalanche, volcanic eruptions
airBlows	An *envirnStrikeAction* that is caused by the air blowing	Blizzard/heavy snowfall; hurricane/cyclone/tropical storm; thunderstorm/lightning/wind/hail; tornado
waterMoves	An *envirnStrikeAction* that is caused by the water moving	Flood/dam failure, tsunami, seiche
fireBurns	An *envirnStrikeAction* that is caused by a fire	Building or countryside fire
systemFails	An *envirnStrikeAction* that is caused by a system failure	Nuclear power plant/hazardous materials/chemical emergency; other man-made or technological disaster
healthConditionAfflicts	An *envirnStrikeAction* that is caused by a health affliction	Famine, epidemic, etc.
obscureVision	An *envirnStrikeAction* that is caused by fog or man-made obscurant that reduces visibility	Includes area target

5.3 C2 Actions

A *C2Action* is an Action that involves command and control (C2). This category of Action is comprised of two subcategories: control Actions and command and control Actions. Figure 5.4 shows the C2 Action part of the Action taxonomy. The "action" class box represents all of the Action classes that are children of the subcategories. The sergeant in the figure represents a command and control Action.

Fig. 5.4 C2 Action taxonomy

Control Actions

A *controlAction* is a *C2Action* that involves control of entities by an entity. This subcategory contains a number of self-referent actions, such as moving one's self and surrendering. Some of them can be instantiated at a granularity of an individual or of an entire nation. These classes are explained in their definitions. Several of these Action classes have been added in the MCO. There are 13 classes in this subcategory (Table 5.4).

Table 5.4 Control Actions

Action class	Definition	Examples
moveSelf	A *controlAction* in which the Element changes location or location distribution; maneuver forces to achieve a position of advantage over the enemy	
changeSelf	A *controlAction* in which the Element changes its own status	
changeElement	A *controlAction* in which the Element changes the status of another element	
controlElement	A *controlAction* in which the Element exerts control – up to full ownership – of another element; control significant areas in the OA whose possession or control provides either side with an operational advantage	

(continued)

Table 5.4 (continued)

Action class	Definition	Examples
changeOwnActivityLevel	A *controlAction* that changes own activity level	
removeObstacle	A *controlAction* that provides mobility to facilitate movement and maneuver without delays caused by terrain or obstacles	
impedeMovementOrAction	A *controlAction* that delays, channels or stops movement and maneuver by enemy formations or impedes any Action by any actor	
searchForAdversary	A *controlAction* that searches for the adversary	
concealSelf	A *controlAction* that conceals self physically, electronically, or from cyber search	
concealEntity	A *controlAction* that conceals another entity physically, electronically, or from cyber search	
defendSelf	A *controlAction* that defends self by any means	
surrenderSelf	A *controlAction* that surrenders self to the adversary	
disengageSelfFromCombat	A *controlAction* that disengages self from combat	

Command and Control Actions

A *commandAndControlAction* is a *C2Action* that involves control of armed forces or other subordinate units. These classes are available to represent the various parts of the standard military command and control processes. However, by analogy, they may also be useful in other organizational situations. Almost all of these Action classes have been added in the MCO. There are nine classes in this subcategory (Table 5.5).

Table 5.5 Command and control Actions

Action class	Definition	Examples
establishCommunicationsProcesses	A *commandAndControlAction* that establishes communications processes	
establishCommandAndControlProcesses	A *commandAndControlAction* that establishes command and control processes or computers	
organizeForce	A *commandAndControlAction* that establishes, organizes, and operates HQ; establishes command authorities among subordinate commanders; assigns tasks, prescribes task performance standards, and designates operational areas; prioritizes, allocates resources	
commandSubordinate	A *commandAndControlAction* that coordinates and controls the employment of lethal and nonlethal capabilities; coordinates, synchronizes, and, when appropriate, integrates operations with operations and activities of other participants	

(continued)

Table 5.5 (continued)

Action class	Definition	Examples
issuePlan	A *commandAndControlAction* that prepares, modifies, and publishes plans, orders, and guidance	
communicate	A *commandAndControlAction* that communicates and maintains the status of information across the staff, force, and public as appropriate; ensures the flow of information; and reports to higher authority	
assessMilitaryProgress	A *commandAndControlAction* that assesses progress toward accomplishing tasks, creating conditions, and achieving objectives	
controlFires	A *commandAndControlAction* that conducts targeting; provides fire support; counters air and missile threats; interdicts enemy capabilities; assesses the results of fires	
coordinateActivity	A *commandAndControlAction* that coordinates an activity among the appropriate Actors	Various parts of the C2 process

5.4 Information Actions

An *informationAction* is an Action that relates to persuasion or observation. This category of Action is comprised of three subcategories: persuasion Actions, monitoring Actions, and intelligence Actions. Figure 5.5 shows the information Action part of the Action taxonomy. The "action" class box represents all of the Action classes that are children of the subcategories. The debate pictured in the figure represents a possible persuasion Action.

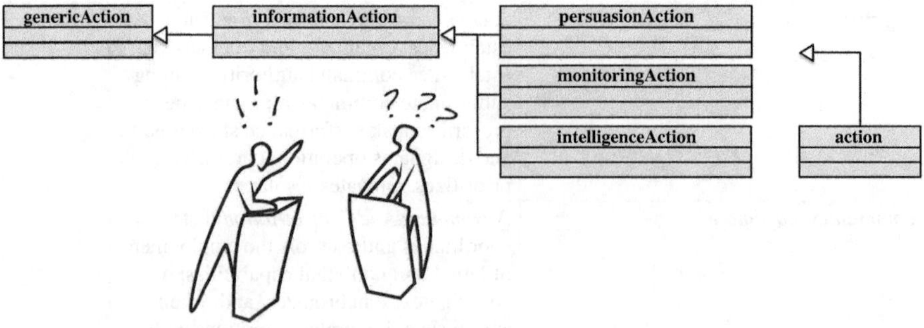

Fig. 5.5 Information Action taxonomy

Persuasion Actions

A *persuasionAction* is an *informationAction* that attempts to change opinions or behaviors of people. The classes range from generic advice and diplomatic action to actions with a specific purpose, such as destabilizing the government and increasing or decreasing the legitimacy of an Actor. Persuasion is currently a domain in which scientific progress is being made toward determining how to do it effectively (Hartley and Jobson 2020). A few of these Action classes have been added in the MCO. There are 18 classes in this subcategory (Table 5.6).

Table 5.6 Persuasion Actions

Action class	Definition	Examples
provideAdvisorsToLeaders	A *persuasionAction* that provides advisors to government or other organizations – other than judicial organizations	
employDiplomaticAction	A *persuasionAction* that implements UN actions/ resolutions, US diplomacy or Host Nation diplomacy	Internal or external, communicate, make alliances, etc.
destabilizeGovt	A *persuasionAction* that destabilizes the Host Nation government	
mediateNegotiateOrPersuadeConflictingParties	A *persuasionAction* that mediates, negotiates or persuades, or supports mediation, negotiation or persuasion of conflicting parties	
provideAdvisorsToJudicialOrganizations	A *persuasionAction* that provides advisors to the police or justice organizations or supports establishment of operations	
coordinateNGOActivities	A *persuasionAction* that coordinates NGO, IO, or IGO activities within Host Nation	
promoteCivicEducation	A *persuasionAction* that promotes civic education	"Civics" classes

(continued)

Table 5.6 (continued)

Action class	Definition	Examples
reduceLikelihoodOfPopulationMovements	A *persuasionAction* that reduces likelihood of population movements	
rebuildSenseOfCommunity	A *persuasionAction* that rebuilds sense of community within Host Nation	
conductBenignPublicInformationOperation	A *persuasionAction* that conducts a positive information operation	Includes PR and counterpropaganda
conductNegativeInformationOperation	A *persuasionAction* that conducts a negative information operation	Includes propaganda
decreaseLegitimacyOfActor	A *persuasionAction* that decreases legitimacy of the Actor in eyes of populace or media	
increaseLegitimacyOfActor	A *persuasionAction* that increases legitimacy of the Actor in eyes of populace or media	
decreaseSenseOfCommunity	A *persuasionAction* that decreases the popular sense that there is a community to belong to	
seekRecruitsByNonNationStateActor	A *persuasionAction* that gains recruits by non-nation-state Actors	
increaseLikelihoodOfPopulationMovements	A *persuasionAction* that increases likelihood of population movements	
seekFundsByNonNationStateActor	A *persuasionAction* that seeks funding by non-nation-state Actor	
seekSupportByNonNationStateActor	A *persuasionAction* that seeks popular support by non-nation-state Actor	

Monitoring Actions

A *monitoringAction* is an *informationAction* that attempts to discover behaviors. These classes cover standard monitoring activities. There are four classes in this subcategory (Table 5.7).

Table 5.7 Monitoring Actions

Action class	Definition	Examples
maintainComplianceWithPeaceAccords	A *monitoringAction* that maintains compliance with peace accord milestones or conditions, with all needed tasks	
monitorPowersharingArrangements	A *monitoringAction* that monitors power sharing arrangements, with all needed tasks	
monitorOrReportOnCorruptionByGovtOfficials	A *monitoringAction* that monitors or reports corruption by government officials	
monitorHumanRightsPractice	A *monitoringAction* that monitors human rights practices	

Intelligence Actions

An *intelligenceAction* is an *informationAction* that relates to the collection and use of intelligence information. The Action classes relating to the formal intelligence process have been added in the MCO. There are 13 classes in this subcategory (Table 5.8).

Table 5.8 Intelligence Actions

Action class	Definition	Examples
IDOrInterdictFundingOfNonNationStateActor	An *intelligenceAction* that identifies or interdicts funding of a non-nation-state Actor or disrupts it	
IDInderdictOrInterruptRecruitmentByNonNationStateActor	An *intelligenceAction* that identifies, interdicts, or interrupts recruitment by a non-nation-state Actor or disrupts it	
IDInstitutionalOrLocalSupportForNonNationStateActor	An *intelligenceAction* that identifies institutional or local support for a non-nation-state Actor or disrupts it	

(continued)

Table 5.8 (continued)

Action class	Definition	Examples
collectData	An *intelligenceAction* that collects information; plans information operations; conducts defensive information operations	On infrastructure, economics, government effectiveness, perceptions, refugees, etc.
establishLiaisonProgramsWithOrganization	An *intelligenceAction* that establishes liaison programs with the government or other organization	
controlInformation	An *intelligenceAction* that controls information – of all kinds	
establishSensorProcess	An *intelligenceAction* that establishes a sensor or intelligence process	
disseminateInformation	An *intelligenceAction* that disseminates and integrates information with operations	All kinds of intelligence
controlIntelligenceProcess	An *intelligenceAction* that plans and directs intelligence activities; evaluates and feeds back intelligence effectiveness and quality	Inform commander; describe OE; identify, define, nominate objectives; support planning, execution of operations; assess effectiveness of operations

(continued)

Table 5.8 (continued)

Action class	Definition	Examples
processAndAnalyse	An *intelligenceAction* that processes and exploits collected data to produce relevant information; analyses information and produces intelligence	
employIRC	An *intelligenceAction* that employs an information-related capability (IRC) to influence disrupt, corrupt, or usurp enemy's decision-making	Tools, techniques, or activities in the information environment
counterDeception	An *intelligenceAction* that counters adversary deception and surprise	
supportDeception	An *intelligenceAction* that supports a friendly deception effort	

5.5 Conflict Actions

A *conflictAction* is an Action that relates to general conflict. This category of Action is comprised of three subcategories: sustainment Actions, security Actions, and general conflict Actions. Figure 5.6 shows the conflict Action part of the Action taxonomy. The "action" class box represents all of the Action classes that are children of the subcategories. The two knights jousting in the figure represents an instantiation of a general conflict Action.

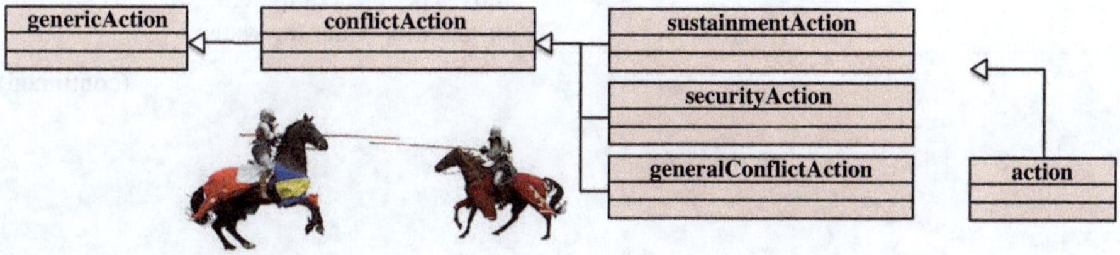

Fig. 5.6 Conflict Action taxonomy

Sustainment Actions

A *sustainmentAction* is a *conflictAction* that sustains forces or sustains others by action of the entity. These classes include logistics actions that support sustainment as well as supply and funding actions. The last three Action classes in the table have been added in the MCO. There are 13 classes in this subcategory (Table 5.9).

Table 5.9 Sustainment Actions

Action class	Definition	Examples
supplyGovtOrShadowGovt	A *sustainmentAction* that provides supplies needed for governing	
rebuildMilitaryInfrastructure	A *civilBuildingAction* that builds, rebuilds, or maintains military structure, including equipment	Bases, structures, etc.
rebuildReplaceMilitaryVehicle	A *civilBuildingAction* that maintains, repairs, or replaces military vehicle	
acquireEquipmentOrMaterial	A *sustainmentAction* that acquires equipment or material	
warehouseEquipmentOrMaterial	A *sustainmentAction* that warehouses equipment or material	
distributeEquipmentOrMaterial	A *sustainmentAction* that distributes equipment or material	
movePeopleEquipmentOrMaterialOnTheGround	A *sustainmentAction* that moves people, equipment, or material on the ground	
movePeopleEquipmentOrMaterialThroughTheAir	A *sustainmentAction* that moves people, equipment, or material through the air	
movePeopleEquipmentOrMaterialOverTheWater	A *sustainmentAction* that moves people, equipment, or material over the water	
movePeopleEquipmentOrMaterialUnderTheWater	A *sustainmentAction* that moves people, equipment, or material under the water	

(continued)

Table 5.9 (continued)

Action class	Definition	Examples
supplyForce	A *sustainmentAction* that supplies part of the force, from a large unit down to an individual military system	Food, operational energy (fuel and other energy requirements), arms, munitions, and equipment, by land, sea, or air (e.g., inflight refueling)
provideGeneralForceSupport	A *sustainmentAction* that coordinates and provides general support for the force	Field services, personnel services support, health services, mortuary affairs, religious support, postal support, morale-welfare-and-recreational support, financial support, and legal services
fundForce	A *sustainmentAction* that acquires, manages, and distributes funds	

Security Actions

A *securityAction* is a *conflictAction* that provides security. This provision of security can be for own forces, for others conducting various kinds of operations, protecting particular people or installations, or providing defense against particular threats. Several Action classes have been added in the MCO. There are 23 classes in this subcategory (Table 5.10).

Table 5.10 Security Actions

Action class	Definition	Examples
establishDemilitarizedZoneSanctionArmsEmbargo	A *securityAction* that establishes a demilitarized zone, sanction, or arms embargo/disarmament	
implementWeaponsControlRegime	A *securityAction* that implement a weapons control regime, including WMD or disarmament	
conductBorderControlBoundarySecurityFreedomOfMovement	A *securityAction* that conducts a border control, boundary security, or freedom of movement operation	
establishConfidenceBuildingOrSecurityMeasure	A *securityAction* that establishes a confidence-building or security measure	
safeguardInstitutionOfGovernanceOrKeyOfficial	A *securityAction* that safeguards Host Nation institutions of governance or Host Nation key officials	
provideSecurityAssistance	A *securityAction* that provides security assistance to the government or others	
conductSecurityCoordination	A *securityAction* that conducts security coordination	
provideForceSecurity	A *securityAction* that provides force security for own forces	Includes prevent friendly fire incidents
provideSecurityForHumanitarianAssistanceActivities	A *securityAction* that provides security for humanitarian assistance activities	
provideSecurityForPeaceOperationActivities	A *securityAction* that provides security for peace operation activities	
provideSecurityForStabilityActivities	A *securityAction* that provides security for stability activities	
mitigatePoliticalOrSocialInstabilityOrIndividualUnrestAction	A *securityAction* that mitigates political or social instability within Host Nation or mitigates individual unrest actions by non-nation-state Actor	
clearMinesIEDs	A *securityAction* that clears mines (or other similar explosives) or other protective Action	
provideRefugeeCampSecurity	A *securityAction* that provides refugee camp security	
provideForceHealthProtection	A *securityAction* that promotes improves, preserves, or restores mental or physical well-being of personnel	
provideAirSpaceMissileDefense	A *securityAction* that provides defense against air, space, and missile attack	

(continued)

Table 5.10 (continued)

Action class	Definition	Examples
protectAttachedCivilian	A *securityAction* that protects civilians and contractors authorized to accompany the force	
provideElectronicAndCyberDefense	A *securityAction* that conducts operations security (OPSEC), cyberspace defense, cybersecurity, defensive electronic attack (EA), and electronic protection activities	
protectMilitaryInfrastructure	A *securityAction* that secures and protects combat and logistics forces, bases, security areas, and lines of communications	
provideCBRNDefense	A *securityAction* that provides chemical, biological, radiological, and nuclear (CBRN) defense; mitigates the effects of CBRN incidents through thorough planning, preparation, response, and recovery	
provideEmergencyManagement	A *securityAction* that provides emergency management and response capabilities and services to reduce the loss of personnel and capabilities due to isolating events, accidents, health threats, and natural disasters	
identifyNeutralizeInsiderThreatSuicideBombingOrSabotage	A *securityAction* that identifies and neutralizes insider threats, suicide bombings, sabotage; conducts identity collection activities, including security screening and vetting in support of identity intelligence	
patrolArea	A *securityAction* that operates within an area to search for or search for and attack an adversary or to deter unwanted actions	Air, land, or sea patrol, military, or law enforcement

General Conflict Actions

A *generalConflictAction* is a *conflictAction* that relates to conducting conflict. There are classes for several categories of conflict, such as stability operations, and for parts of operations, such as deploying, redeploying, and holding in place. More than half of these Action classes have been added in the MCO. There are 34 classes in this subcategory (Table 5.11).

Table 5.11 General conflict Actions

Action class	Definition	Examples
conductPeacekeepingOperation	A *generalConflictAction* that is part of or supports a general intervention peacekeeping operation	UN Chapter VI, preventive diplomacy, delegatory peacekeeping, peacemaking, military support to (domestic) civil authorities (MSCA)
conductConventionalWarOperation	A *generalConflictAction* that is part of or supports a conventional war operation	By the Host Nation or intervenors against others or against the Host Nation by intervenors
conductIrregularWarOperation	A *generalConflictAction* that is part of or supports an irregular war operation	Of any type
establishObserverMissionOrInterposeForces	A *generalConflictAction* that establishes or supports an observer mission or interposes forces	
conductMilitaryExercise	A *generalConflictAction* that is part of or supports the execution of a military exercise with or without Host Nation military	
conductStabilityOperation	A *generalConflictAction* that is part of or supports an intervention stability operation	Of any type
conductPersonnelRecoveryOperation	A *generalConflictAction* that is or supports a personnel recovery operation	Search and rescue
conductTerrorismOperation	A *generalConflictAction* that is part of or supports a terrorism operation within Host Nation (or nearby areas)	
provideConsequenceManagementSupport	A *generalConflictAction* that provides or supports consequence management support	As a result of use of chemical, biological, or radiological weapons
conductPiracyOperation	A *generalConflictAction* that is part of or supports a piracy operation	

(continued)

Table 5.11 (continued)

Action class	Definition	Examples
conductHumanitarianAssistanceOperation	A *generalConflictAction* that is part of or supports an intervention humanitarian assistance operation	Humanitarian or civic assistance
conductNonCombatantEvacuationOperation	A *generalConflictAction* that is or supports a non-combatant evacuation operation within Host Nation (or nearby areas)	Permissive or nonpermissive NEO
executeCivilDefensePlan	A *generalConflictAction* that executes or supports part of a Host Nation civil defense plan	
conductAntiOrCounterterrorismOp	A *generalConflictAction* that conducts part of or supports an anti- or counter-terrorism operation within the Host Nation (and nearby areas)	
conductAntiPiracyOperation	A *generalConflictAction* that is part of or supports an anti-piracy operation	
shapeSituation	A *generalConflictAction* that shapes the PMESII environment in a way beneficial for the side	
deployIntoTheater	A *generalConflictAction* that deploys or supports deployment of part of the personnel and materiel into the theater	
redeploy	A *generalConflictAction* that redeploys or supports redeployment of part of the personnel and materiel within or out of the theater	
engageInCombat	A *generalConflictAction* that engages in combat (no matter what kind of overall operation is being conducted)	
holdInPlace	A *generalConflictAction* in which not much is done	Garrison operations, waiting, etc.

(continued)

Table 5.11 (continued)

Action class	Definition	Examples
conductCBRNOperation	A *generalConflictAction* that is part of or supports a chemical, biological, radiological, or nuclear war operation	
conductEconomicWarOperation	A *generalConflictAction* that is part of or supports an economic war operation	State-driven, organized crime, drug trade, trade war, financial actions, etc.
conductInformationWarOperation	A *generalConflictAction* that is part of or supports an information war operation	Cyberwar, propaganda, etc.
conductShowOfForceOperation	A *generalConflictAction* that is part of or supports a show of force operation	
conductSanctionMIOQuarantineOperation	A *generalConflictAction* that is part of or supports an enforcement of sanctions, maritime intercept operation (MIO), or quarantine operation	Sanctions, MIO, quarantine, exclusion zone
conductFONOverflightOperation	A *generalConflictAction* that is part of or supports a freedom of navigation (FON) or freedom of overflight operation	
conductCounterdrugOperation	A *generalConflictAction* that is part of or supports a counterdrug operation	
conductCounterinsurgencyOperation	A *generalConflictAction* that is part of or supports a counterinsurgency operation	
conductInsurgencyOperation	A *generalConflictAction* that is part of or supports an insurgency operation	Guerilla war, etc.
conductDisasterReliefOperation	A *generalConflictAction* that is part of or supports a disaster relief operation	Domestic disaster relief, international disaster relief, disaster control
conductPeaceEnforcementOperation	A *generalConflictAction* that is part of or supports a peace enforcement operation	UN Chapter VI 1/2, aggravated peace support operation (APSO-), UN Chapter VII

(continued)

Table 5.11 (continued)

Action class	Definition	Examples
conductNationAssistanceOperation	A *generalConflictAction* that is part of or supports a nation assistance operation	Security assistance, nation building, foreign internal defense
conductOtherPeaceOperation	A *generalConflictAction* that is part of or supports other peace operations	Pre-conflict peace building, post-conflict peace building, arms control, deterrence, disarmament, counterproliferation
conductRelocationOperation	A *generalConflictAction* that is part of or supports relocation related to displaced persons and other movement of people	Relocation of refugees or illegal immigrants or illegal emigrants

5.6 Conflict Organization or Personnel Actions

A *conflictOrgOrPersAction* is an Action that changes conflict organizations and personnel. This category of Action is comprised of two subcategories: conflict organizational Actions and conflict personnel Actions. Figure 5.7 shows the conflict organizational and personnel Action part of the Action taxonomy. The "action" class box represents all of the Action classes that are children of the subcategories. The planning cycle in the figure contains an organizing action that represents a possible conflict organization Action.

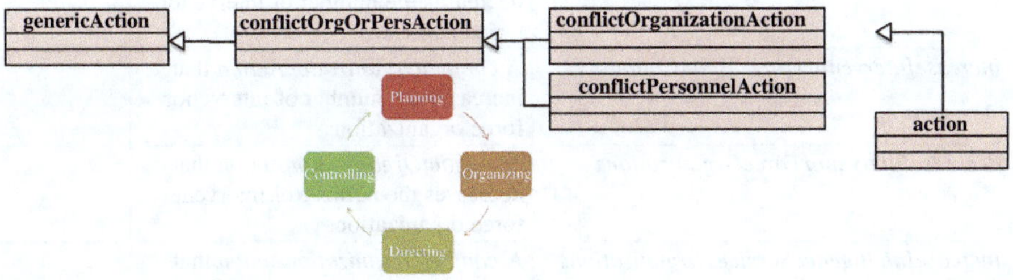

Fig. 5.7 Conflict organization or personnel Action taxonomy

Conflict Organization Actions

A *conflictOrganizationAction* is a *conflictOrgOrPersAction* that relates to changes in the conflict organization. Two of these classes cover creating and reforming military and intelligence services. The other classes cover increasing and decreasing the number of armed organizations (which is different from changing the number of personnel in an organization, contained in the next subcategory). There are 22 classes in this subcategory (Table 5.12).

Table 5.12 Conflict organization Actions

Action class	Definition	Examples
createOrReformOrMonitorMilitary	A *conflictOrganizationAction* that creates, reforms, or monitors military or armed group; staffs, funds, arms, and/or supplies it	
createOrReformOrMonitorIntelligence Services	A *conflictOrganizationAction* that creates, reforms, or monitors an intelligence service; provides staff for it; funds it; supplies it	
increaseExternalForceOrganizations AdvocatingConflict	A *conflictOrganizationAction* that increases the number of external force organizations advocating conflict	
decreaseExternalForceOrganizations AdvocatingConflict	A *conflictOrganizationAction* that decreases the number of external force organizations advocating conflict	
decreaseIntelligenceServicesOrganizations	A *conflictOrganizationAction* that decreases the number of intelligence services by all means	
increaseIntervenorOrganizations	A *conflictOrganizationAction* that increases the number of intervenor organizations	
decreaseIntervenorOrganizations	A *conflictOrganizationAction* that decreases the number of intervenor organizations	
increaseIntervenorForceOrganizations	A *conflictOrganizationAction* that increases the number of intervenor force organizations	
decreaseIntervenorForceOrganizations	A *conflictOrganizationAction* that decreases the number of intervenor force organizations	
increaseIntelligenceServicesOrganizations	A *conflictOrganizationAction* that increase the number of intelligence services organization	
increaseMilitaryOrganizations	A *conflictOrganizationAction* that increases the number of military organizations	

(continued)

Table 5.12 (continued)

Action class	Definition	Examples
decreaseMilitaryOrganizations	A *conflictOrganizationAction* that decreases the number of military organizations	
increaseRegimeSponsoredNonMilitary ArmedForceOrganizations	A *conflictOrganizationAction* that increases the number of regime-sponsored, non-military armed force organizations	
decreaseRegimeSponsoredNonMilitary ArmedForceOrganizations	A *conflictOrganizationAction* that decreases the number of regime-sponsored, non-military armed force organizations	
increaseParamilitaryOrganizations	A *conflictOrganizationAction* that increases the number of paramilitary organizations	
decreaseParamilitaryOrganizations	A *conflictOrganizationAction* that decreases the number of paramilitary organizations	
increasePrivateSecurityOrganizations	A *conflictOrganizationAction* that increases the number of private security organizations	
decreasePrivateSecurityOrganizations	A *conflictOrganizationAction* that decreases the number of private security organizations	
increaseInsurgentOrganizations	A *conflictOrganizationAction* that increases the number of insurgent organizations	
decreaseInsurgentOrganizations	A *conflictOrganizationAction* that decreases the number of insurgent organizations	
increaseTerroristOrganizations	A *conflictOrganizationAction* that increases the number of terrorist organizations	
decreaseTerroristOrganizations	A *conflictOrganizationAction* that decreases the number of terrorist organizations	

Conflict Personnel Actions

A *conflictPersonnelAction* is a *conflictOrgOrPersAction* that relates to changes in conflict personnel. Two of these classes relate to training military and intelligence services. The remainder of the classes cover increasing and decreasing the number of personnel in various armed organizations. The only new Action classes added in the MCO relate to increasing or decreasing the number of military systems and animals (categorizing these as if they were personnel for convenience). The reason for a decrease is unspecified in the class. For example, it could be the result of combat attrition or administrative reorganization. There are 26 classes in this subcategory (Table 5.13).

Table 5.13 Conflict personnel Actions

Action class	Definition	Examples
trainMilitaryForces	A *conflictPersonnelAction* that trains military forces	
trainIntelligenceService	A *conflictPersonnelAction* that trains an intelligence service	
increaseIntervenorDiplomaticPersonnel	A *conflictPersonnelAction* that increases the number of diplomats	
decreaseIntervenorDiplomaticPersonnel	A *conflictPersonnelAction* that decreases the number of diplomats	
increaseTheInterventionForcesPersonnel	A *conflictPersonnelAction* that increases the number of intervention forces personnel	
decreaseTheInterventionForcesPersonnel	A *conflictPersonnelAction* that decreases the number of intervention forces personnel	
decreaseGovtMilitaryForcesPersonnel	A *conflictPersonnelAction* that decreases the number of government military forces personnel	
increaseRegimeSponsoredNonMilitary ArmedForcesPersonnel	A *conflictPersonnelAction* that increases regime-sponsored, non-military armed forces personnel	
decreaseRegimeSponsoredNonMilitary ArmedForcesPersonnel	A *conflictPersonnelAction* that decreases regime-sponsored, non-military armed forces personnel	
increaseParamilitaryForcePersonnel	A *conflictPersonnelAction* that increases number of paramilitary force personnel	
decreaseParamilitaryForcePersonnel	A *conflictPersonnelAction* that decreases number of paramilitary force personnel	
increasePrivateSecurityForcesPersonnel	A *conflictPersonnelAction* that increases number of private security forces personnel	
decreasePrivateSecurityForcesPersonnel	A *conflictPersonnelAction* that decreases number of private security forces personnel	
increaseTerroristPersonnel	A *conflictPersonnelAction* that increases number of terrorists	
decreaseTerroristPersonnel	A *conflictPersonnelAction* that decreases number of terrorists	
increaseInsurgentPersonnel	A *conflictPersonnelAction* that increases number of insurgents	
decreaseInsurgentPersonnel	A *conflictPersonnelAction* that decreases number of insurgents	

(continued)

Table 5.13 (continued)

Action class	Definition	Examples
decreaseIntelligenceServicePersonnel	A *conflictPersonnelAction* that decreases number of intelligence service personnel	
increaseIntervenorSupportPersonnel	A *conflictPersonnelAction* that increases number of intervenor support personnel	To intervenors or government or proto-government or judicial systems
decreaseIntervenorSupportPersonnel	A *conflictPersonnelAction* that decreases number of intervenor support personnel	To intervenors or government or proto-government or judicial systems
increaseGovtMilitaryForcesPersonnel	A *conflictPersonnelAction* that increases the number of government military personnel	
increaseIntelligenceServicesPersonnel	A *conflictPersonnelAction* that increases the number of intelligence services personnel	
increaseMilitarySystems	A *conflictPersonnelAction* that increases the number of a military system Actor	
decreaseMilitarySystems	A *conflictPersonnelAction* that decreases the number of a military system Actor	
increaseAnimals	A *conflictPersonnelAction* that increases the number of an animal Actor or members of a group	
decreaseAnimals	A *conflictPersonnelAction* that decreases the number of an animal	

5.7 Human Affairs Actions

A *humanAffairsAction* is an Action that affects human affairs. This category of Action is comprised of four subcategories: civil training Actions, civil personnel Actions, social aid Actions, and Actions to change the civil situation. Figure 5.8 shows the human affairs Action part of the Action taxonomy. The "action" class box represents all of the Action classes that are children of the subcategories. The fire-fighting training picture in the figure represents a possible civil training Action.

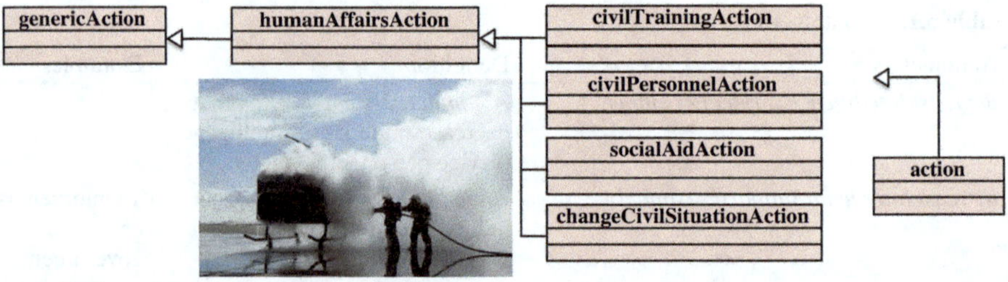

Fig. 5.8 Human affairs Action taxonomy

Social Aid Actions

A *socialAidAction* is a *humanAffairsAction* that involves social aid. There are classes that address providing supplies and services. There are also classes that address changing the number of nongovernmental organizations (NGOs) that provide aid and the numbers of their personnel. There are 20 classes in this subcategory (Table 5.14).

Table 5.14 Social aid Actions

Action class	Definition	Examples
importFood	A *socialAidAction* that imports food	
distributeFood	A *socialAidAction* that distributes food	
distributeWater	A *socialAidAction* that distributes water	
provideSanitationOrWasteWaterManagement	A *socialAidAction* that provides sanitation or wastewater management	
reducePollution	A *socialAidAction* that reduces pollution	
distributeDurableGoodsRelief	A *socialAidAction* that distributes durable goods relief	Other than food and water
prepositionHumanitarianReliefStocks	A *socialAidAction* that prepositions humanitarian relief stocks in or near Host Nation	
provideEducationSupplies	A *socialAidAction* that provides education supplies	
provideMedicalTreatment	A *socialAidAction* that provides medical treatment	
supportHealthcare	A *socialAidAction* that supports general healthcare	Other than medical treatment

(continued)

Table 5.14 (continued)

Action class	Definition	Examples
resettlePeople	A *socialAidAction* that resettles people	Both benign or malicious motives
provideSocialProtectionProgram	A *socialAidAction* that provides a social protection program	Of various kinds
reduceDrugDemand	A *socialAidAction* that reduces use or abuse of or demand for drugs or psychotropic substances	
increaseNGOOrganizations	A *socialAidAction* that increases the number of NGOs, IOs, or IGOs	
decreaseNGOOrganizations	A *socialAidAction* that decreases the number of NGOs, IOs, or IGOs	
increaseNGOWorkers	A *socialAidAction* that increases number of NGO, IO, or IGO workers	
decreaseNGOWorkers	A *socialAidAction* that decreases number of NGO, IO, or IGO workers	
removeWaste	A *socialAidAction* that removes trash, waste, etc.	
provideHealthcareSupplies	A *socialAidAction* that provides healthcare supplies or equipment	
producePotableWater	A *socialAidAction* that produces potable water	Through reverse osmosis, importation, or other means

Civil Training Actions

A *civilTrainingAction* is a *humanAffairsAction* that performs training or education in a civil setting. These classes address training and education for civilians, rather than for active armed forces. No classes were added to the MCO in this subcategory. There are nine classes in this subcategory (Table 5.15).

Table 5.15 Civil training Actions

Action class	Definition	Examples
educateGovtAndShadowGovtPersonnel	A *civilTrainingAction* that educates government or shadow government personnel in how various government processes work	
trainFirstResponders	A *civilTrainingAction* that trains first responders	Fire, medical, etc.
trainNewPoliticalLeaders	A *civilTrainingAction* that trains new political leaders in responsibilities, processes, etc.	
trainLawEnforcementPersonnel	A *civilTrainingAction* that trains law enforcement personnel	
provideJobTrainingOrEmploymentForDischargedMilitaryPersonnel	A *civilTrainingAction* that provides job training or employment for discharged military personnel	
trainEducators	A *civilTrainingAction* that trains educators	
provideJobTraining	A *civilTrainingAction* that provides general job training	
sponsorMediaTrainingOrProfessionalization	A *civilTrainingAction* that sponsors media training or professionalization	Fair, honest, and complete reporting, etc.
educateStudents	A *civilTrainingAction* that educates students	

Civil Personnel Actions

A *civilPersonnelAction* is a *humanAffairsAction* that relates to changes in personnel in a civil setting. Most of these classes address changes to the number of personnel in civilian organizations (rather than armed organizations). Other classes address changes to the number of people in certain categories, such as migrants and workers. The only new Action classes added in the MCO relate to increasing or decreasing the number of civilian vehicles (categorizing these as if they were personnel for convenience). There are 34 classes in this subcategory (Table 5.16).

Table 5.16 Civil personnel Actions

Action class	Definition	Examples
provideProgramOrPersonnelSupport	A *civilPersonnelAction* that provides program design or learning or personnel to support FASP programs	Foreign assistance standardized program
changeWorkersJobStatus	A *civilPersonnelAction* that hires, fires, retires, or changes job of a worker	
decreaseFirstRespondersPersonnel	A *civilPersonnelAction* that decreases number of first responders	
increasePoliticalPopulation	A *civilPersonnelAction* that increases the number of politically active people	
decreasePoliticalPopulation	A *civilPersonnelAction* that decrease the number of politically active people	
increaseWorkers	A *civilPersonnelAction* that increases number of workers	
decreaseWorkers	A *civilPersonnelAction* that decreases number of workers	
increaseIDPs	A *civilPersonnelAction* that increases number of internally displaced persons	
decreaseIDPs	A *civilPersonnelAction* that decreases number of internally displaced persons	
increaseMigrants	A *civilPersonnelAction* that increases number of migrants from outside the country (due to pull factors)	
decreaseMigrants	A *civilPersonnelAction* that decreases number of migrants from outside the country (due to pull factors)	
increaseRefugees	A *civilPersonnelAction* that increases number of refugees from outside the country (due to push factors)	
decreaseRefugees	A *civilPersonnelAction* that decreases number of refugees from outside the country (due to push factors)	
increaseExpatriates	A *civilPersonnelAction* that increases number of expatriates	

(continued)

Table 5.16 (continued)

Action class	Definition	Examples
decreaseExpatriates	A *civilPersonnelAction* that decreases number of expatriates	
increaseGeneralPopulation	A *civilPersonnelAction* that increase number in general population	
decreaseGeneralPopulation	A *civilPersonnelAction* that decreases number in general population	
decreaseLawEnforcementPersonnel	A *civilPersonnelAction* that decreases number of law enforcement personnel	
increaseGovtPersonnel	A *civilPersonnelAction* that increases number of government personnel	
decreaseGovtPersonnel	A *civilPersonnelAction* that decreases number of government personnel	
increaseContractorPersonnel	A *civilPersonnelAction* that increases number of contractor personnel	
decreaseContractorPersonnel	A *civilPersonnelAction* that decreases number of contractor personnel	
increaseInternationalOrLocalMediaPersonnel	A *civilPersonnelAction* that increases number of international or local media personnel	
decreaseInternationalOrLocalMediaPersonnel	A *civilPersonnelAction* that decreases number of international or local media personnel	
decreaseEducators	A *civilPersonnelAction* that decrease number of education personnel	
increaseHealthcarePersonnel	A *civilPersonnelAction* that increases the number of healthcare workers	
decreaseHealthcarePersonnel	A *civilPersonnelAction* that decreases the number of healthcare workers	
increaseCriminalPopulation	A *civilPersonnelAction* that increases the size of the criminal population	
decreaseCriminalPopulation	A *civilPersonnelAction* that decreases the size of the criminal population	
increaseEducators	A *civilPersonnelAction* that increases number of education personnel	

(continued)

Table 5.16 (continued)

Action class	Definition	Examples
increaseFirstRespondersPersonnel	A *civilPersonnelAction* that increases the number of first responder personnel	
increaseLawEnforcementPersonnel	A *civilPersonnelAction* that increases the number of law enforcement personnel	
increaseCivilianVehicles	A *civilPersonnelAction* that increases the number of a civilian vehicle Actor	
decreaseCivilianVehicles	A *civilPersonnelAction* that decreases the number of a civilian vehicle Actor	

Change Civil Situation Actions

A *changeCivilSituationAction* is a *humanAffairsAction* that relates to changes in the civil situation. These include changes in the composition of various factions and populations, changes in leadership, and response to emergencies. There are eight classes in this subcategory (Table 5.17).

Table 5.17 Change civil situation Actions

Action class	Definition	Examples
negotiateWithBureaucracies ToGetRelief	A *changeCivilSituationAction* that negotiates with bureaucracies (Host Nation or international) to get relief for Host Nation populace	
changePoliticalFactions	A *changeCivilSituationAction* that changes the composition of political factions	
changeCulturalPopulation	A *changeCivilSituationAction* that changes the makeup of cultural population	
changeReligiousPopulation	A *changeCivilSituationAction* that changes the makeup of religious population	
changeKeyLeaderIdentities	A *changeCivilSituationAction* that adds, subtracts, or changes names of various types of key leaders	
respondToCivilEmergencies	A *changeCivilSituationAction* that responds to emergencies	Needing fire, medical, police, etc.
changeSocialFactions	A *changeCivilSituationAction* that changes the composition of social factions	
changeReligiousFactions	A *changeCivilSituationAction* that changes the composition of religious factions	

5.8 Economic Actions

An *economicAction* is an Action that is economic in nature. This category of Action is comprised of four subcategories: business economic Actions, consume or produce Actions, civic building Actions, and business organizational Actions. Figure 5.9 shows the economic Action part of the Action taxonomy. The "action" class box represents all of the Action classes that are children of the subcategories. The picture of a building under construction in the figure represents a possible civil building Action.

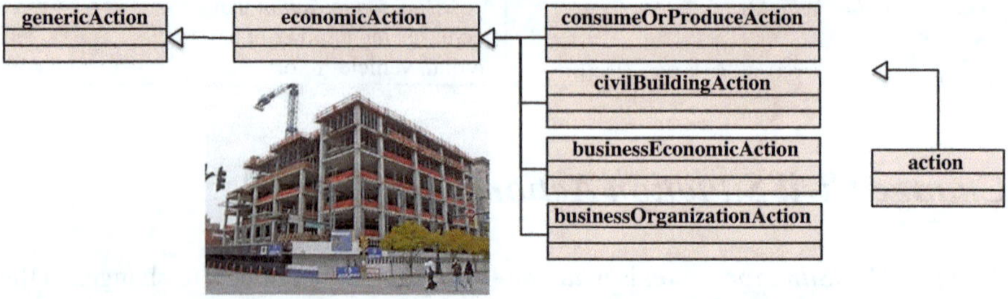

Fig. 5.9 Economic Action taxonomy

Business Economic Actions

A *businessEconomicAction* is an *economicAction* that relates to business. These are general business actions, rather than those relating to a specific business. The last two classes have been added as natural opposites to two preceding classes. There are eight classes in this subcategory (Table 5.18).

Table 5.18 Business economic Actions

Action class	Definition	Examples
seekInvestment Capital	A *businessEconomicAction* that seeks investment capital for use in Host Nation from within Host Nation or from foreign sources	
conductLabor Strikes	A *businessEconomicAction* that conducts labor strikes against business or government organizations	
conductTradeIn GoodsOrServices	A *businessEconomicAction* that imports or exports goods or services	
conductCultural Event	A *businessEconomicAction* that conducts cultural event	Theater, museum, or sports
conductBusiness Management	A *businessEconomicAction* that identifies companies to purchase; reorganizes companies; obtains financing; manages operations	
obtainFunds	A *businessEconomicAction* that acquires funds in various ways	
inhibitInvestment Capital	A *businessEconomicAction* that inhibits investment capital for use in the Host Nation from within or from foreign services	
disburseFunds	A *businessEconomicAction* that spends funds for various purposes	

Consume or Produce Actions

A *consumeOrProduceAction* is an *economicAction* that relates to consumption or production of a good or service. The last four classes have been added to the MCO, in part to allow for the consumption of power, supplies, and money in the operation of the infrastructure. There are 13 classes in this subcategory (Table 5.19).

Table 5.19 Consume or produce Actions

Action class	Definition	Examples
conductAgricultureOperation	A *consumeOrProduceAction* that produces food and other agricultural products (operate part of business infrastructure)	Farming, cattle raising, etc.
consumeFood	A *consumeOrProduceAction* that consumes food	
produceGoodsOrEquipment	A *consumeOrProduceAction* that produces goods or equipment (other than food) (operates part of business infrastructure)	Mining, manufacturing, building materials, etc.
consumeGoodsOrEquipment	A *consumeOrProduceAction* that consumes goods or equipment (other than food)	
produceWaste	A *consumeOrProduceAction* that produces waste	Trash, garbage, human waste, etc.
consumeNaturalResources	A *consumeOrProduceAction* that consumes natural resources	
produceEnergy	A *consumeOrProduceAction* that produces energy of any type (operates part of energy infrastructure)	Electricity, oil, gas, etc.
operateFishingBusiness	A *consumeOrProduceAction* that produces food and other fishing-related products (operates part of business infrastructure)	Fishing, aquaculture, etc.
provideServices	A *consumeOrProduceAction* that provides services of various kinds (operates part of business infrastructure)	Shops, financial services, media, etc.
operateWaterInfrastructure	A *consumeOrProduceAction* that operates part of the water infrastructure	Potable water, sewage treatment, recreation, etc.
operateCivilTransportInfrastructure	A *consumeOrProduceAction* that operates part of the civil (non-military) transportation infrastructure	Road, rail, water, air transportation
operateSocialInfrastructure	A *consumeOrProduceAction* that operates part of the social infrastructure	Living in houses, running healthcare center, etc.
operateNonMilGovtInfrastructure	A *consumeOrProduceAction* that operates part of the non-military government infrastructure	Runs prisons, government buildings, etc.

Civil Building Actions

A *civilBuildingAction* is an *economicAction* that builds, rebuilds, or repairs civil infrastructure. Each class focuses on a different type of infrastructure. There are 29 classes in this subcategory (Table 5.20).

Table 5.20 Civil building Actions

Action class	Definition	Examples
provideTemporaryShelterHousingRefugeeCamps	A *civilBuildingAction* that provides temporary shelter/ housing/refugee camps (in or near Host Nation)	
rebuildCivilianHousing	A *civilBuildingAction* that builds or rebuilds civilian housing	
rebuildReligiousFacility	A *civilBuildingAction* that builds or rebuilds religious facilities	
rebuildManufacturingInfrastructure	A *civilBuildingAction* that builds or rebuilds manufacturing structures or equipment	
rebuildAgricultureInfrastructureLivestock	A *civilBuildingAction* that builds or rebuilds agriculture structures, equipment, or livestock	
rebuildShopsOrCommercialInfrastructure	A *civilBuildingAction* that builds or rebuilds shops, commercial structures, or their equipment	
rebuildInformationAndMediaInfrastructure	A *civilBuildingAction* that builds or rebuilds media infrastructure or equipment	
rebuildSchoolOrEducationalInfrastructure	A *civilBuildingAction* that builds or rebuilds schools or educational structures, including equipment	
repairHealthcareInfrastructure	A *civilBuildingAction* that builds or rebuilds health infrastructure, including equipment	
rebuildElectricityProductionPlant	A *civilBuildingAction* that builds or rebuilds electricity production plants, including equipment	
rebuildElectricityDistributionInfrastructure	A *civilBuildingAction* that builds or rebuilds electricity distribution system, including equipment	

(continued)

Table 5.20 (continued)

Action class	Definition	Examples
rebuildExtractiveEnergyProductionInfrastructure	A *civilBuildingAction* that builds or rebuilds extractive energy (oil, coal, etc.) production, including equipment	
rebuildExtractiveEnergyTransportationInfrastructure	A *civilBuildingAction* that builds or rebuilds extractive energy transportation (e.g., oil pipeline), including equipment	
rebuildGovtInfratructure	A *civilBuildingAction* that builds or rebuilds government structure (including police), including equipment	
rebuildRoadInfrastructure	A *civilBuildingAction* that builds or rebuilds road	
rebuildRailroadInfrastructure	A *civilBuildingAction* that builds or rebuilds railroad, including equipment	
rebuildBridgeOrTunnelInfrastructure	A *civilBuildingAction* that builds or rebuilds bridges or tunnels, including equipment	
rebuildSeaportInfrastructure	A *civilBuildingAction* that builds or rebuilds seaports, including equipment	
rebuildAirportInfrastructure	A *civilBuildingAction* that builds or rebuilds airports, including equipment	
rebuildReplaceVehicle	A *civilBuildingAction* that repairs or replaces non-military vehicle	
rebuildWaterwaysInfrastructure	A *civilBuildingAction* that builds or rebuilds waterways, including equipment	
rebuildWaterDistributionInfrastructure	A *civilBuildingAction* that builds or rebuilds water distribution, including equipment	
rebuildWaterOrSewageTreatmentFacilities	A *civilBuildingAction* that builds or rebuilds water or sewage treatment facilities, including equipment	
rebuildDamInfrastructure	A *civilBuildingAction* that builds or rebuilds dams, including equipment	

(continued)

Table 5.20 (continued)

Action class	Definition	Examples
assistInMISOrItsUse	A *civilBuildingAction* that assists in creating, protecting, or using MIS systems	
buildPrisonInfrastructure	A *civilBuildingAction* that builds or rebuilds prison or jail	
rebuildMiningInfrastructure	A *civilBuildingAction* that builds or rebuilds mining infrastructure	
rebuildCulturalInfrastructure	A *civilBuildingAction* that builds or rebuilds cultural infrastructure	Theater, museum, or sports
rebuildFinancialInfrastructure	A *civilBuildingAction* that builds or rebuilds financial infrastructure	

Business Organization Actions

A *businessOrganizationAction* is an *economicAction* that relates to changes in a business organization. These classes allow for the change in number of business organizations of each type. Changes in the number of personnel in business organizations are provided by the two classes, *increaseWorkers* and *decreaseWorkers* in the *civilPersonnelAction* subcategory of the *humanAffairsAction* category. There are 34 classes in this subcategory (Table 5.21).

Table 5.21 Business organization Actions

Action class	Definition	Examples
increaseEnergyBusinesses	A *businessOrganizationAction* that increases energy businesses	
decreaseEnergyBusinesses	A *businessOrganizationAction* that decreases energy businesses	
increaseFinancialServicesIndustryBusinesses	A *businessOrganizationAction* that increases size of financial services industry	
decreaseFinancialServicesIndustryBusinesses	A *businessOrganizationAction* that decreases size of financial services industry	
increaseWorkerOrganizations	A *businessOrganizationAction* that increases number of worker organizations	
decreaseWorkerOrganizations	A *businessOrganizationAction* that decreases number of worker organizations	

(continued)

Table 5.21 (continued)

Action class	Definition	Examples
increaseMarkets	A *businessOrganizationAction* that increases number or effectiveness of markets	
decreaseMarkets	A *businessOrganizationAction* that decreases number or effectiveness of markets	
increaseCommercialSector	A *businessOrganizationAction* that increases the commercial sector	
decreaseCommercialSector	A *businessOrganizationAction* that decreases the commercial sector	
increaseMediaBusinesses	A *businessOrganizationAction* that increases number of media businesses	
decreaseMediaBusinesses	A *businessOrganizationAction* that decreases number of media businesses	
increaseManufacturingBusinesses	A *businessOrganizationAction* that increases manufacturing businesses	
decreaseManufacturingBusinesses	A *businessOrganizationAction* that decreases manufacturing businesses	
increaseServiceBusinesses	A *businessOrganizationAction* that increases number or diversity of service businesses	
decreaseServiceBusinesses	A *businessOrganizationAction* that decreases number or diversity of service businesses	
increaseTransportationBusinesses	A *businessOrganizationAction* that increases transport businesses	
decreaseTransportationBusinesses	A *businessOrganizationAction* that decreases transport businesses	
increaseTourismIndustryBusinesses	A *businessOrganizationAction* that increases tourism industry	
decreaseTourismIndustryBusinesses	A *businessOrganizationAction* that decreases tourism industry	
increaseMiningBusinesses	A *businessOrganizationAction* that increases mining businesses	
decreaseMiningBusinesses	A *businessOrganizationAction* that decreases mining businesses	
increaseFishingBusinesses	A *businessOrganizationAction* that increases fishing businesses	
decreaseFishingBusinesses	A *businessOrganizationAction* that decreases fishing businesses	
increaseTimberBusinesses	A *businessOrganizationAction* that increases timber businesses	
decreaseTimberBusinesses	A *businessOrganizationAction* that decreases timber businesses	

(continued)

Table 5.21 (continued)

Action class	Definition	Examples
increaseCulturalBusinesses	A *businessOrganizationAction* that increases cultural businesses	Theater, museum, or sports
decreaseCulturalBusinesses	A *businessOrganizationAction* that decreases cultural businesses	Theater, museum, or sports
increaseAgricultureBusinesses	A *businessOrganizationAction* that increases agriculture businesses	
decreaseAgricultureBusinesses	A *businessOrganizationAction* that decreases agriculture businesses	
increaseCriminalOrganizations	A *businessOrganizationAction* that increases criminal organizations	
decreaseCriminalOrganizations	A *businessOrganizationAction* that decreases criminal organizations	
increaseContractorBusinesses	A *businessOrganizationAction* that increases contractor businesses	
decreaseContractorBusinesses	A *businessOrganizationAction* that decreases contractor businesses	

5.9 Policing or Criminal Actions

A *policingOrCriminalAction* is an Action that relates to police work or security. This category of Action is comprised of two subcategories: criminal Actions and policing Actions. Figure 5.10 shows the policing or criminal Action part of the Action taxonomy. The "action" class box represents all of the Action classes that are children of the subcategories. The robber making off with the loot in the figure represents a criminal Action.

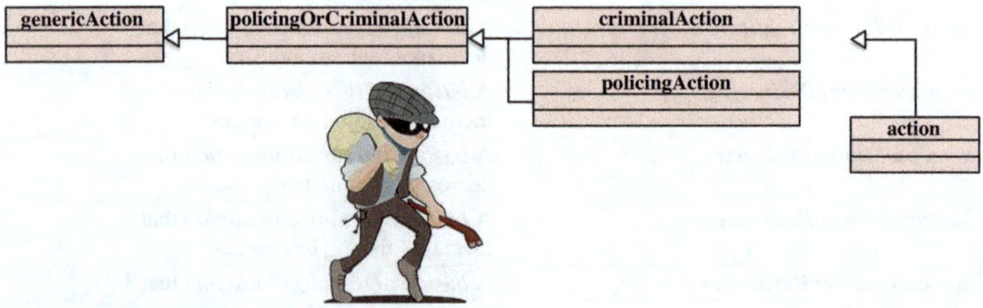

Fig. 5.10 Policing or criminal Action taxonomy

Criminal Actions

A *criminalAction* is a *policingOrCriminalAction* that is one type of criminal Action. The classes that have been defined are based on observed activities in conflict situations. The class on human trafficking has been added to match the opposing class in the *policingAction* subcategory. There are nine classes in this subcategory (Table 5.22).

Table 5.22 Criminal Actions

Action class	Definition	Examples
extortSuppressPopulationOpposition	A *criminalAction* that conducts extortion or suppression actions against the general population or opposition groups	
conductExtrajudicialAction	A *criminalAction* that conducts various extrajudicial actions	Killings, intimidation, including corrupt activities
engageInCriminalOrCorruptAction	A *criminalAction* that engages in criminal or corrupt activities	Intimidation, kidnapping, murder, smuggling, drug trafficking, bribery, "protection," illicit services, self-dealing, prostitution, etc.
conductDrugTrade	A *criminalAction* that engages in drug production and movement or trafficking activities	
conductFinancialCrimeOrMoneyLaundering	A *criminalAction* that conducts financial crimes or money laundering	
conductIntellectualPropertyTheft	A *criminalAction* that conducts intellectual property theft, corporate espionage, or cyber crimes	
engageInOrganizedOrGangRelatedCrime	A *criminalAction* that creates organizations or gangs and engages in crime	
createCivilDisturbance	A *criminalAction* that creates civil disturbances that may lead to illegal actions	Parades, demonstrations, peaceful protests, riots
conductTraffickingInPersons	A *criminalAction* that conducts trafficking in persons	

Policing Actions

A *policingAction* is a *policingOrCriminalAction* that is one type of policing Action. The classes that have been defined are based on observed activities in conflict situations. The class on white collar crime has been added to oppose the white collar crime classes in the *criminalAction* subcategory. There are six classes in this subcategory (Table 5.23).

Table 5.23 Policing Actions

Action class	Definition	Examples
conductPolicingOperation	A *policingAction* that conducts a law enforcement type operation	Patrol, arrest, etc.
provideAntiTraffickingInPersons	A *policingAction* that provides anti-trafficking in persons	
interdictDrugs	A *policingAction* that prevents, interrupts, captures, or eliminates drug production and movement or trafficking activities	
reduceOrganizedOrGangRelatedCrime	A *policingAction* that reduces organized or gang-related crime	
quellCivilDisturbance	A *policingAction* that reduces or stops civil disturbances	Parades, demonstrations, peaceful protests, riots
conductAntiWhiteCollarCrimeOp	A *policingAction* that conducts operations against white collar crime	Financial, money laundering, intellectual property theft, etc.

5.10 Civil Government Actions

A *civilGovernmentAction* is an Action that is governmental in nature, excluding purely nongovernment economic actions and military actions. This category of Action is comprised of three subcategories: policy or legal Actions, government economic Actions, and government organizational Actions. Figure 5.11 shows the civil government Action part of the Action taxonomy. The "action" class box represents all of the Action classes that are children of the subcategories. The veto stamp in the figure represents a possible policy or legal Action.

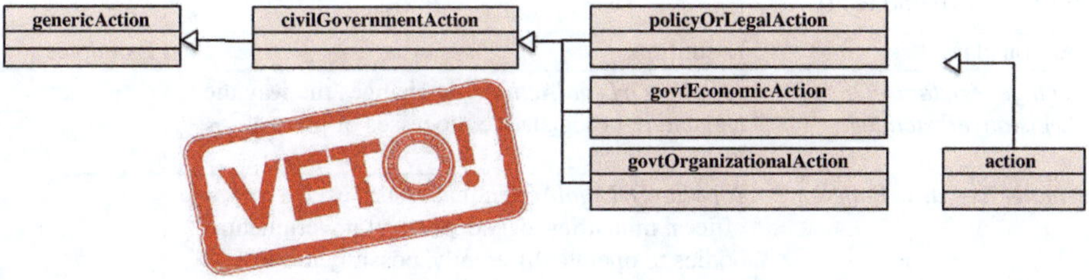

Fig. 5.11 Civil government Action taxonomy

Policy or Legal Actions

A *policyOrLegalAction* is a *civilGovernmentAction* that relates to government policy and legal actions. The policy change classes categorize the types of policies that a government makes. The legal actions reflect the activities that have been observed in conflict situations. There are 26 classes in this subcategory (Table 5.24).

Table 5.24 Policy or Legal Actions

Action class	Definition	Examples
developConstitution	A *policyOrLegalAction* that writes constitution; advocates for or establishes constitutional reform process	
transferControlOfGovt Functions	A *policyOrLegalAction* that transfers control from one governing body to another, in part or in toto	
changeThePenalSystem	A *policyOrLegalAction* that creates or changes the penal systems	
changeTheLegalSystem	A *policyOrLegalAction* that creates or changes the legal system	
conductWarCrimes Investigation	A *policyOrLegalAction* that conducts war crimes investigations, tribunals, etc.	
changePropertyProcedure	A *policyOrLegalAction* that changes property law, regulations, enforcement, etc.	
changeAgriculturalPolicy	A *policyOrLegalAction* that changes agricultural policy	
supportReductionOfDrug Crops	A *policyOrLegalAction* that supports reduction of drug crops	Eradication, replacement crops, etc.
changeSocialSafetyNet	A *policyOrLegalAction* that changes Host Nation social safety net	
reduceFinancialCrimes OrMoneyLaundering	A *policyOrLegalAction* that assists in drafting legislation or implementing regulations or training to reduce financial crimes or money laundering	
reduceIntellectual PropertyTheft	A *policyOrLegalAction* that builds capacity to detect, investigate, prosecute, or prevent intellectual property theft and corporate espionage or to increase cyber security	

(continued)

Table 5.24 (continued)

Action class	Definition	Examples
changeLegislatureOr LegislativePractices	A *policyOrLegalAction* that changes the way the legislature or legislative processes or procedures work	
changeExecutiveFunction	A *policyOrLegalAction* that assists executive offices, ministries, or independent governmental bodies to operate differently, possibly more efficiently or effectively or democratically	
promoteAntiCorruption Reforms	A *policyOrLegalAction* that promotes anti-corruption institutions, processes or policies regarding all sources of corruption	
promoteCivilControlOf SecuritySector	A *policyOrLegalAction* that supports civil control of military or other elements of the security sector	
changeInformationAnd MediaPolicies	A *policyOrLegalAction* that promotes policies for media freedom or freedom of information	
conductJudicialAction	A *policyOrLegalAction* that holds courts; hears evidence; makes decisions, etc.	
conductLegislativeAction	A *policyOrLegalAction* that proposes or debates laws; passes some, etc.	
conductExecutiveAction	A *policyOrLegalAction* that executes laws, performs administrative tasks, establishes regulations, etc.	
changeTransportation Policy	A *policyOrLegalAction* that changes Host Nation transportation policy	
changeEnergyPolicy	A *policyOrLegalAction* that changes Host Nation energy policy	
changeNaturalResources Policy	A *policyOrLegalAction* that changes Host Nation natural resources policy	
changeLaborPolicy	A *policyOrLegalAction* that changes Host Nation labor policy	
changeEducationPolicy	A *policyOrLegalAction* that changes Host Nation education policy	
changeHealthcarePolicy	A *policyOrLegalAction* that changes Host Nation healthcare policy	
changeSocialOr CulturalPolicy	A *policyOrLegalAction* that changes Host Nation social or cultural policy	

Government Economic Actions

A *govtEconomicAction* is a *civilGovernmentAction* that relates to government economic actions. These classes relate to government economic actions, as opposed to business economic actions. However, depending on the economic system (capitalist, socialist, etc.), the line between these may shift. The class on nationalizing a business has been added as the antithesis of the class on privatizing a business. There are 16 classes in this subcategory (Table 5.25).

Table 5.25 Government economic Actions

Action class	Definition	Examples
supportAgricultureDirectly	A *govtEconomicAction* that supports agriculture directly	Buy or sell produce/food, support planting or conduct agriculture support programs, etc.
importEnergy	A *govtEconomicAction* that imports energy into Host Nation (petroleum, electricity, coal, etc.)	
createInsuranceSystem	A *govtEconomicAction* that creates or changes the Host Nation insurance system	
createInterbanksPaymentSystem	A *govtEconomicAction* that creates or changes the Host Nation interbank payment system	
createNewCurrency	A *govtEconomicAction* that creates new currency -or other similarcurrencychangeactions-	
developMicrofinanceSystem	A *govtEconomicAction* that develops or changes microfinance systems in Host Nation	
createStockMarket	A *govtEconomicAction* that creates or changes the Host Nation stock market	
privatizeBusiness	A *govtEconomicAction* that privatizes a Host Nation or external business operating in Host Nation	
changeGovtEconomicOrFinancialPolicy	A *govtEconomicAction* that changes the Host Nation government economic or financial policy	
assistEconomicIntegrationOrCooperation	A *govtEconomicAction* that assists in Host Nation economic integration or cooperation across private, government, international sectors	Strategy/ assessment, prices or subsidies, debt management, arears clearance
changeCommercialLaw	A *govtEconomicAction* that strengthens the Host Nation commercial law	
changeTaxOrTradePolicy	A *govtEconomicAction* that changes the Host Nation tax or trade policy	

(continued)

Table 5.25 (continued)

Action class	Definition	Examples
createPublicWorksProgram	A *govtEconomicAction* that creates a public works program within Host Nation to generate jobs	
manageNaturalResources	A *govtEconomicAction* that manages Host Nation natural resources	Energy reserves, water, raw materials, land, etc.
participateDirectlyInEconomy	A *govtEconomicAction* that consists of direct external participation in economy	Buying or selling commodities, other than agricultural
nationalizeBusiness	A *govtEconomicAction* that nationalizes a Host Nation or external business operating in the Host Nation	

Government Organization Actions

A *govtOrganizationAction* is a *civilGovernmentAction* that attempts to change a government organization. These classes address changes to the number of government organizations of various types and other changes to the government, such as conducting elections. There are 22 classes in this subcategory (Table 5.26).

Table 5.26 Government organization Actions

Action class	Definition	Examples
createGovt	A *govtOrganizationAction* that creates or reforms government organizations or whole government	
conductElection	A *govtOrganizationAction* that consists of election planning or execution, elections monitoring, elections outreach	
establishStaffOrFundTransitionGovt	A *govtOrganizationAction* that creates a transition (substitute Host Nation) government; provides staff for it; funds it; supplies it	
rebuildOrMonitorLawEnforcementOrganizations	A *govtOrganizationAction* that creates, reforms, or monitors a law enforcement organization, including provision of staff, funds, or supplies	

(continued)

Table 5.26 (continued)

Action class	Definition	Examples
demobilizeReduceReintegrateMilitaryOrParamilitaryUnits	A *govtOrganizationAction* that demobilizes, reduces, or reintegrates military or paramilitary units	
assistInCreatingSocialServices	A *govtOrganizationAction* that promotes good social service policies or social service institutions	
increaseHealthcareOrganizations	A *govtOrganizationAction* that increases the number of healthcare organizations	
decreaseHealthcareOrganizations	A *govtOrganizationAction* that decreases the number of healthcare organizations	
increaseEducationOrganizations	A *govtOrganizationAction* that increases the number of education organizations	
decreaseEducationOrganizations	A *govtOrganizationAction* that decreases the number of education organizations	
increaseLawEnforcementOrganizations	A *govtOrganizationAction* that increases the number of law enforcement organizations	
decreaseLawEnforcementOrganizations	A *govtOrganizationAction* that decreases the number of law enforcement organizations	
increaseBureaucracyOrganizations	A *govtOrganizationAction* that increases the number of bureaucracy organizations	
decreaseBureaucracyOrganizations	A *govtOrganizationAction* that decreases the number of bureaucracy organizations	
increaseGovtOrganizations	A *govtOrganizationAction* that increases the number of government organizations	
decreaseGovtOrganizations	A *govtOrganizationAction* that decreases the number of government organizations	
increaseJudicialOrganizations	A *govtOrganizationAction* that increases the number of judicial organizations (including alternative courts)	
decreaseJudicialOrganizations	A *govtOrganizationAction* that decreases the number of judicial organizations (including alternative courts)	
increaseFirstResponderOrganizations	A *govtOrganizationAction* that increases the number of first responder organizations	

(continued)

Table 5.26 (continued)

Action class	Definition	Examples
decreaseFirstResponderOrganizations	A *govtOrganizationAction* that decreases the number of first responder organizations	
increaseSocialServicesOrganizations	A *govtOrganizationAction* that increases the number of social services organizations	
decreaseSocialServicesOrganizations	A *govtOrganizationAction* that decreases the number of social services organizations	

5.11 Action Relationships

The only relationships used in this chapter are the *is-a* and *hasMetric* relationships, explained as follows:

is-a: Each instance of A is an instance of B, and each property of B is a property of A. Its inverse relationship is *superClassOf.*

hasMetric: A has Metric B (also shown as *described by*). Its inverse relationship is *metricOf.*

5.12 Review of Actions

We often refer to "Actions" as shorthand for "Action classes." Each Action class is a thing that defines the instantiations of the class to be a particular subset of all of the things that are the direct causes of changes. The Action categories and subcategories are an aid to this definition process: each narrows the subset in a meaningful way.

The verbs of the unconventional conflict domain are represented as Actions. Despite the content of a verb, there is no assurance that the Action is or will be successful. These Actions are initiated by the Actors of Chap. 4 and affect the Actors and the Objects, as discussed in Chap. 6. Actions may also be categorized by the DIME paradigm, as discussed in Chap. 9.

Chapter 6
MCO Object Ontology

Objects are the recipients of actions in the Modern Conflict Ontology (MCO): they don't cause things to happen; they are passive and acted upon. Generally, Objects are not people, but may contain people in certain cases. Figure 3.8 showed how the Object Ontology fits into the overall MCO structure (with Fig. 6.1 illustrating its icon as used in the Context Diagram, Fig. 3.5).

Fig. 6.1 Object icon

6.1 Overview of Objects

There are seven categories of Objects: infrastructure Objects, needed thing Objects, natural Objects, conflict Objects, governing Objects, conceptual Objects, and composite Objects. Figure 6.2 shows the top level of the Object taxonomy. The abstract sculpture in the figure illustrates the wide variety of Objects that may be represented in the ontology.

D. S. Hartley III, *An Ontology of Modern Conflict*, Understanding Complex Systems, https://doi.org/10.1007/978-3-030-53214-7_6

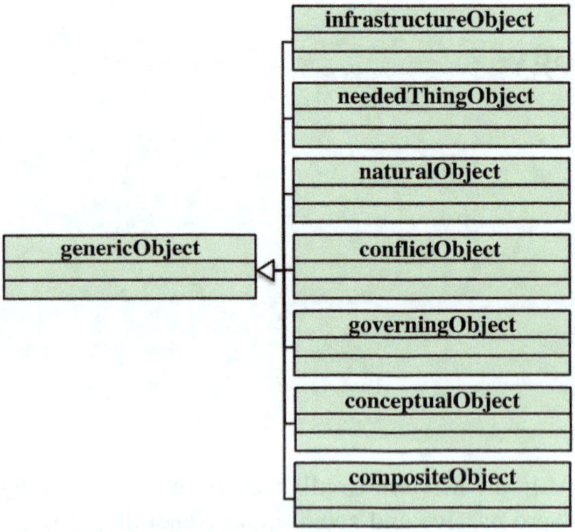

Fig. 6.2 Object taxonomy overview

Changes to Objects from the UCO

The first change is that, in the Unconventional Conflict Ontology (UCO) terminology, these things were environment elements, referring to their status as passive parts of the environment. That terminology was adequate, but clumsy. The term "object" is also perhaps only adequate, but less clumsy. This change is also reflected in the category and subcategory names. In Chap. 3, we discussed the general changes from the UCO to the Modern Conflict Ontology (MCO). Most of these changes are formal, e.g., changing spelling conventions and definition standards. For the most part, the tables of the Environment Elements contained in the book about the UCO describe the content adequately (Hartley D. S., An Ontology for Unconventional Conflict, 2018). However, there have been changes at the category and subcategory level of the Object Elements in the MCO, which make cross-referencing difficult. For example, the *ShelterInfrastucture* subcategory of the UCO has been eliminated and the two classes within it moved to other subcategories in the MCO. In addition, a new category, *compositeObject*, is introduced in the MCO.

An Insight into Objects

The Infrastructure Objects category is the most easily described category because so much of it is concrete (pun intended). The infrastructure subcategories are easily understood: water, transport, government, energy, business, and social. Each of these is decomposed into natural classes. As with the other elements of the ontology, these classes may be instantiated in various ways. For example, the *airportInfrastructure* class might be instantiated as the infrastructure for a particular airport or it might be instantiated as the airport infrastructure for all of the airports in a province or it might be instantiated as a particular kind of airport infrastructure at a particular airport. The choices depend upon the needs of the user.

The other categories of Objects become less obvious as they become more abstract; however, the nature of the decomposition and instantiation remains the same.

6.2 Infrastructure Objects

An *infrastructureObject* is an Object that is a portion of infrastructure. This category of Object is comprised of six subcategories: water infrastructure, transport infrastructure, government infrastructure, energy infrastructure, business infrastructure, and social infrastructure. Figure 6.3 shows the infrastructure Object part of the Object taxonomy. The "object" class box represents all of the Object classes that are children of the subcategories. The railroad marshalling yard in the figure represents an instantiation of a transportation infrastructure Object.

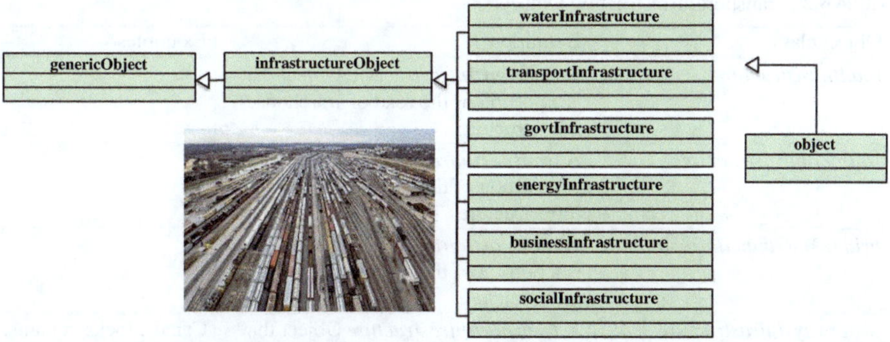

Fig. 6.3 Infrastructure Object taxonomy

Water Infrastructure

A *waterInfrastructure* Object is an *infrastructureObject* that is a portion of the water supply infrastructure. Each class identifies a different part of the infrastructure. There are three classes in this subcategory (Table 6.1).

Table 6.1 Water infrastructure Objects

Object class	Definition	Examples
waterDistributionInfrastructure	A *waterInfrastructure* Object that is a portion of the water distribution infrastructure	Pipes, pumping station
waterAndSewageTreatmentInfrastructure	A *waterInfrastructure* Object that is a portion of the water and sewage treatment infrastructure	Pipes, settling pond
damInfrastructure	A *waterInfrastructure* Object that is a portion of the dam infrastructure	Dam and equipment

Transport Infrastructure

A *transportInfrastructure* Object is an *infrastructureObject* that is a portion of the transportation infrastructure. Each class identifies a different part of the infrastructure. A general transportation infrastructure Object in the UCO has been modified and moved to the *compositeObject* category in the MCO. There are seven classes in this subcategory (Table 6.2).

Table 6.2 Transport infrastructure Objects

Object class	Definition	Examples
roadInfrastructure	A *transportInfrastructure* Object that is a portion of the road or warehouse infrastructure	
railroadInfrastructure	A *transportInfrastructure* Object that is a portion of the railroad infrastructure	
bridgeAndTunnelInfrastructure	A *transportInfrastructure* Object that is a portion of the bridge or tunnel infrastructure	
waterwaysInfrastructure	A *transportInfrastructure* Object that is a portion of the waterways infrastructure	Canals, locks at dams, equipment
seaportInfrastructure	A *transportInfrastructure* Object that is a portion of the seaport infrastructure	Docks, buildings, cranes, etc.
airportInfrastructure	A *transportInfrastructure* Object that is a portion of the airport infrastructure	Buildings, runways, equipment
vehicleNonMilitary	A *transportInfrastructure* Object that is a non-military vehicle	Autos, planes, ships, etc. that aren't being considered actors

Government Infrastructure

A *govtInfrastructure* Object is an *infrastructureObject* that is a portion of the government infrastructure. Each class identifies a different part of the infrastructure. This subcategory has been extensively reworked from the UCO subcategory. A general government infrastructure Object has been modified and moved to the *compositeObject* category and several Objects relevant to combat have been added. There are nine classes in this subcategory (Table 6.3).

Table 6.3 Government infrastructure Objects

Object class	Definition	Examples
prisonInfrastructure	A *govtInfrastructure* Object that is a portion of the prison infrastructure	
adminInfrastructure	A *govtInfrastructure* Object that is a portion of the government -including police- administrative and ceremonial infrastructure	
militaryInfrastructure	A *govtInfrastructure* Object that is a portion of the military -including intervention- infrastructure	
militaryVehicle	A *govtInfrastructure* Object that is a military -including intervention- vehicle	Vehicles that aren't being considered actors
interventionC4I	A *govtInfrastructure* Object that is a portion of the intervention force C4I	Command, control, communications, computers, and intelligence
HNMilitaryC4I	A *govtInfrastructure* Object that is a portion of the Host Nation's military command, control, communications, computers, and intelligence	Internet, computer systems, etc.
otherC4I	A *govtInfrastructure* Object that is a portion of the C4I for non-intervention, non-Host Nation military	Command, control, communications, computers, and intelligence
munition	A *govtInfrastructure* Object that is a munition	Bomb, tank round, rocket fired from reusable launcher
weapon	A *govtInfrastructure* Object that is a weapon	Gun tube on towed or self-propelled howitzer, gun on ship

Energy Infrastructure

An *energyInfrastructure* Object is an *infrastructureObject* that is a portion of the energy system infrastructure. Each class identifies a different part of the infrastructure. A general energy infrastructure Object in the UCO has been modified and moved to the *compositeObject* category. There are four classes in this subcategory (Table 6.4).

Table 6.4 Energy infrastructure Objects

Object class	Definition	Examples
electricityProductionPlant	An *energyInfrastructure* Object that is a portion of the electricity production infrastructure	Nuclear plant, coal-fired plant, generator at dam, etc.
electricityDistributionInfrastructure	An *energyInfrastructure* Object that is a portion of the electricity distribution infrastructure	Power line, substation, etc.
extractiveEnergyProductionInfrastructure	An *energyInfrastructure* Object that is a portion of the extractive energy production infrastructure	Oil, coal, etc.
extractiveEnergyTransportationInfrastructure	An *energyInfrastructure* Object that is a portion of the extractive energy transportation infrastructure	Oil, gas pipeline, storage tank, etc.

Business Infrastructure

A *businessInfrastructure* Object is an *infrastructureObject* that is a portion of the business infrastructure. Each class identifies a different part of the infrastructure. A general business infrastructure Object in the UCO has been modified and moved to the *compositeObject* category. There are six classes in this subcategory (Table 6.5).

Table 6.5 Business infrastructure Objects

Object class	Definition	Examples
shopAndCommercialInfrastructure	A *businessInfrastructure* Object that is a portion of the shop or commercial infrastructure	In a town or a chain such as McDonald's
manufacturingInfrastructure	A *businessInfrastructure* Object that is a portion of the manufacturing infrastructure	
agricultureInfrastructure	A *businessInfrastructure* Object that is a portion of the agriculture, timber or fishing structure	
informationAndMediaInfrastructure	A *businessInfrastructure* Object that is a portion of the media and information infrastructure	TV and radio stations, transmission towers, business computer systems, Internet equipment
miningInfrastructure	A *businessInfrastructure* Object that is a portion of the mining and associated infrastructure	
financialInfrastructure	A *businessInfrastructure* Object that is a portion of the financial infrastructure	Banks, stock exchanges, computer systems, insurance, etc.

Social Infrastructure

A *socialInfrastructure* Object is an *infrastructureObject* that is a portion of the social infrastructure. Each class identifies a different part of the infrastructure. The *civilianHousing* class has been moved here from the discontinued *shelterInfrastructure* subcategory of the UCO. There are six classes in this subcategory (Table 6.6).

Table 6.6 Social infrastructure Objects

Object class	Definition	Examples
civilianHousing	A *socialInfrastructure* Object that is a portion of the civilian housing	
religiousBuilding	A *socialInfrastructure* Object that is a religious building	
educationInfrastructure	A *socialInfrastructure* Object that is a portion of the education infrastructure	
healthcareInfrastructure	A *socialInfrastructure* Object that is a portion of the healthcare infrastructure	
culturalInfrastructure	A *socialInfrastructure* Object that is a portion of the cultural infrastructure	Theater, museum, or sports
computerInfrastructure	A *socialInfrastructure* Object that is a portion of the Host Nation MIS infrastructure	Internet, computer systems, etc.

6.3 Needed Thing Objects

A *neededThingObject* is an Object that is something needed by people. This category of Object is comprised of three subcategories: business Objects, immediate needs, and service needs. Figure 6.4 shows the needed thing Object part of the Object taxonomy. The "object" class box represents all of the Object classes that are children of the subcategories. The tent city in the figure represents an instantiation of an immediate need Object.

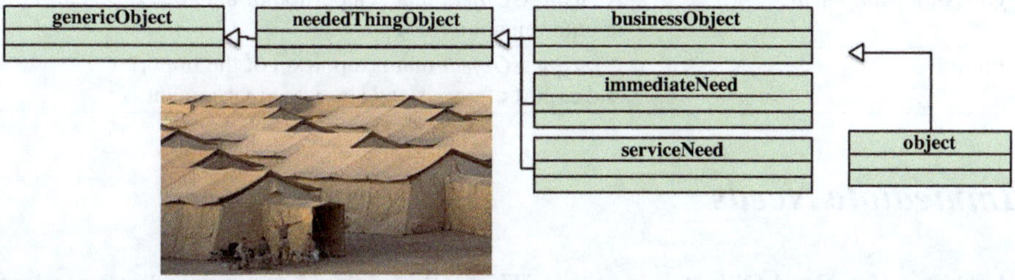

Fig. 6.4 Needed thing Object taxonomy

Business Objects

A *businessObject* is a *neededThingObject* that is part of the business environment that makes business possible. Each class identifies a different part of the environment. This subcategory has been extensively reworked from the UCO subcategory. For example, the Objects relating to drugs have been moved to the *criminalObject* subcategory, so that they have only one parent, and Objects that were in the UCO *Service* subcategory have been moved in. There are 13 classes in this subcategory (Table 6.7).

Table 6.7 Business Objects

Object class	Definition	Examples
energySupplyAndDistribution	A *businessObject* that is a portion of the energy supply and distribution	
financialSystem	A *businessObject* that is a portion of the Host Nation financial system	
insuranceSystem	A *businessObject* that is a portion of the Host Nation insurance system	
foreignAndLocalInvestment	A *businessObject* that is a portion of the foreign or local investment pool in the country	
acceptableJobs	A *businessObject* that is a portion of the acceptable job availability environment – employment from worker point of view	
employment	A *businessObject* that is a portion of the employment environment -from an economic point of view-	
market	A *businessObject* that is a portion of the market in goods and services, including stock market	
criticalIndustries	A *businessObject* that is a portion of the Host nation critical industries environment	
livestockAndAgricultureEquipment	A *businessObject* that is a portion of the livestock or agricultural equipment	
commercialEquipment	A *businessObject* that is a portion of the commercial equipment	
bankAccountAndFunds	A *businessObject* that is a portion of the money on deposit and available for use and allocations of funding	
goodsAndEquipment	A *businessObject* that is a portion of the goods or equipment on hand	
trade	A *businessObject* that is a portion of the import and export of goods and services environment	

Immediate Needs

An *immediateNeed* Object is a *neededThingObject* that is an immediate need of people. Each class identifies a different need. Some classes that had multiple parents in the UCO have been removed from this subcategory. There are four classes in this subcategory (Table 6.8).

Table 6.8 Immediate needs Objects

Object class	Definition	Examples
foodSupply	An *immediateNeed* Object that is a portion of the food supply	
potableWaterSupply	An *immediateNeed* Object that is a portion of the potable water supply	
IDPRefugeeCampAndTemporaryShelter	An *immediateNeed* Object that is an internally displaced persons or refugee camp or other temporary shelter	
healthcareSupplies	An *immediateNeed* Object that is a portion of the healthcare supplies or equipment	

Service Needs

A *serviceNeed* Object is a *neededThingObject* that is a service. Each class identifies a different service. Several classes have been removed from the corresponding subcategory in the UCO. There are seven classes in this subcategory (Table 6.9).

Table 6.9 Service needs Objects

Object class	Definition	Examples
trashDisposal	A *serviceNeed* Object that is a portion of the trash disposal environment	
wasteWaterTreatment	A *serviceNeed* Object that is a portion of the waste water treatment environment	
educationalSystem	A *serviceNeed* Object that is a portion of the educational system	Job-worthy graduates – general knowledge and skills level; elite/expert knowledge and skills level; civics education
healthTreatment	A *serviceNeed* Object that is a portion of the public health requirements satisfaction environment	
informationAndEntertainment	A *serviceNeed* Object that is a portion of the public information and entertainment environment	
culturalActivity	A *serviceNeed* Object that is a portion of the cultural events and opportunities to participate environment	Theater, museum, or sports
educationSupplies	A *serviceNeed* Object that is a portion of the education supplies or equipment	

6.4 Natural Objects

A *naturalObject* is an Object that is a portion of the natural environment. This category of Object is comprised of two subcategories: condition and geographical Object. Figure 6.5 shows the natural Object part of the Object taxonomy. The "object" class box represents all of the Object classes that are children of the subcategories. The photograph of mesas in a storm represents both a condition Object and a geographical Object.

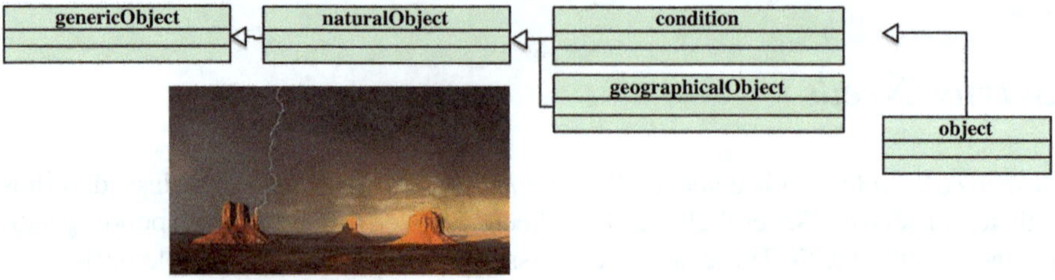

Fig. 6.5 Natural Object taxonomy

Conditions

A *condition* Object is a *naturalObject* that is a manmade or natural disaster or serious condition. Each class identifies a different condition. The four subcategories, *Disaster-Manmade*, *Disaster-Natural*, *Condition-Manmade*, and *Condition-Natural*, in the UCO have been combined and simplified in the MCO as part of defining each class to have only one parent. The questions of whether the Object is a disaster or a condition and whether the disaster or condition is manmade or natural are answered by the (instantiation) values of state variables attached to the (instantiated) Objects. There are 13 classes in this subcategory (Table 6.10).

Table 6.10 Conditions Objects

Object Class	Definition	Examples
pollution	A *condition* Object that is a portion of the pollution situation	Individual, agricultural, industrial pollution
illHealthResults	A *condition* Object that is a portion of ill health results other than health emergencies	Death and illness from disease or other health issues
healthEmergency	A *condition* Object that is a portion of the health emergency situation	Famine, epidemic, etc.

(continued)

Table 6.10 (continued)

Object Class	Definition	Examples
dayNightSeason	A *condition* Object that is a portion of the time of day, day/night, season environment	
obscurant	A *condition* Object that is a portion of the fog or man-made obscurant situation	Fog or manmade obscurant
temperatureHeatOrColdWave	A *condition* Object that is a portion of the temperature, heat/cold wave environment	
tempest	A *condition* Object that is a portion of the storm situation	Blizzard/heavy snowfall; hurricane/cyclone/tropical storm; thunderstorm/lightning/wind/hail; tornado
precipitationDrought	A *condition* Object that is a portion of the precipitation, drought/flood situation	
seastate	A *condition* that is a portion of the seastate situation	Choppy water; tides
shiftingEarth	A *condition* Object that is a portion of the earth movement situation	Earthquake; landslide/mudslide/ avalanche, volcanic eruptions
conflagration	A *condition* Object that is a portion of the fire situation	Building or countryside fire
flood	A *condition* Object that is a portion of the water movement situation	Flood/dam failure, tsunami, seiche
manmadeDisasterSituation	A *condition* Object that is a portion of the man-made disaster situation	Nuclear power plant/hazardous materials/chemical emergency; other man-made or technological disaster

Geographical Objects

A *geographicalObject* is a *naturalObject* that is a geographical Object. Each class identifies a different type of geographical Object. The *GeographicalSubdivision* of the UCO has been modified and moved to the *compositeObject* category, and a new *humanPhysicalFeature* class has been added. Also, a *boundary* class has been added to support *compositeObject* classes that need the concept. There are five classes in this subcategory (Table 6.11).

Table 6.11 Geographical Objects

Object class	Definition	Examples
landCharacterization	A *geographicalObject* that is a portion of a terrain characterization	Trafficability, cover, vegetation type, etc.
naturalFeature	A *geographicalObject* that is a portion of a natural feature	River, mountain, etc.
naturalResource	A *geographicalObject* that is a portion of a natural resource	Arable land, mineral deposit, forest, etc.
humanPhysicalFeature	A *geographicalObject* that is a portion of a human-built physical feature	Non-infrastructure things: mine, man-made lake, mine, park, etc.
boundary	A *geographicalObject* that is 0-, 1-, or 2-dimensional thing separating 1-, 2-, or 3-dimensional things	A part of a one-dimensional separation of two geographicalSubdivisions, the surface separating the ground from the air

6.5 Conflict Objects

A *conflictObject* is an Object that is a portion of the conflict environment. This category of Object is comprised of three subcategories: hot conflict environment, warm conflict environment, and cool conflict environment. The temperature adjectives are related to the intensity of conflict. The corresponding UCO category had two subcategories, which have been re-divided into these three. Figure 6.6 shows the conflict Object part of the Object taxonomy. The "object" class box represents all of the Object classes that are children of the subcategories. The ambulance scene in the figure represents an instantiation of the *deathAndInjuryOfCiviliansFromConflict* Object in a hot conflict environment.

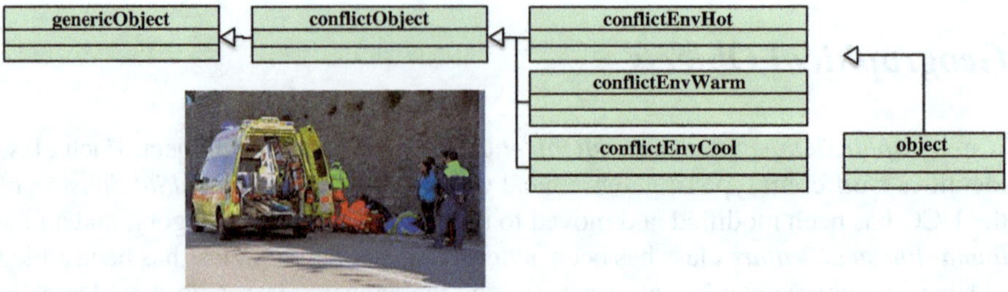

Fig. 6.6 Conflict Object taxonomy

Conflict Environments: Hot

A *conflictEnvHot* Object is a *conflictObject* that relates to a high-danger environment. Each class identifies a different type of environment. There are nine classes in this subcategory (Table 6.12).

Table 6.12 Conflict environment: hot Objects

Object class	Definition	Examples
deathAndInjuryOfCiviliansFromConflict	A *conflictEnvHot* Object that involves death and injury of civilians from conflict	
deathAndInjuryOfCombatantsFromConflict	A *conflictEnvHot* Object that involves death and injury of combatants from conflict	
propertyDestructionFromConflict	A *conflictEnvHot* Object that involves property destruction from conflict	
foreignConflict	A *conflictEnvHot* Object that is a foreign conflict environment that involves the Host Nation	
militaryOperation	A *conflictEnvHot* Object that is a portion of a military operation	Combat operation, peace operation, exercise, terrorism, etc.
terrorism	A *conflictEnvHot* Object that is a portion of terrorism environment	
mineAndIED	A *conflictEnvHot* Object that is a portion of minefields and IEDs environment	
piracy	A *conflictEnvHot* Object that is a portion of piracy environment	
violence	A *conflictEnvHot* Object that is a portion of environment of violence in Host Nation	

Conflict Environments: Warm

A *conflictEnvWarm* Object is a *conflictObject* that relates to a dangerous environment. Each class identifies a different type of environment. There are 12 classes in this subcategory (Table 6.13).

Table 6.13 Conflict environments: warm Objects

Object class	Definition	Examples
oppositionPartyUseOfForce	A *conflictEnvWarm* Object that is a portion of the use of force environment by opposition to Host Nation	
factionalDispute	A *conflictEnvWarm* Object that is a portion of a factional dispute	
civil_Internal_Unrest	A *conflictEnvWarm* Object that is a portion of the civil -internal- unrest environment	
nonNationStateActorFunding	A *conflictEnvWarm* Object that is a portion of the non-nation-state Actor funding environment	
nonNationStateActorRecruiting	A *conflictEnvWarm* Object that is a portion of the non-nation-state actor recruiting environment	
nonNationStateActorSupport	A *conflictEnvWarm* Object that is a portion of the non-nation-state Actor popular support environment	
stressMigration	A *conflictEnvWarm* Object that is a portion of the stress migration environment	
forcedPopulationMovement	A *conflictEnvWarm* Object that is a portion of the forced population movement -whether into refugee camps or ethnic cleansing- environment	
perceptionOfASafeAndSecureEnvironment	A *conflictEnvWarm* Object that is the perception of a safety and security of environment	
forceAndOperationsSecurity	A *conflictEnvWarm* Object that is a portion of the security environment for military forces and for their operations	
civilDisturbance	A *conflictEnvWarm* Object that is a portion of the civil disturbance environment	Parade, demonstration, peaceful protest, riot
demobilizedArmedForcePresence	A *conflictEnvWarm* Object that is the presence of a portion of the demobilized armed force	

Conflict Environments: Cool

A *conflictEnvCool* Object is a *conflictObject* that relates to a low-danger environment. Each class identifies a different type of environment. There are eight classes in this subcategory (Table 6.14).

Table 6.14 Conflict environments: cool Objects

Object class	Definition	Examples
resolutionOfDifferencesByCompetingGroups	A *conflictEnvCool* Object that is a portion of the environment of resolving differences between pairs of groups	Ranging from violent conflict to peaceful negotiation
civilStabilityAndDurablePeace	A *conflictEnvCool* Object that is a portion of the environment of civil stability and durable peace	
returnOfExpatriates	A *conflictEnvCool* Object that is a portion of the returning of expatriates environment	
changeInPopulationComposition	A *conflictEnvCool* Object that is a portion of the change in the relative composition of mobile population categories	
positiveAndNegativeImpactOfIntervention	A *conflictEnvCool* Object that is a portion of the influence of positive and negative events coming from the intervention	More security, rapes, etc.
complianceWithPeaceAccords	A *conflictEnvCool* Object that is a portion of the compliance environment of peace accords and conditions	
politicalPowersharing	A *conflictEnvCool* Object that is a portion of the political power-sharing arrangements environment	
securityInRefugeeCamp	A *conflictEnvCool* Object that is a portion of the security environment in refugee camp or temporary shelters	

6.6 Governing Objects

A *governingObject* is an Object that describes the governance situation. This new category of Object is comprised of four subcategories: government Object, economic Object, criminal Object, and intervention Object. The classes of these subcategories had been part of two subcategories in the *ConceptualEnvironment* category of the UCO. Figure 6.7 shows the governing Object part of the Object taxonomy. The "object" class box represents all of the Object classes that are children of the subcategories. The government building in the figure represents the legislature, which is a government Object.

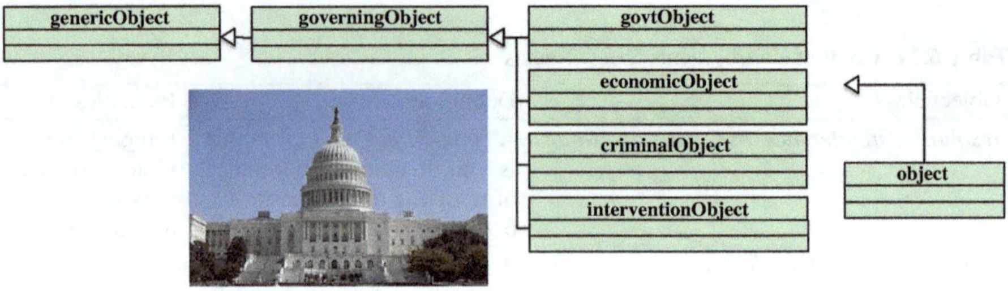

Fig. 6.7 Governing Object taxonomy

Government Objects

A *govtObject* is a *governingObject* that relates to aspects of the government. Each class identifies a different aspect of the government. There are 17 classes in this subcategory (Table 6.15).

Table 6.15 Government Objects

Object class	Definition	Examples
governance	A *govtObject* that is the Host Nation governance	
changeInGovtLeadership	A *govtObject* that is the Host Nation leadership change environment	
govtDecisionMaking	A *govtObject* that is the distribution of power environment	
legalSystemTradition	A *govtObject* that is the legal system tradition	Common law, civil law -French-, religious, Asian, no law, etc.
connectionBetweenLawEnforcementAndTheMilitary	A *govtObject* that is the environment of connection between law enforcement and the military	Law enforcement as part of military or not, multiple levels of police, border guards, etc.

(continued)

Table 6.15 (continued)

Object class	Definition	Examples
government	A *govtObject* that is the environment generated by the entire government	
publicRecordsTransparency	A *govtObject* that is the transparency of government information	Records exist and are available, public reporting -push-, visibility of actions, etc.
borderControl	A *govtObject* that is the Host Nation border-control environment	
typeGovt	A *govtObject* that is the type of the government	Autocratic, democracy, etc.
otherGovtPolicy	A *govtObject* that is the Host Nation non-economic policy	Health, education, labor, information and media, social and cultural, energy, natural resources, agriculture, transportation, or trade policy
generalGovtPolicy	A *govtObject* that is the general Host Nation government policy	
penalSystem	A *govtObject* that is the penal system	
socialServicesSystem	A *govtObject* that is the social services system at any level of government or part	Child or health services
waterAndWasteSystem	A *govtObject* that is the Host Nation waste handling system	W, sewage, or trash, etc.
bureaucracy	A *govtObject* that is the bureaucracy at any level of government	Executive bureaucracy
legislativeBranchPresence	A *govtObject* that is the presence of the legislative branch	
executiveBranchPresence	A *govtObject* that is the presence of the executive branch	

Economic Objects

An *economicObject* is a *governingObject* that relates to aspects of the economy. This is a new subcategory. Each class identifies a different aspect of the economy. There are five classes in this subcategory (Table 6.16).

Table 6.16 Economic Objects

Object class	Definition	Examples
monetaryHealth	An *economicObject* that is the Host Nation money	Inflation, deflation, etc.
govtEconomicAndFinancialPolicy	An *economicObject* that is the government's economic policy	Economic, financial, budget
economicStatistics	An *economicObject* that is encompasses the various economic statistics other than monetary health	
naturalResourceMgmntEnvironment	An *economicObject* that is the Host Nation basic natural resources management environment	
taxationStructuresAndPolicy	An *economicObject* that is the Host Nation taxation structure or policy	

Criminal Objects

A *criminalObject* is a *governingObject* that is related to crime. Each class identifies a different aspect of crime. There are 13 classes in this subcategory (Table 6.17).

Table 6.17 Criminal Objects

Object class	Definition	Examples
crimeCommon	A *criminalObject* that is the common crime environment	
crimeOrganized	A *criminalObject* that is the organized crime environment	
corruptionInCulture	A *criminalObject* that is the corruption environment in the culture	
corruptionInSocialServices	A *criminalObject* that is the corruption environment in the social services	
corruptionInLawEnforcement	A *criminalObject* that is the corruption environment in law enforcement organizations	
corruptionInCentralAuthority	A *criminalObject* that is the corruption environment in the authority	

(continued)

Table 6.17 (continued)

Object class	Definition	Examples
corruptionInMilitary	A *criminalObject* that is the corruption environment in the military	
drugUse	A *criminalObject* that is the drug use environment	
drugCultivation	A *criminalObject* that is the drug cultivation business environment	
drugManufacture	A *criminalObject* that is the drug manufacture business environment	
drugTransshipment	A *criminalObject* that is the drug transshipment business environment	
blackAndGrayMarket	A *criminalObject* that is the black and gray market business environment	
corruptionInBusiness	A *criminalObject* that is the corruption environment in business	

Intervention Objects

An *interventionObject* is a *governingObject* that relates to the intervention (if there is one). This is a new subcategory. Each class identifies a different aspect of the relations between the intervention and the government. There are three classes in this subcategory (Table 6.18).

Table 6.18 Intervention Objects

Object class	Definition	Examples
relationshipWithIntervenors	An *interventionObject* that is the relationship environment between Host Nation and intervenors	
cooperationBetweenGovtMilitaryAndIntervenors	An *interventionObject* that is the cooperation environment between the Host Nation military and the intervenors	
interventionOrganizationEnvironment	An *interventionObject* that is the organizational composition of the intervention	Organization of diplomats, advisors, support personnel, etc.

6.7 Conceptual Objects

A *conceptualObject* is an Object that is not tangible. This category of Object is comprised of three subcategories: rights Object, cognitive Object, and document. Its precursor, *ConceptualEnvironment* in the UCO, lost two subcategories to the *governingObject* category. The remaining subcategories have been joined with the new document subcategory. Figure 6.8 shows the conceptual Object part of the Object taxonomy. The "object" class box represents all of the Object classes that are children of the subcategories. The Bill of Rights in the figure represents an instantiation of a rights Object.

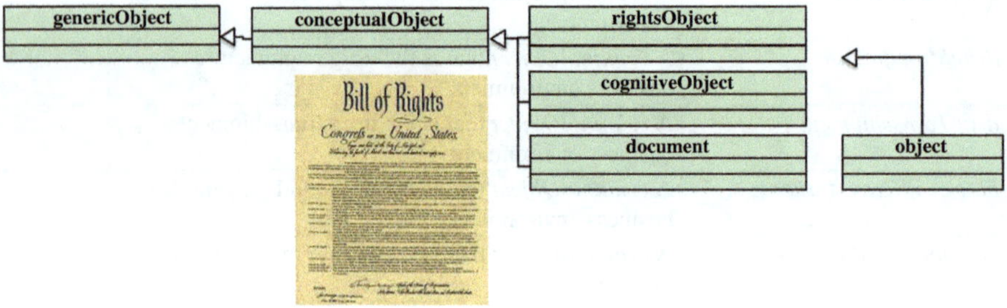

Fig. 6.8 Conceptual Object taxonomy

Rights Objects

A *rightsObject* is a *conceptualObject* that is related to the rights of people. Each class identifies a different right (or its antithesis). There are ten classes in this subcategory (Table 6.19).

Table 6.19 Rights Objects

Object class	Definition	Examples
rightsAndFreedoms	A *rightsObject* that is the general rights and freedoms environment	
crimePolitical	A *rightsObject* that is the "political" crime environment, as defined by the government	
protectionOfHumanRights	A *rightsObject* that is the human rights protection environment	
politicalPersecution	A *rightsObject* that is the political persecution environment	
propertyRightsAndAccess	A *rightsObject* that is the property rights and access environment	
freedomOfMovement	A *rightsObject* that is the freedom of movement of the populace environment	Politically restricted, tied to the land, free to move, etc.

(continued)

Table 6.19 (continued)

Object class	Definition	Examples
satisfactionOfPeoplesSpiritualNeeds	A *rightsObject* that is the satisfaction of people's spiritual needs environment	
observationOfCulturalAndSocialInterest	A *rightsObject* that is the cultural rights environment	Observation of social anniversaries, other cultural events and interests
freedomOfDomesticMedia	A *rightsObject* that is the domestic media freedom environment	
freedomOfInternationalMedia	A *rightsObject* that is the international media freedom environment within the country	

Cognitive Objects

A *cognitiveObject* is a *conceptualObject* that is related to cognitive activities. These include issues of legitimacy, ideas and norms, perceptions and opinions, and data/ information. There are 16 classes in this subcategory (Table 6.20).

Table 6.20 Cognitive Objects

Object class	Definition	Examples
domesticLegitimacyOfActor	A *cognitiveObject* that is the domestic legitimacy environment of the Actor	
internationalLegitimacyOfActor	A *cognitiveObject* that is the international legitimacy environment of the Actor	
socialIssueDecisionMaking	A *cognitiveObject* that is the social decision making environment	
keyIdea	A *cognitiveObject* that is a key idea	
socialNorm	A *cognitiveObject* that is a social norm	
perceptionByPeopleThatTheirInterestsAreRepresented	A *cognitiveObject* that is the popular perception that their interests are represented	

(continued)

Table 6.20 (continued)

Object class	Definition	Examples
perceptionByPeopleOfChangesInTheirSocialStatus	A *cognitiveObject* that is the popular perception of change in social status	
toleranceByPeopleOfTheStatusQuo	A *cognitiveObject* that is the tolerance by people of the situation	
opinionOfPopulation	A *cognitiveObject* that is the opinion of a population	
opinionOfSignificantGroup	A *cognitiveObject* that is the opinion of a significant group	
opinionOfSignificantLeader	A *cognitiveObject* that is the opinion of a significant leader	
opinionChangeOfPopulation	A *cognitiveObject* that is the opinion change for the population	
opinionChangeOfSignificantGroup	A *cognitiveObject* that is the opinion change for a significant group	
opinionChangeOfSignificantLeader	A *cognitiveObject* that is the opinion change for a significant leader	
popularSenseOfCommunity	A *cognitiveObject* that is the sense that there is a community to belong to	
dataInformation	A *cognitiveObject* that consists of data or information	Data or information of any kind

Documents

A *document* Object is a *conceptualObject* that is a document, whether electronic or physical. This is a new subcategory. There are three classes in this subcategory. These are the constitution, civil defense plans and military plans (Table 6.21).

Table 6.21 Document Objects

Object class	Definition	Examples
constitution	A *document* Object that is the Host Nation constitution	
civilDefensePlan	A *document* Object that is a portion of the Host Nation civil defense plan	
militaryPlan	A *document* Object of an armed force embodying a portion of the plans for action of the force	An operations plan, fires plan, communications plan, etc.

6.8 Composite Objects

A *compositeObject* is an Object that is composed of other objects. This is a new category in the MCO. This category of Object is comprised of three subcategories: simple composite Object, hierarchical Object, and complex composite Object. A separate chapter (Chap. 7) is provided to supply examples of instantiations of these new classes. Figure 6.9 shows the composite Object part of the Object taxonomy. The "object" class box represents all of the Object classes that are children of the subcategories. The *drugCrimeOverall* part of the figure, with a transportation network, growing marijuana, a drug submarine, and injectable drugs (for drug use), represents a simple composite Object.

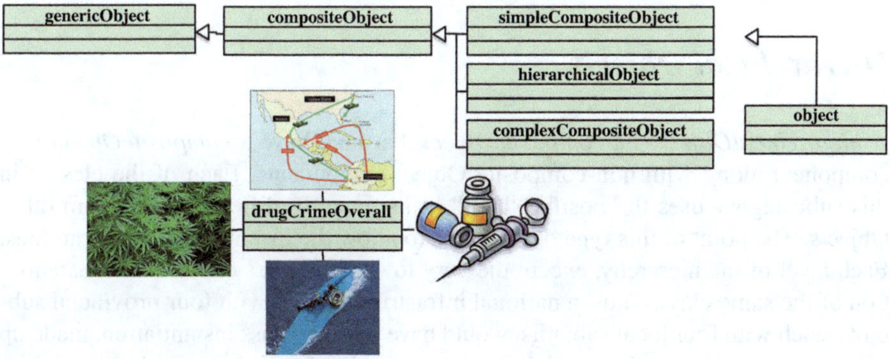

Fig. 6.9 Composite Object taxonomy

Simple Composite Objects

A *simpleCompositeObject* is a *compositeObject* that has only non-composite Objects as components. There are four classes in this subcategory. These are predefined to address drug crime, the general economy, the overall immediate needs of the people, and a *geographical* subdivision (Table 6.22).

Table 6.22 Simple composite Objects

Object class	Definition	Examples
drugCrimeOverall	A *simpleCompositeObject* that has *drugUse*, *drugCultivation*, *drugManufacture*, and *drugTransshipment* as components	
generalEconomy	A *simpleCompositeObject* that has a portion of all *economicObject* objects and all *businessObject* objects as components	
overallImmediateNeedsOfThePeople	A *simpleCompositeObject* that has *foodSupply*, *potableWaterSupply*, *healthcareSupplies*, and IDP_*RefugeeCampAndTemporaryShelter* as components	
geographicalSubdivision	A *simpleCompositeObject* that is a portion of a geographical subdivision of the world with *geographicalObject* components	Country, province, city, arbitrary geographic area

Hierarchical Objects

A *hierarchicalObject* is a *compositeObject* that may have a *compositeObject* as a component along with non-composite Object components. Each of the classes in this subcategory uses the postfix "Unit" in its name to differentiate it from other Objects. The point of this type of Object is to allow the instantiation of hierarchies. Each level of the hierarchy, except the very lowest, requires one or more instantiation of the same class. Thus, a national infrastructure unit with four provincial subunits, each with four local subunits, would have a "unit" class instantiation, made up of four different "unit" class instantiations, each of which is made up of four different instantiations of either a the same "unit" class or some other type of class. (See Chap. 7 for examples.) There are 12 classes in this subcategory. These are predefined to address various situations. Note that the *compositeGeographicalUnit* is different from the *geographicalSubdivision*. The "unit" class has the latter class as components (Table 6.23).

Table 6.23 Hierarchical Objects

Object class	Definition	Examples
crimeOverallUnit	A *hiearchicalObject* that has *commonCrime*, *organizedCrime*, *blackAndGrayMarkets*, *drugCrimeOverall*, *corruptionOverallUnit*, and possibly a *crimeOverallUnit* class as components	
commercialSectorUnit	A *businessObject* that is a portion of the overall commercial sector – all *businessObject* objects and possibly a *commercialSectorUnit* class as components	
compositeInfrastructureUnit	A *hiearchicalObject* that has the six infrastructure units and a possible *compositeInfrastructureUnit* class as components	
economicFoundationUnit	A *hiearchicalObject* that is composed of a portion of the *generalEconomy* and the *generalInfrastructure* and a possible *economicFoundationUnit* class as components	
energyInfrastructureUnit	A *hiearchicalObject* that is a portion of the four energy infrastructure components and a possible *energyInfrastructureUnit* class as components	
transportInfrastructureUnit	A *transportInfrastructure* Object that is a portion of the seven transportation infrastructure components and a possible *transportInfrastructureUnit* class as components	
corruptionOverallUnit	A *hiearchicalObject* that is a portion of six corruption Object components and a possible *corruptionOverallUnit* class as components	
compositeGeographicalUnit	A *hiearchicalObject* composed of one or more *geographicalSubdivision* classes and a possible *compositeGeographicalUnit* class	
waterInfrastructureUnit	A *hiearchicalObject* that is a portion of the three water infrastructure components and a possible *waterInfrastructureUnit* class as components	
govtInfrastructureUnit	A *hiearchicalObject* that is a portion of the nine government infrastructure components and a possible *govtInfrastructureUnit* class as components	
businessInfrastructureUnit	A *hiearchicalObject* that is a portion of the six business infrastructure components and a possible *businessInfrastructureUnit* class as components	
socialInfrastructureUnit	A *hiearchicalObject* that is a portion of the six social infrastructure components and a possible *socialInfrastructureUnit* class as components	

Complex Composite Objects

A *complexCompositeObject* is a *compositeObject* that has both Actor and Object components. There are two classes in this subcategory. These are pre-defined to address geopolitical units and facility units. Note that the *geopoliticalUnit* may include both *compositeGeographicalUnit* classes and *geographicalSubdivision* classes as components (Table 6.24).

Table 6.24 Complex composite Objects

Object class	Definition	Examples
geopoliticalUnit	A *complexCompositeObject* that has one or more geographical subdivisions, a geographical population, infrastructure, govt leadership, and a sub *geopoliticalUnit*	A town, province, etc.
facilityUnit	A *complexCompositeObject* that has one or more geographical subdivisions, one or more organizations, infrastructure, leadership, and a sub *facilityUnit*	A military airbase, civilian seaport, manufacturing facility, etc.

6.9 Object Relationships

The only relationships used in this chapter are the *is-a* and *hasMetric* relationships, explained as follows:

is-a: Each instance of A is an instance of B and each property of B is a property of A. Its inverse relationship is *superClassOf.*

hasMetric: A has Metric B (also shown as *described by*). Its inverse relationship is *metricOf.*

6.10 Review of Objects

We often refer to "Objects" as shorthand for "Object classes." Each Object class is a thing that defines the instantiations of the class to be a particular subset of all of the things that are passive and acted upon. The Object categories and subcategories are an aid to this definition process: each narrows the subset in a meaningful way.

The nouns of the unconventional conflict domain that are not active elements are represented as Objects. Some of the Objects are concrete things (sometimes literally concrete), such as infrastructure elements. Some are more abstract, such as conceptual objects. The others have both concrete and abstract characteristics and include "needed thing" elements, natural objects, and conflict objects. These Objects are affected by the Actions of Chap. 5.

The descriptions of composite Objects and composite Actors will be amplified by examples of instantiations in Chap. 7.

The Metrics of Chap. 8 can be compared to adjectives that modify the nouns (Actors and Environment Elements) and adverbs that modify the verbs (Actions). However, the Metrics provide more information than standard adjectives and adverbs.

Chapter 7
MCO Composite Actors and Objects

The essentials of the element ontologies become clearer when applied. We are going to be looking at some modeling alternatives. We are using the word "modeling" in its broader sense of abstracting the important parts of something into a representation of the thing. This is useful in creating computer models; however, it is also useful in reducing clutter to focus on the most important aspects of something. In this sense, an ontology is a model of its domain. One important part of modeling lies in deciding what is important and should be retained in the model and what is not important and can be discarded. Commonly, the intended use for the model is the major deciding factor in such decisions. The Modern Conflict Ontology (MCO) is a fairly general ontology, meaning that its users still have the latitude to make modeling choices.

The Actors, Actions, and Objects of the ontology (other than the composite classes) appear relatively straightforward and, at first glance, complete. However, as we shall see, there are parts of the domain that require more complicated constructions. It is easy to see how one can represent *thisLion* as an instance of the *genericAnimal* class. But we know that lions are not just solitary animals, they are parts of prides of lions (or, crucially, not parts of one). We could say that *thisPride* is an instance of the *animalGroup* class. But we would be omitting what we know about the social structure of lion prides. Lions are not important in the domain of human conflict; but humans have more complex societies and have more complex relationships that are important in human conflict. The composite classes allow us to represent some of these complexities. In particular, they offer alternative ways to represent a complex thing, ranging from the very simple to the very detailed.

We will start simply and build complexity in small steps. Then we will look at a generalization of what we have done and apply it.

D. S. Hartley III, *An Ontology of Modern Conflict*, Understanding Complex Systems, https://doi.org/10.1007/978-3-030-53214-7_7

7.1 Military Composite Classes and Instantiations

The military provides a classic case for hierarchies. We will use the military to explore the value of composite classes, which then can be used in understanding their general value.

Simple Implementation

First, consider the Actor class named *landSystem*. This is a class that is used for military vehicles and systems that are land-based. In this example, we will consider a tank, shown in Fig. 7.1.

Fig. 7.1 *thisTank* instantiation

We could have divided the Actor class into tanks, artillery, trucks, missile silos, and other land systems that can act on the world. However, the metrics supply the identifying information to differentiate the systems, so we will proceed directly to an instance, called *thisTank*. (Remember an instance is a particular real-world item, such as a particular pie in a store. You would expect to have many instances of the *landSystem* class, such as *blueTank14*, *redTank72*, and *syrianTruck1205*.) Some of the metrics are shown in the figure, including an identification, the number of members (in this case, personnel), and the weapons that are mounted on the system. (Each instance has its own metric values, although similar instances may have the same values for some metric types, with different values for the identification metric.)

However, most use cases will probably involve units larger than single tanks. ("Use case" is a term of art that permits segregation of different uses for a system into different cases.) Consider the *significantGroupActor* category, *armedForce* subcategory class named *interventionForceOrganization* that could address this need. This Actor class can be instantiated as any size armed force. Here we instantiate it as a tank platoon, named *thisTankPlatoon* in Fig. 7.2. This Actor class also has metrics, some of which are shown in the figure. We learn that its name tells us about its place in a hierarchy. We also see that there could be two kinds of members, the tank crews and the tanks themselves. The weapons are also accumulated by type. For many uses, this Actor class is entirely adequate. However, in hierarchical organizations, the structure of the hierarchy is important, and in some uses it is important to be able to refer to the unit as a whole and to each of its components separately.

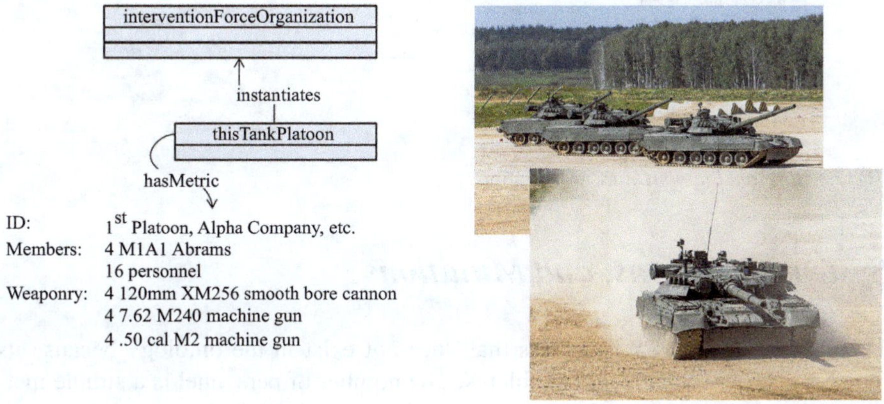

Fig. 7.2 *thisTankPlatoon* instantiation

Composite Actor Implementation

Consider a hypothetical *compositeActor* class named *armedForceUnitX* (not the class contained in the ontology) that could address this need, as shown in Fig. 7.3. This class is composed of other classes. Because we are going to model a tank platoon, the classes are *landSystem* classes, and there are four of them. We instantiate each of these as *thisTank1* through *thisTank4*, representing different tanks. The metrics for each of these are identical except for the serial numbers. There is also an instantiation of the *armedForceUnitX* class into *thisTankPlatoonUnit*. Because it is composed of *thisTank1* through *thisTank4*, shown by the red connections, its metrics reflect the combined metrics of its components. This model of the tank platoon permits independent actions by each tank, while allowing combined actions by the platoon as a whole. The cost for this flexibility is increased complexity over the *armedForce* class-based model in Fig. 7.2.

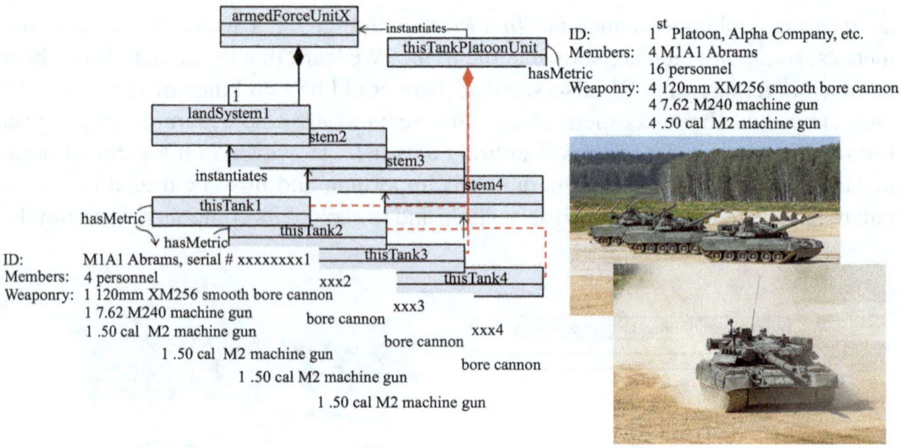

Fig. 7.3 *thisTankPlatoonUnitX* instantiation

Systems, Weapons, and Munitions

The class *armedForceX* is a class that does not exist in the ontology because its member metrics would cause problems. The number of personnel is a simple metric; however, the number of tanks refers to a count of Actor classes (or their instantiations as objects). The second part of the metric would require identifying (referencing) the class, as well as the number. If you look carefully at the weaponry metric in Fig. 7.2, you will see that the weaponry metric has numbers and identifications for the weapon type. Why is this structure allowed for weapons, but not for tanks and such? The reason is structural, but the structure is based on a modeling rationale. Some things in the real world are regarded as too unimportant to include in the ontology. Other things are important enough to include but not worth expending too much effort on (think of the crowd scene "extras" in movies). Still other things are very important and deserve more effort (think of the leading parts in movies).

Our Actors are the leading parts. While not each Actor is a single human being, all may be thought of as embodying extensions of this concept. A group of people may consider a situation, decide to act, and then perform the action. (Not all actions a person might perform are possible or appropriate for all Actors. For example, a person can punch another person in the face; a committee cannot.) Similarly, some inanimate objects are listed as Actors because they act like people: consider the tornado that is on track to hit an open field but then abruptly changes course to smash into a trailer park. However, most of the inanimate actors will actually contain people who are the controllers of the action. The fighter jet is considered the Actor; however, it is the pilot who makes the decisions.

The military inanimate actors are of the *militarySystem* class, whether *landSystem*, *airSystem*, *seaSystem*, or *spaceSystem*. Mobile inanimate systems with a human in the driver's seat are easily determined to be Actors. However, some are more difficult to determine. For example, consider a "rocket." (1) This might be an RPG carried and fired by a person. (2) It might be a large missile fired by a multiple launch rocket system (MLRS) carried on and fired from a tracked vehicle. (3) It might be an unguided or a guided rocket fired from an aircraft. (4) It might also be an intercontinental ballistic missile (ICBM). Case (1) is a personal weapon. Case (2) is a weapon that is "attached" to an *airSystem* (technically, it is the munition for the weapon). Similarly, case (3) is a weapon that is "attached" to a *landSystem*. Case (4) is a *spaceSystem*. Weapons are defined in an Actor's metrics and are carried on or by the Actor. Munitions are the expendable parts of weapons. If the rocket is expected to be replaced in the launcher after firing, cases 1–3, then it is a weapon/munition. If the rocket is the major item and the launcher is a one-use item, such as an ICBM in its silo, the rocket may be labeled as the *spaceSystem* or the rocket and silo together may be labeled as the system, with the rocket as a weapon. A missile launching submarine poses a similar problem. Clearly, the submarine is a *seaSystem*. However, are the missiles separate *spaceSystems* as the aircraft on an aircraft carrier *seaSystem* are separate *airSystems* or are the missiles weapons/munitions?

Adding Complexity

Now we can see where some generalizations can be useful. In Fig. 7.4, we introduce and name the two subcategories of the *compositeActor* category. The *simpleCompositeActor* subcategory *is-a compositeActor* that has all of its components of the same type, but all are non-composite Actors, as was the case in the more complicated model above. We introduce and name the *milSystemGroup* class to be a *simpleCompositeActor* that is composed of *militarySystem* classes, such as *landSystem*. Note that the *militarySystem* classes (from 1 to n in number) are not subclasses of the *milSystemGroup* but are components, represented by the closed diamond arrowhead.

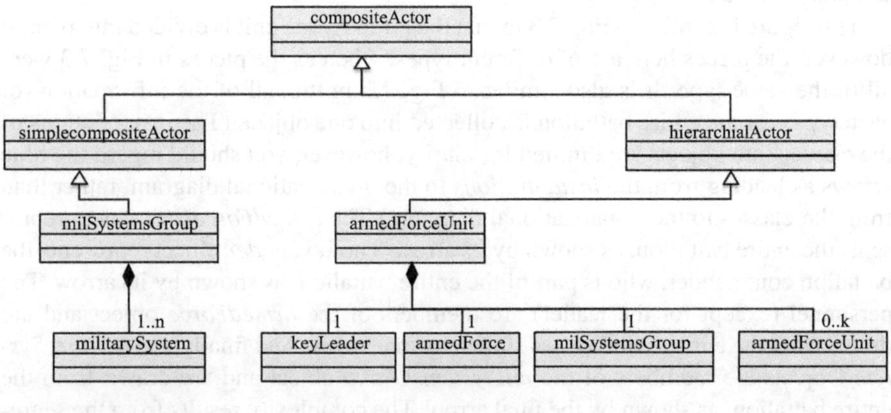

Fig. 7.4 Modeling military units with composite actors

We also introduce and name the *hierarchicalActor* subcategory, which *is-a compositeActor*. Like the *simpleCompositeActor*, the *hierarchicalActor* has multiple components; however, some components are non-composite Actors and some are composite Actors. The *armedForceUnit is-a hierarchicalActor* and has the "standard" composition of one *keyLeader* class, one *armedForce* class, and one *milSystemGroup* class and has zero to some integer *k armedForceUnit* classes. As with the *milSystemsGroup*, the classes of the *armedForceUnit* are not subclasses but components, as represented by the closed arrowhead in the figure.

The reasoning behind this composition derives from a frequent pattern in describing military units, the table of organization and equipment (TOE). The personnel hierarchy is described with the major equipment (vehicles, weapon systems, etc.) attached. Generally, an organization is divided into a headquarters (HQ) organization and several operational organizations. These sub-organizations are normally one size level (echelon) lower than the organization itself, although some specialized units may be at a still lower echelon.

The *armedForce* model in Fig. 7.2 combined all of the personnel and all of the major equipment items (which coincides with our *militarySystem* concept) into its metrics. The *armedForceUnit* model in Fig. 7.3 separated these by unit, but not by personnel and major equipment. The classes of Fig. 7.4 support the desired separation. The *milSystemGroup* class is composed of *militarySystem* classes, which still contain information about personnel; however, this information is retained to support a larger goal, ensuring that the number of all personnel associated with a military system is captured. The *armedForce* model is defined to include only the personnel of the armed force, together with their personal weapons. This captures all of the personnel in the unit. It is often necessary to identify the leader of the unit (and sometimes more than one leader). Thus the minimal definition of the *armedForceUnit*, as shown in Fig. 7.4, includes a *keyLeader* class, an *armedForce* class (containing all of the personnel, except the leader), and a milSystemGroup class (containing all of the military systems).

Figure 7.5 applies this to a nominal infantry battalion, consisting of a headquarters and headquarters company (HHC), three rifle companies (CO), and a heavy weapons company.

This figure is similar to Fig. 7.3 in that the armed force unit is divided into pieces; however, the pieces here are of different types, whereas the pieces in Fig. 7.3 were all of the same type. It is also similar to Fig. 7.2 in that all of the information (of each type) for the entire battalion is collected into one object. (The instantiations of the classes into objects are omitted for clarity; however, you should regard the blue arrows as leading from the *instantiations* to the organizational diagram, rather than from the classes to the organizational diagram.) The *armedForceUnit* object represents the entire battalion, as shown by its arrow. The *keyLeader* object represents the battalion commander, who is part of the entire battalion, as shown by its arrow. The personnel (except for the leader) are members of the *armedForce* object and are drawn from the entire battalion, as shown by that arrow. And finally, the *militarySystem* Objects are members of the *milSystemsGroup* object and are drawn from the entire battalion, as shown by the final arrow. The complexity results from the segre-

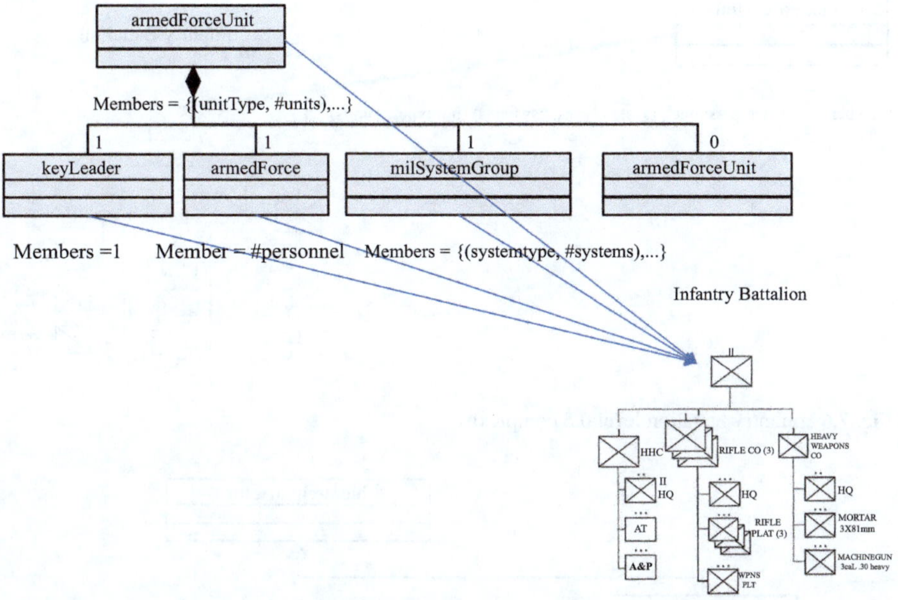

Fig. 7.5 Infantry battalion level 1 complexity

gation of the leader from the rest of the personnel and from the military systems (major equipment), as desired.

Table 7.1 recapitulates the representation of the infantry battalion at the battalion echelon or level 1 complexity. The first column identifies the type of Actor. The second column shows the number of each component type. The third column identifies the thing being represented. The last entry, the second *armedForceUnit* entry, represents the subordinate commands of the first entry.

Table 7.1 Battalion representation – level 1 complexity

armedForceUnit		Whole battalion
keyLeader	1	Battalion commander, in the battalion
armedForce	1	Battalion personnel and personal weapons (minus battalion commander)
milSystemGroup	1	Battalion *militarySystems*
armedForceUnit	0	

Note that there is a simpler picture. Figure 7.6 shows the *armedForceUnit* object standing alone, with the total information contained in its metrics. This picture is actually just a shortcut, not an actual model. All of the complexity of Fig. 7.5 is required, so that the metric information shown in Fig. 7.6 can be derived (by external computation), but the complexity is just not shown.

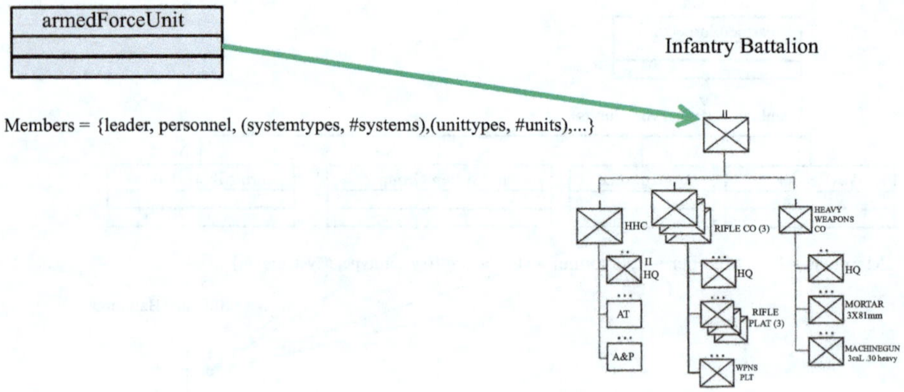

Fig. 7.6 Infantry battalion level 0.5 complexity

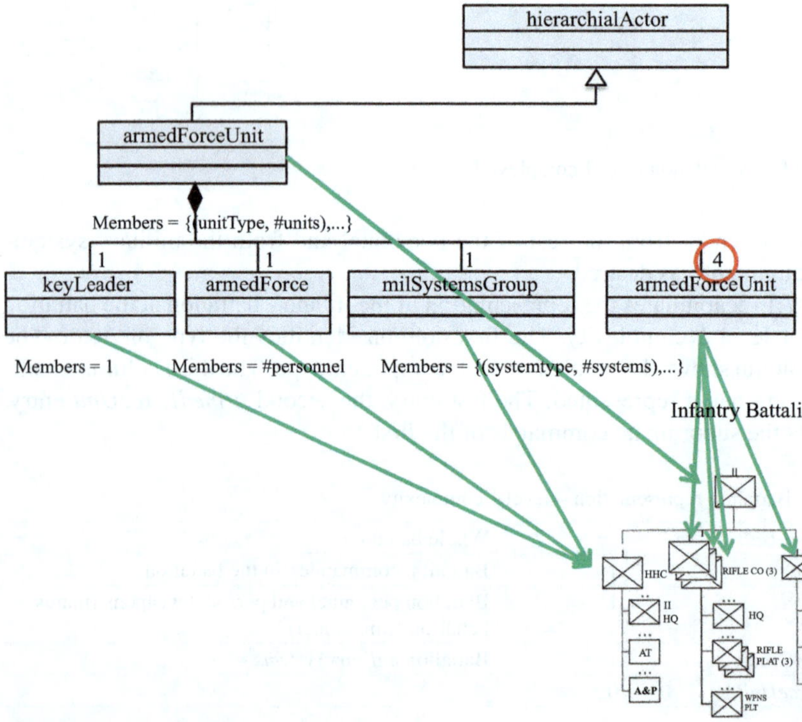

Fig. 7.7 Infantry battalion level 1.5 complexity

Figure 7.7 applies to the same infantry battalion; however, it is analogous to Fig. 7.3 (only) in that it models the next lower echelon of the battalion.

Four new *armedForceUnit* objects are added to the components of the original *armedForceUnit* object. The original *armedForceUnit* object still represents the entire battalion, as shown by its arrow. The (battalion) leader is now found in the Headquarters & Headquarters Company (HHC). The *armedForce* only represents the personnel in the HHC (minus the leader) and the *milSystemGroup* only represents the *militarySystem* objects in the HHC. The reason for this is that each of the four new *armedForceUnit* Objects represents one of the four other companies in the battalion. As shown, the new *armedForceUnit* objects are pictured simply as in Fig. 7.6, hence the designation as Level 1.5, rather than 2.0.

Table 7.2 recapitulates the representation of the infantry battalion at the company echelon or level 2 complexity, which includes a representation of each company at level 1 complexity. (The HHC company is taken care of in the battalion-level representations.) The first column identifies the type of Actor. The second column shows the number of each component type. The third column identifies the thing being represented. The last entry, the second *armedForceUnit* entry, in each group of rows represents the subordinate commands of the first entry in the group of rows.

Table 7.2 Battalion representation – level 2 complexity

armedForceUnit		Whole battalion
keyLeader	2	Battalion commander, in the battalion HHC plus company commander HHC
armedForce	1	Battalion HHC personnel and personal weapons (minus battalion commander and HHC company commander)
milSystemGroup	1	Battalion HHC *militarySystems*
armedForceUnit	4	Battalion companies (at level 1 for each company)
Representative level 1 company		
armedForceUnit		Whole company
keyLeader	1	Company commander, in the company
armedForce	1	Company personnel and personal weapons (minus company commander)
milSystemGroup	1	Company *militarySystems*
armedForceUnit	0	

Figure 7.8 carries the decomposition of the battalion one step further. The original *armedForceUnit* object still represents the entire battalion, as shown by its arrow. However, now the leader is found in the HQ of the HHC; the *armedForce* and the *milSystemGroup* only represent the HQ, and an extra *armedForceUnit* has been added to represent the rest of the HHC. Each *armedForceUnit* at this echelon is expanded to identify the leader, personnel, and military systems that it represents, together with the appropriate *armedForceUnit* classes to represent the platoon echelon within it. Note that connections are shown only for one of the platoons to minimize complexity in the figure. The complexity level of this figure is shown as 2.5 because the platoons are represented in the manner of Fig. 7.6, rather than the expansion of Fig. 7.5.

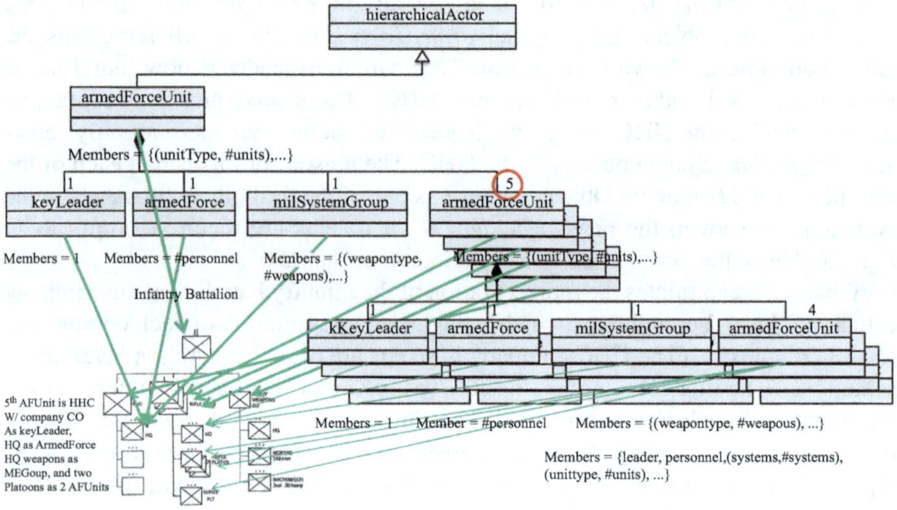

Fig. 7.8 Infantry battalion level 2.5 complexity

Table 7.3 recapitulates the representation of the infantry battalion at the platoon echelon or level 3 complexity, which includes a representation of each company at level 2 complexity and a representation of each platoon at level 1 complexity. (The HHC is the fifth company. The HQ platoon (assuming the HHC is structured to have

Table 7.3 Battalion representation – level 3 complexity

armedForceUnit		Whole battalion
keyLeader	3	Battalion commander, in the battalion HQ of the HHC and HQ company commander and HQ platoon leader [only 2 if no HQ platoon]
armedForce	1	Battalion HQ personnel and personal weapons (minus battalion commander, HQ platoon leader)
milSystemGroup	1	Battalion HQ *militarySystems*
armedForceUnit	5	Battalion companies, incudes HHC (at level 2 for each company)
Representative level 2 company		
armedForceUnit		Whole company
keyLeader	2	Company commander, in the company HQ and HQ platoon leader [only 1 if no HQ platoon]
armedForce	1	Company HQ personnel and personal weapons (minus company commander, HQ platoon leader)
milSystemGroup	1	Company HQ *militarySystems*
armedForceUnit	4	Company platoons (at Level 1 for each platoon)
Representative level 1 platoon		
armedForceUnit		Whole platoon
keyLeader	1	Platoon leader, in the platoon
armedForce	1	Platoon personnel and personal weapons (minus platoon leader)
milSystemGroup	1	Platoon *militarySystems*
armedForceUnit	0	

one) is taken care of in the battalion level representations.) The first column identifies the type of Actor. The second column shows the number of each component type. (Note the reduction in the whole battalion *keyLeader* count because the HHC company commander is covered in that company.) The third column identifies the thing being represented. The last entry, the second *armedForceUnit* entry, in each group of rows represents the subordinate commands of the first entry in the group of rows.

Obviously, the number of echelons that are represented is not determined by the ontology. It is determined by the ontology's use. However, the ontology supports the need to represent the hierarchical structure of a military force. Military forces are not the only organizations that use a hierarchical structure.

Non-land Examples

In Fig. 7.9 we see an instantiation of an *airSystem* unit, rather than a *landSystem*. (Note that the instantiations of the *armedForceUnit* were not actually pictured in the figures but alluded to by the green arrows. The instances were only specified in the third columns of the tables as the names of the things being represented. In Fig. 7.9, they are pictured and have names beginning with "this.") The class structure is the same; however, the details at the instantiation level are different. Note that the five lower-echelon *armedForceUnit* classes are instantiated by four flights (one possible

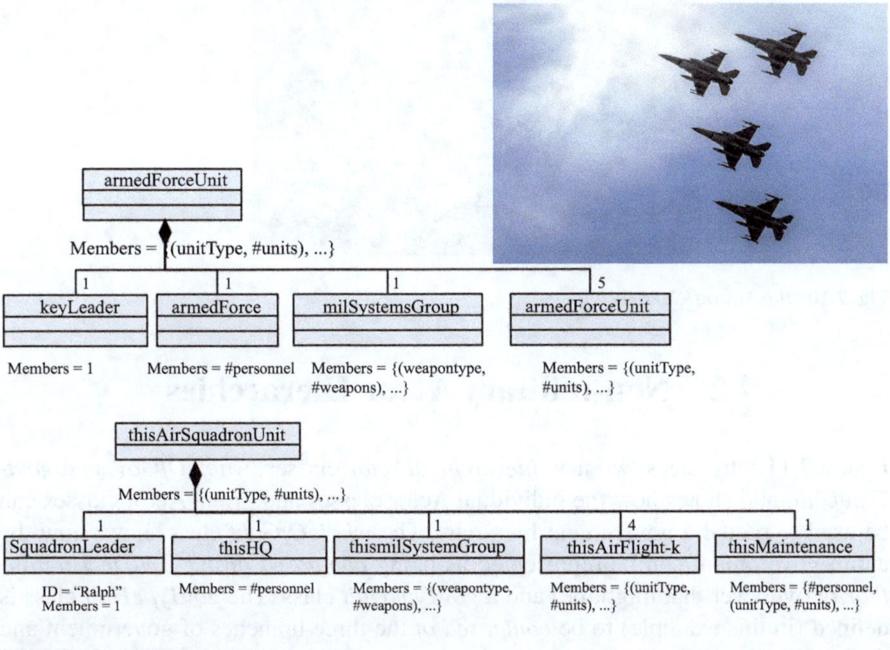

Fig. 7.9 *thisAirSquadronUnit*

air squadron subdivision) and one maintenance unit. Some military services include organic maintenance units at lower levels and some include the maintenance units at the higher echelon, so this figure should be regarded as illustrative, not prescriptive.

For the sea system example, we skip the class structure and just show the instantiations. Figure 7.10 illustrates the instantiation of the *armedForceUnit* class into the ship's commander, the commander's HQ unit, the aircraft carrier itself as the sole member of the *milSystemGroup* (at this echelon), and two *armedForceUnit* objects at the lower echelon, the air wing and the carriers operating departments as a group. If the carrier has an admiral and staff aboard (in naval parlance, a higher-echelon commander and staff embarked), they would not show up at this echelon because the carrier is subordinate to the admiral's (higher) echelon. In fact, the figure for that echelon would show the entire carrier and complement as one of the *armedForceUnit* objects, each representing a ship under the command of the admiral.

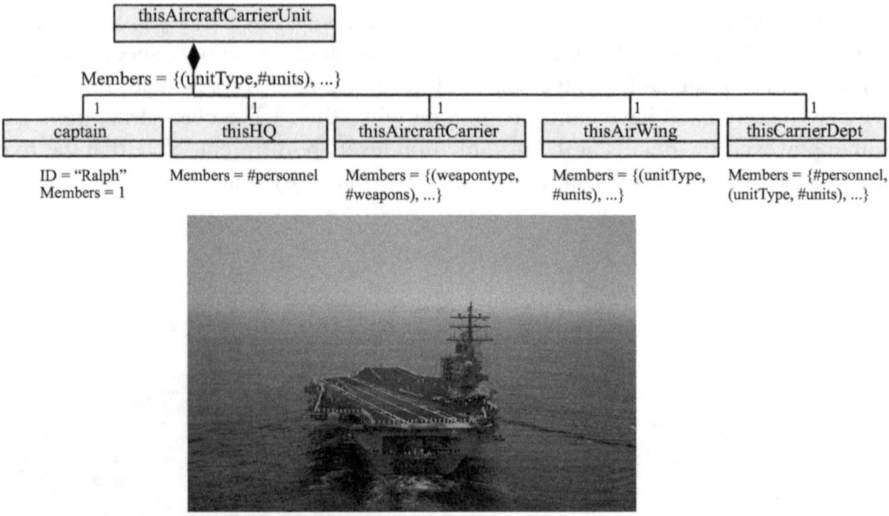

Fig. 7.10 *thisAircraftCarrierUnit*

7.2 Non-military Actor Hierarchies

Figure 7.11 introduces two new *hierarchicalActor* classes, *wholeOfGovt* and *govtTypeUnit*, and shows how the individual Actor classes and group Actor classes can be used to model a government hierarchy. The *wholeOfGovt* class shows how the entire government can be represented as being *composed of* the *govtDecisionAuthority* (whoever that might be) and a *govTypeUnit* class. The *govTypeUnit* class is defined (in this example) to be *composed of* the three branches of government and the bureaucracy, plus the key leaders from each (shown to be parts of each, respectively), plus zero to *n govTypeUnit* classes. (Note, in the United States, most of the

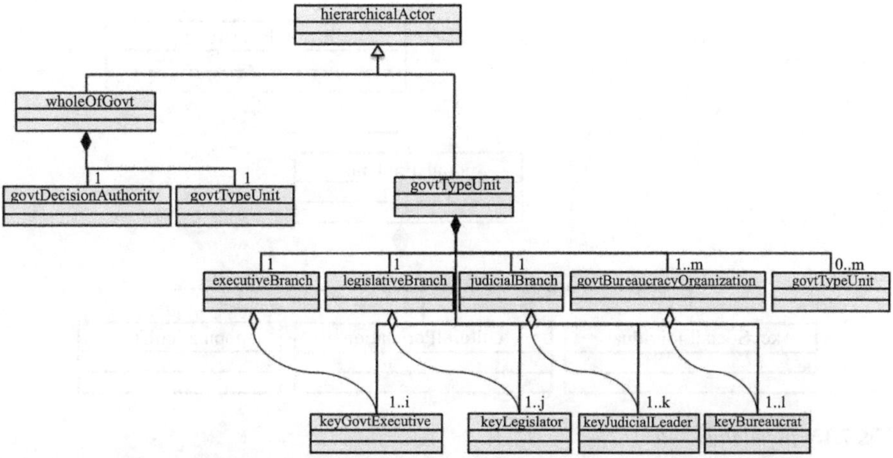

Fig. 7.11 Government hierarchies

bureaucracies are part of the Executive branches, but some are not. This complexity can be instantiated; however, it is too complicated to put into a readable figure.) These optional *govTypeUnit* classes support representing the next lower echelon/level of government. In the United States, these would be the 50 states, plus perhaps the governments of the District of Columbia (Washington, DC) and the territories. (In some other country, these might be called provinces.) Because each of these *n* *govTypeUnit* classes contains zero to *n* *govTypeUnit* classes, the next lower echelon can be represented (perhaps called counties or districts).

Figure 7.12 introduces another new *hierarchicalActor* class, the *organizationUnit* class. This class can be used to define the hierarchy of any organization, such as a company or an NGO, in the same recursive fashion as shown above. This allows for a significant differentiation between some religious organizations. For example, the Roman Catholic Church is distinctly hierarchical, whereas the Islamic sects are not.

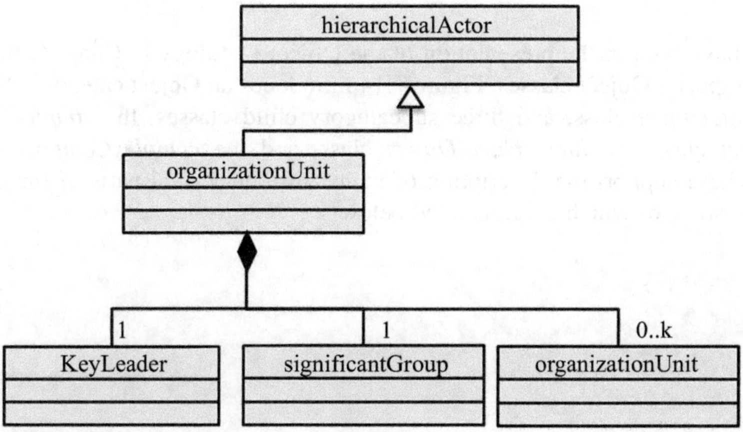

Fig. 7.12 *organizationUnit*

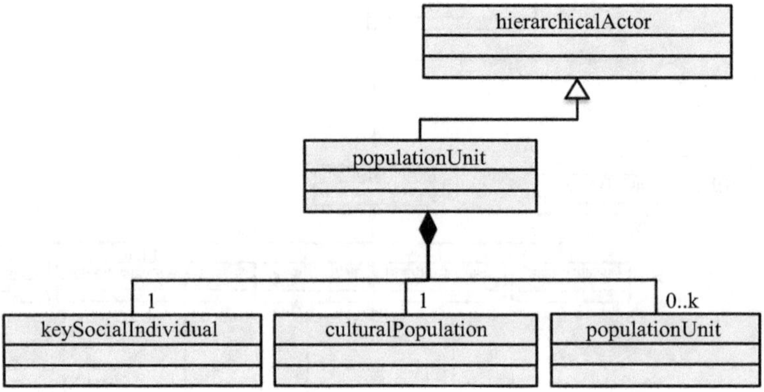

Fig. 7.13 *populationUnit*

Figure 7.13 introduces the *populationUnit* class. This is a particularly important hierarchy within the cultural population in many countries. Note that the instantiation of the *culturalPopulation* class is not necessarily the entire population of a country, but the population of a particular culture. If there are subordinate *populationUnit* classes, the *culturalPopulation* class at the highest level would only contain that part of the particular population that is not contained in the subunits. For example, the tribes of some countries consist of organized units, with distinct areas of influence (the subunits), and people who are only loosely affiliated with the tribe (perhaps assigned [by the modeler] to the general, or highest level, population).

These examples explore some ways in which the composite Actor classes allow us to express what we know about parts of the conflict domain.

7.3 Non-actor Hierarchies

As we have seen in the presentation of the Object Ontology in Chap. 6, there are also composite Object classes. Figure 7.14 introduces an Object category class, the *compositeObject* class, and three subcategory child classes, the *simpleCompositeObject* class, the *hierarchicalObject* class, and the *complexCompositeObject* class. These support the description of many different useful parts of the conflict domain, some of which are illustrated below.

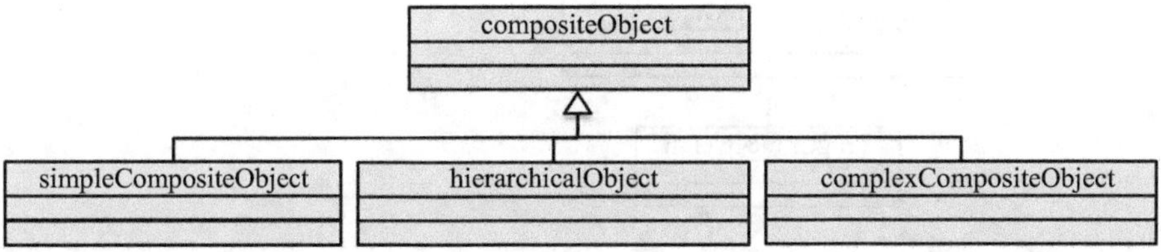

Fig. 7.14 Composite object category and subcategories

Simple Object Examples

One *simpleCompositeObject* is the *overallImmediateNeedsOfThePeople*, which naturally is meant to represent the immediate needs of food, water, healthcare, and shelter in a humanitarian emergency situation. Figure 7.15 illustrates the *is-a* relationship between the first two classes and the component relationship between the individual needs and the overall need. An instantiation of these class relationships might include the needs within an entire theater or within some portion of a country or might concern a particular delivery of the needs. Especially in the last instance, one or more of the individual needs might be missing, as either not needed in that instance or not provided for some reason. (The use of the open diamond as the arrowhead in this figure emphasizes this. The open diamond arrowhead indicates aggregation, not composition.)

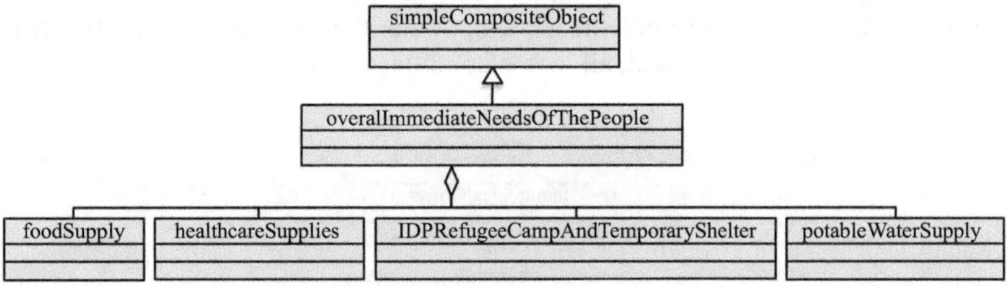

Fig. 7.15 *overallImmediateNeedsOfThePeople*

Figure 7.16 illustrates a second *simpleCompositeObject*, the *drugCrimeOverall* class. It is also the aggregation of four Objects and can have various types of instantiations. Having a composite Object allows for discussion of a consolidated topic, such as drug crime, as well as discussion of its individual component crimes. Remember that a class can be instantiated in many ways. For example, the *drugCrimeOverall* class could be instantiated as *opioidCrimeOverall*. In this instance (pun intended), there is no instance of the *drugCultivation* class. This explains the use of aggregation rather than composition.

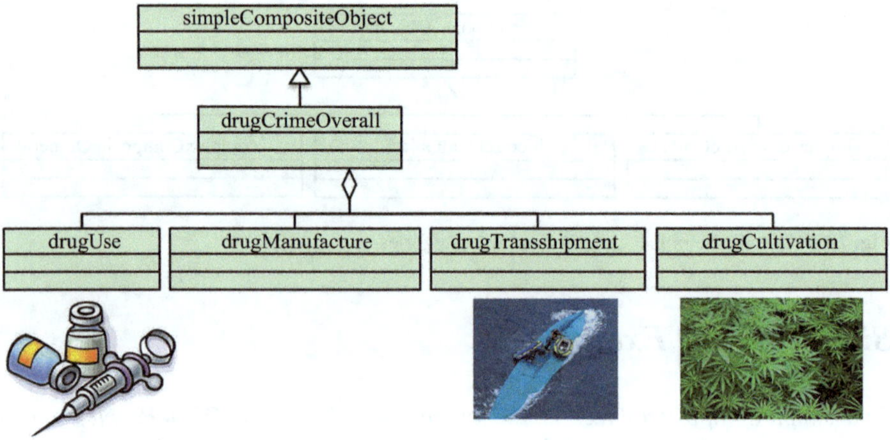

Fig. 7.16 *drugCrimeOverall*

Hierarchical Object Examples

The existence of drug crimes naturally leads to corruption. Several corruption Object classes are contained in the criminal object subcategory, which leads to the need for a collective Object. Rather than a simple composite Object, this is implemented as a hierarchical Object, with a post-fix of "Unit," as shown in Fig. 7.17. Each of the individual corruption Object classes is shown as a component. And another copy of the *corruptionOverallUnit* class is also shown as a component. This permits an instantiation to implement corruption at the national level, at the province level, and at the local level, all in one complicated instance.

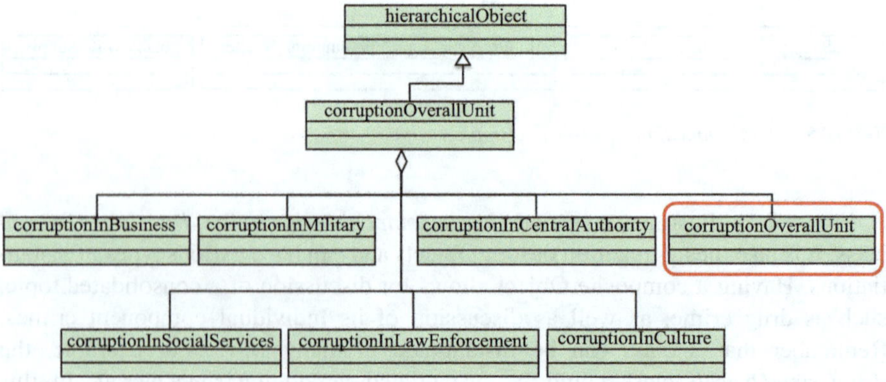

Fig. 7.17 *corruptionOverallUnit*

This hierarchical representation permits a note about naming of classes. The *corruptionInCentralAuthority* class could be read as meaning only the country's central authority. However, the class definition actually says of the "authority." Thus, in the instantiation, the copy for the country level would refer to the national authority; each copy at each province level would refer to that province's authority; and each copy at each local level would refer to the local authority. Thus, as with all classes, the instantiation may consist of a subset of the entire possible class meaning. Also note that an instantiation might omit one or more of the component classes as required by the situation – including the recursive class when no recursion is desired.

The second hierarchical representation example uses both of the preceding examples. Figure 7.18 shows the *crimeOverallUnit* class as composed of three regular Object classes (common crime, organized crime, and black and gray market), one simple composite class (overall drug crime), a different hierarchical Object class (overall corruption), and a recursion *crimeOverallUnit* class. The possible instantiations begin to be complicated.

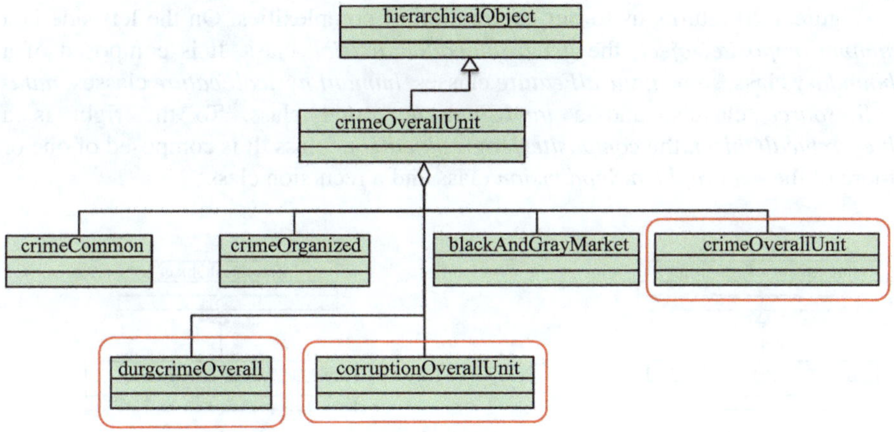

Fig. 7.18 *crimeOverallUnit*

The *commercialSectorUnit* is illustrated to provide a little relief from complexity. There are 13 classes in the *businessObject* subcategory. Together, they describe the commercial sector. Together with another copy of the *commercialSectorUnit* class, they are its components. This is a good point to reiterate a modeling note about hierarchical classes: the contents of lower levels should not be contained in the upper levels. Thus, if we are representing the country level commercial sector and the province's commercial sectors, the *employment* class (one of the 13 classes) figures at the national level should not include the *employment* class figures for any of the included provinces. One instantiation might single out only a few provinces for greater detail, combining all the rest of the provinces into a national lump. Another instantiation might include all of the provinces and only include those national classes for which there are actual non-province contents (Fig. 7.19).

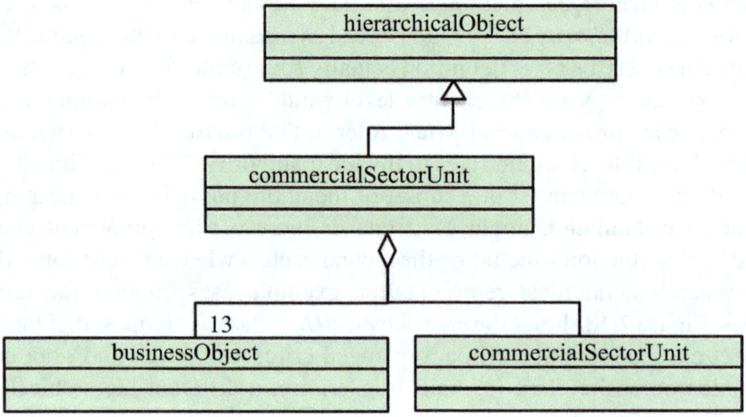

Fig. 7.19 *commercialSectorUnit*

Figure 7.20 returns us to our exploration of complexities. On the left side is a *simpleCompositeObject*, the *geographicalSubdivision* class. It is composed of a *boundary* class, some *naturalFeature* classes, *humanPhysicalFeature* classes, *naturalResource* classes, and a *landCharacterization* class. To the right is a *hierarchicalObject*, the *compositeGeographicalUnit* class. It is composed of one or more of the *geographicalSubdivision* class and a recursion class.

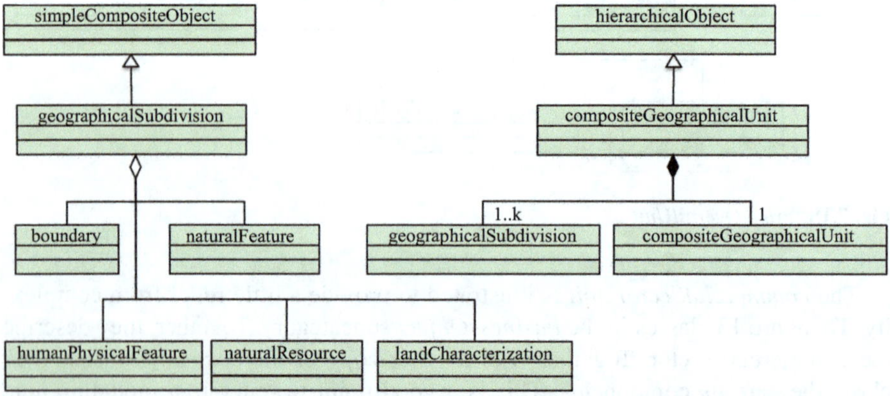

Fig. 7.20 *geographicalSubdivision* and *compositeGeographicalUnit*

Logically, the recursion is not necessary, because as many *geographicalSubdivision* classes as are desired can be included as components at the top level. However, if it is desired to subdivide one or more of the physical geographical subdivisions, it might be appropriate to segregate them. Again, if this is done, care should be taken to avoid double-counting.

Infrastructure is a natural subject for hierarchical objects, and they are provided for each of the six types of infrastructure. Figure 7.21 illustrates the one for water infrastructure. The *waterInfrastructureUnit* class includes the three *waterInfrastructure* classes (*waterDistributionInfrastructure, waterAndSewageTreatmentInfrastructure*, and *damInfrastructure*), plus a recursion class.

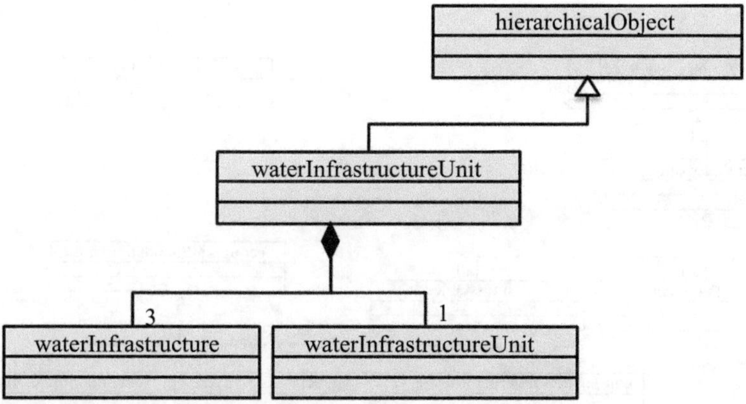

Fig. 7.21 *waterInfrastructureUnit*

Naturally, there is also an object to represent all of the infrastructure, the *compositeInfrastructureUnit* (Fig. 7.22). Because each of the components is a hierarchical class, there is no obvious need for a separate recursion class. However, some instance might require it, so it is included.

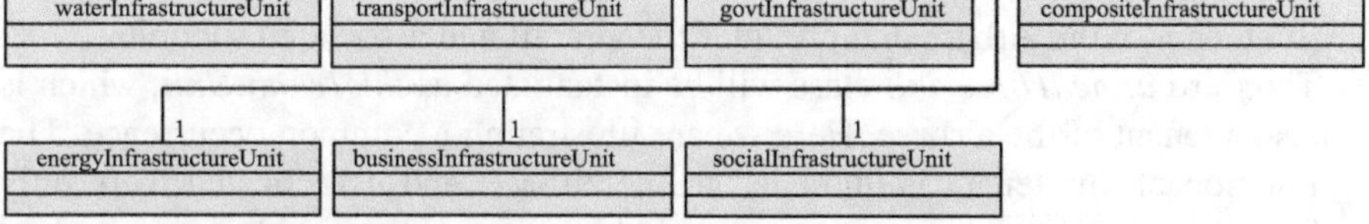

Fig. 7.22 *compositeInfrastructureUnit*

The apex of pure Object complexity lies in representing the economy (Fig. 7.23). On the left side is a *simpleCompositeObject*, the *generalEconomy* class. This class is composed of all of the *economicObject* classes and all of the *businessObject* classes. On the right side is a *hierarchicalObject*, the *economicFoundationUnit* class. This is composed of the *generalEconomy* class, the *compositeInfrastructureUnit* class, and a recursion class. Thus, the entire economy can be represented by instantiations of this class.

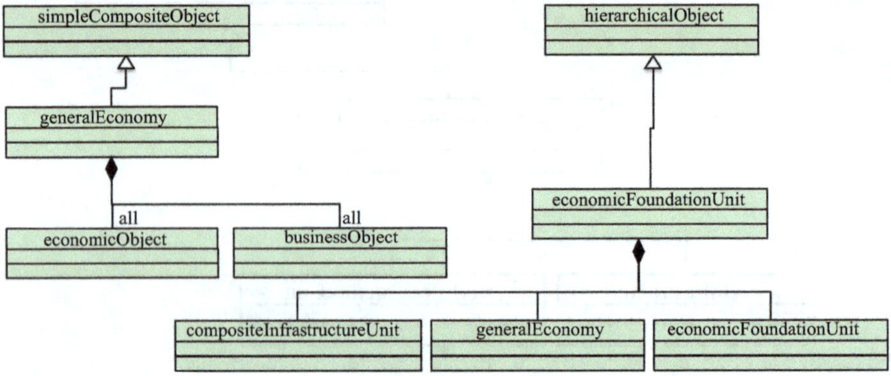

Fig. 7.23 *generalEconomy* and *economicFoundationUnit*

Complex Composite Objects

Complex composite Objects contain both Objects and Actors. There are two entity types that require this level of complexity, geopolitical entities (*geopoliticalUnit* class) and facilities (*facilityUnit* class). Only the facilityUnit class will be illustrated – the other will be left to the reader's imagination.

We describe this class with an instantiation of a military airbase (Fig. 7.24). The components of the *facilityUnit* class are three *armedForceUnit* classes, a *compositeInfrastructureUnit* class, and a *compositeGeographicalUnit* class.

- The first *armedForceUnit* class will be instantiated as *thisAirbaseUnit*, which owns the airbase and provides its overall leader.
- The second *armedForceUnit* class will be instantiated as *thisAirWingUnit*, which is a tenant of the airbase and provides the aircraft and associated subunits.
- The third *armedForceUnit* class will be instantiated as *thisTenantUnit*, which is also a tenant of the airbase. Here we are illustrating a common occurrence. The function of this tenant is unrelated to the airbase, and it is located there only because it needs to be somewhere and *thisAirbaseUnit* provides security and utilities cheaper than other options.

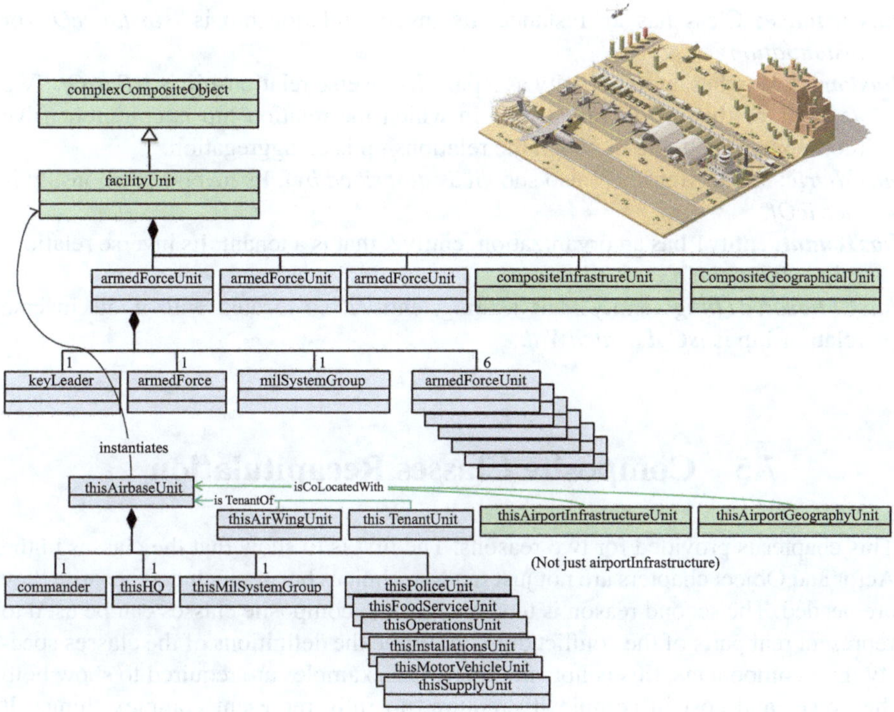

Fig. 7.24 Airbase facility

- The *compositeInfrastructureUnit* is instantiated as *thisAirportInfrastructureUnit* and is co-located with *thisAirbaseUnit* and represents its infrastructure.
- The *compositeGeographicalUnit* is instantiated as *thisAirportGeographyUnit* and is also co-located with *thisAirbaseUnit* and provides the geographical representation of the airbase.

The instantiation of *thisAirbaseUnit* is partially fleshed-out to indicate some of the next level of complexity. The other parts of the instantiation are left at the highest level, with only the note that the infrastructure will be more complicated than just that found in the *airportInfrastructure* class.

7.4 Composite Relationships

Several relationships are used in this chapter and are explained as follows:

is-a: Each instance of A is an instance of B, and each property of B is a property of A. Its inverse relationship is *superClassOf*.

hasInstance: Class has an instance. Its inverse relationship is *isInstanceOf* (or *instantiation*).

hasPart: Entity has another entity as a part. Its inverse relationship is *isPartOf*. Two versions of this are included: one in which the relationship is comprehensive (component) and one in which the relationship is an aggregation.

hasMetric: A has Metric B (also shown as *described by*). Its inverse relationship is *metricOf*.

hasTenant: entity1 has an organization, entity2, that is a tenant. Its inverse relationship is *isTenantOf*.

hasCoLocatedThing: entity1 has a thing, entity2, co-located with it. Its inverse relationship is *isCoLocatedWith*.

7.5 Composite Classes Recapitulation

This chapter is provided for two reasons. The first is to show that the classes in the Actor and Object chapters are not just random choices but are included because they are needed. The second reason is to show how the composite classes can be used to represent real parts of the conflict domain. While the definitions of the classes specify their components, this is not sufficient. The examples are required to show both the power and cost in complexity required to fully represent complex things. It should be noted that the choice of using component classes and the level of their complexity that is instantiated is a **modeling** choice, not an ontology requirement.

Chapter 8
MCO State Variable Ontology

In a sense, state variables or Metrics are the adjectives and adverbs of the Modern Conflict Ontology (MCO). They describe the status of the nouns (Actors and Objects) and verbs (Actions). The MCO contains two Metrics structures, the State Variable or MType Metric Ontology and the PMESII Ontology, as shown in Fig. 3.8. This chapter describes the Metrics in terms of the MType structure. Fig. 8.1 illustrates its icon as used in the Context Diagram, Fig. 3.5. Chap. 9 will discuss the (alternate) PMESII structure.

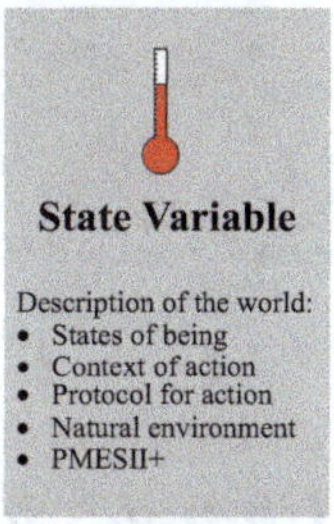

Fig. 8.1 State variable icon

8.1 Overview of Metrics

There are 5930 Metric classes. Each has an ID number, a class name, a defining question (to aid in understanding the meaning of the Metric), a list of typical variables (for which its instances have values), an *MType* value, and key fields for con-

D. S. Hartley III, *An Ontology of Modern Conflict*, Understanding Complex Systems, https://doi.org/10.1007/978-3-030-53214-7_8

necting to the *metricVectorKey*, described below. Each Metric is a subclass of one of the subcategories described below. The particular subcategory for the Metric is identified by the *MType* value.

The *MType* values support the creation of the MCO State Variable Ontology, described in this chapter. MTypes define attributes or properties. The Metric classes are also connected to the PMESII taxonomy in a PMESII Ontology, described in Chap. 9.

There are five MType categories of Metrics: vector key Metrics, physical Metrics, flow Metrics, relational Metrics, and Metrics of Human, Social, Cultural Behavior (HSCB). Figure 8.2 shows the top level of the MType metric taxonomy. The "metric" class box represents all of the Metric classes that are children of the subcategories. The ruler in the figure represents the concept of measuring things (Fig. 8.2).

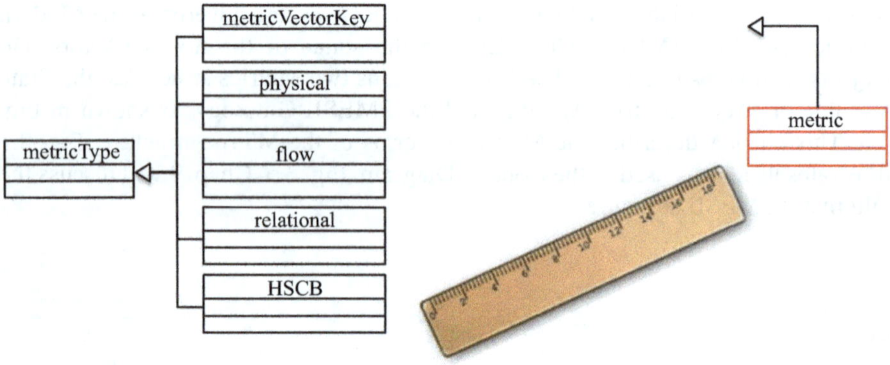

Fig. 8.2 MType metric taxonomy overview

Changes to Metrics from the UCO

The Unconventional Conflict Ontology (UCO) positioned the PMESII Ontology as the principal Metric Ontology and the MType Metric Ontology as the subordinate ontology. The MCO reverses this ordering. The MType Metric Ontology is the more defensible ontology: it derives from an earlier concept of "properties" and "property types." Each element has various properties; however, it took a while to realize that "properties" were actually categories of Metrics. "Property types" became Metric Type (or MType) categories and "properties" became MType subcategories. The PMESII structure already existed, so it was the default principal Metric Ontology. The PMESII Metric Ontology was appealing because it attached the overall ontology to the PMESII paradigm. However, in practice, the decision to declare a particular Metric to relate to a particular part of the PMESII structure was difficult, leading to relations to multiple parts of the PMESII structure for most Metrics and the realization that different people would decide on somewhat different sets of relations.

The major change in the MType Metric Ontology from that of the UCO is the creation of the *metricVectorKey* category and placing of three Metric subcategories in this category. The *flow* category has also been added. The *relationship* category has been renamed the *relational* category to reduce confusion between it and the relationships. The other category-subcategory relations have been modified and some of the subcategory names have been changed. Finally, the miscellaneous category has been eliminated.

The major change in the PMESII Ontology is the connection of the Metric classes to the ontology. In the UCO, this connection was an *is-a* relationship. In the MCO, this connection is an *isPartOf* relationship (inverse of *hasPart*), with an aggregation meaning, rather than a component meaning. This change allows a Metric class to be a part of multiple PMESII subcategories, without violating the one-parent rule for the *is-a* relationship. It also better expresses the concept of a Metric contributing to a PMESII subcategory, rather than defining it. Thus, a Metric could contribute to several PMESII subcategories and multiple Metrics could contribute to a single PMESII subcategory.

An Insight into Metrics

State variables or Metrics are simple in concept, but somewhat confusing in detail. The concept is that each element in the MCO, whether Actor, Action, or Object, has a set of Metrics that describe the state of the element at any given time. Thus, each element class has a defined set of Metric classes that are connected to it. The set of Metric classes for a given element form a single vector of classes.

The description above is almost correct, but not quite. A Metric class that is attached to, for example, an Actor class says nothing about the state of the class. The instantiation of the Metric class that is attached to the instantiation of that Actor class says something about the state of the instantiated Actor. The Metric classes that are attached to the element classes describe the *kind of information* that is available/required concerning that type of element. The version in the preceding paragraph is easier to say, so if some of that language slips in, just remember that it is really at the instance level that Metrics contain state variable values. Each Metric instantiation is a vector consisting of the instantiations of all of the relevant Metric classes. The number and MType identities depend on the nature of the element.

With the exception of four types of metric, each element is assigned its own Metric for each type of metric that is applicable to the element's subcategory. This permits the assignment of the Metrics for each element to PMESII subcategories that are appropriate for the type of Metric and for the meaning of the element. The four types that are not differentiated by element are the *situation*, *identity*, *time*, and *location* types. For each of these, there is only one Metric to which each element is assigned. The reason is that these Metrics should have no impact on the PMESII values, no matter the element. Accordingly, these four are assigned to the *null* PMESII category with one subcategory, *nullPrimitive*, which is not added to any of the PMESII categories. This will become clearer in Chap. 9, which discusses the PMESII structure.

Each MType subcategory has an MType value that is given in the descriptions below. Essentially these are internal variables to permit sorting and cross-referencing. As such, they are not of interest to the reader. However, they were originally used in tables as abbreviations for the MType subcategory name. Providing these values ensures a way for the reader to figure out the reference. Later versions of the tables substitute the names.

8.2 Metric Vector Key State Variables

The *metricVectorKey* MType category is a *metricType* that contains the key information identifying the entire metric vector. This category has three subcategories: *situationKey*, *identityKey*, and *timeKey*. Figure 8.3 shows the *metricVectorKey* part of the State Variable taxonomy. The "metric" class box represents all of the Metric classes that are children of the subcategories. The finger print in the figure represents the concept of identification.

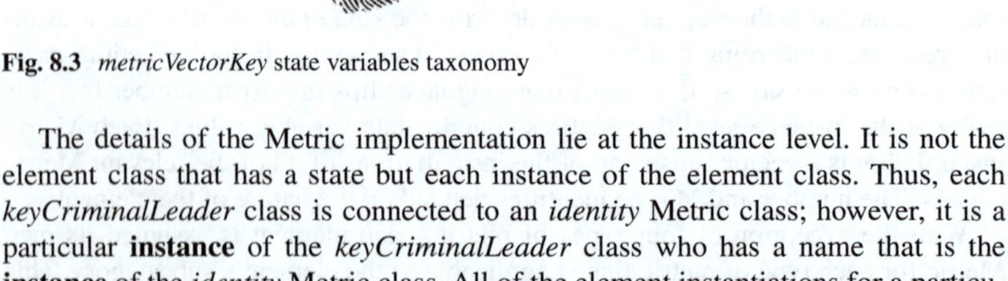

Fig. 8.3 *metricVectorKey* state variables taxonomy

The details of the Metric implementation lie at the instance level. It is not the element class that has a state but each instance of the element class. Thus, each *keyCriminalLeader* class is connected to an *identity* Metric class; however, it is a particular **instance** of the *keyCriminalLeader* class who has a name that is the **instance** of the *identity* Metric class. All of the element instantiations for a particular situation (defined by the user, but generally defining the entire operational environment of interest, say the current Syria situation) are connected to the same instantiation of the *situation* Metric class. This allows information about different situations to be stored in the same database/knowledge base. The *time* Metric class permits the storage of different instantiations of the non-*metricVectorKey* classes, based on the time at which the instantiation values hold true.

Each Metric class has three empty fields (*instanceScenarioID*, *instanceIdentificationID*, and *instanceTimeID*). These fields are filled with the appropriate data at instantiation. (The *situation* instantiation fills only the *instanceScenarioID* field and retains the other empty fields; the *identity* instantiation fills only the *instanceScenarioID* and *instanceIdentificationID* fields; and the time instantiation fills all the fields.) Thus, the entire Metric vector has a set of vector instantiations, keyed by the *situation*, *identity*, and *time* instantiations.

Situation Key

The *situationKey* subcategory is a *metricVectorKey* that identifies the situation, which is the same for all entities for a given situation. Its *MType* value is 100 and its variable is the situation name. That is, there is only one Metric in this subcategory. Table 8.1 shows the single *situationKey* Metric.

Table 8.1 *situationKey* metric

ClassName	Defining questions	Variables in vector
currentSituation	What is the identifying information on the current situation?	situation name

Identity Key

The *identityKey* subcategory is a *metricVectorKey* that uniquely identifies the entity and does not change. Its *MType* value is 120 and its typical variables are a name, a unit-ID, a unit type, etc. There is only one Metric in this subcategory. Table 8.2 shows the single *identityKey* Metric.

Table 8.2 *identityKey* metric

ClassName	Defining questions	Variables in vector
identity	What is the name or other unique identification of the entity?	Name, unit-ID, unit-type, etc.

Time Key

The *timeKey* subcategory is a *metricVectorKey* that identifies the current time data of the metric vector for the entity. Its *MType* value is 110 and its typical variables are a date/time group and possibly a duration or frequency of event. There is only one Metric in this subcategory. Table 8.3 shows the single *timeKey* Metric.

Table 8.3 *timeKey* metric

ClassName	Defining questions	Variables in vector
time	What is the time data?	Date/time group, time duration, frequency of event

8.3 Physical State Variables

The *physical* MType category is a *metricType* that describes physical attributes. This category has seven subcategories: *locationData*, *quantity*, *members*, *disaster-Indicators*, *movable*, *weaponry*, and *damage*. Figure 8.4 shows the *physical* part of the State Variable taxonomy. The "metric" class box represents all of the Metric classes that are children of the subcategories. The location indications on the map in the figure represents the location data MType subcategory.

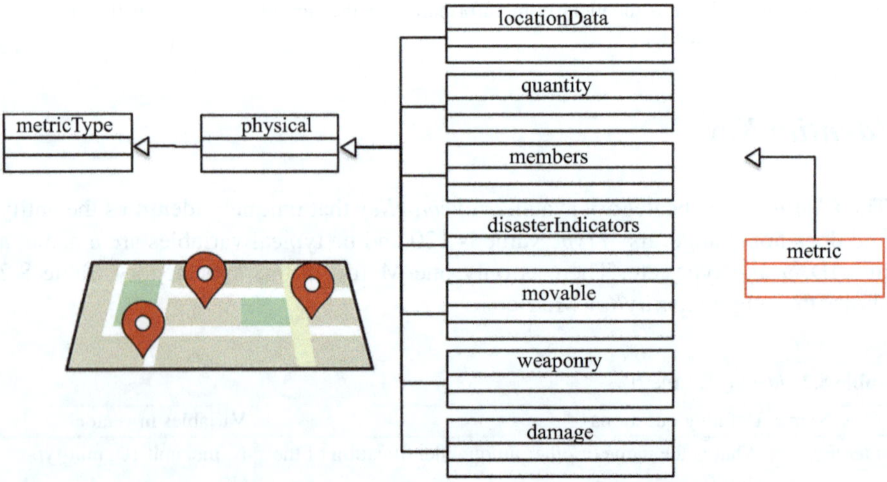

Fig. 8.4 *physical* state variables taxonomy

Location

The *locationData* subcategory is a *physical metricType* that designates the current location of the entity. Its *MType* value is 210, and its variable is a set of points or an area description, defining the x-y or latitude-longitude, depending on the type of entity, along with elevation or depth, as appropriate. It may be null, where location is not applicable. There is only one Metric in this subcategory. Table 8.4 shows the single *locationData* Metric.

Table 8.4 *locationData* metric

ClassName	Defining questions	Variables in vector
location	What is the current location description?	One or more points, on a network, as area, or density over area; with elevation/depth. May be null.

Quantity

The *quantity* subcategory is a *physical metricType* that designates the quantity or capacity affected by the action. Its *MType* value is 230 and its variable depends on the particular element. Table 8.5 shows a sample of the *quantity* Metrics.

Table 8.5 *quantity* metrics sample

ClassName	Defining questions	Variables in vector
externalGovtAdvisorsMoP	What is the number of external advisors provided to the government (not counting Rule of Law advisors)?	
govtPersonnelEducatedChangeMoP	What is the number of government personnel educated?	Civil service, legislative staff, local government staff, etc.
firstRespondersTrainedMoP	What is the number of first responders trained?	
firstRespondersJobsCreatedMoP	How many first responder jobs have been created?	

Members

The *members* subcategory is a *physical metricType* that designates the number and kind of the entity's members. Its *MType* value is 231 and its variable gives number of members by type. Table 8.6 shows a sample of the *members* Metrics.

Table 8.6 *members* metrics sample

ClassName	Defining questions	Variables in vector
govtDecisionAuthorityMembers	What is the number of members by type?	
firstResponderOrganizationMembers	What is the number of members by type?	
socialServicesOrganizationMembers	What is the number of members by type?	

Disaster Indicators

The *disasterIndicators* subcategory is a *physical metricType* that indicates whether the entity is a disaster or a condition and whether it is man-made or natural. Its *MType* value is 240 and its variables indicate whether it is a disaster or condition and whether man-made or natural. Table 8.7 shows a sample of the *disasterIndicators* Metrics.

Table 8.7 *disasterIndicators* metrics sample

ClassName	Defining questions	Variables in vector
conflagrationDisasterIndicators	Is the entity a disaster or a condition; is it man-made or natural?	Disaster or condition, man-made or natural
shiftingEarthDisasterIndicators	Is the entity a disaster or a condition; is it man-made or natural?	Disaster or condition, man-made or natural
floodDisasterIndicators	Is the entity a disaster or a condition; is it man-made or natural?	Disaster or condition, man-made or natural

Movable

The *movable* subcategory is a *physical metricType* that indicates whether the entity can be moved or not and its current speed of movement. (Naturally, a non-movable thing *should* have a zero speed.) Its *MType* value is 250 and its variables indicate movable indicator, current speed, and possibly max speed. Table 8.8 shows a sample of the *movable* Metrics.

Table 8.8 *movable* metrics sample

ClassName	Defining questions	Variables in vector
workerPopulationMovableMetric	Can the entity be moved or not? What is its current speed of movement?	Movable indicator, current speed, possibly max speed
govtDecisionAuthorityMovableMetric	Can the entity be moved or not? What is its current speed of movement?	Mindicator, current speed, possibly max speed
governanceMovableMetric	Can the entity be moved or not? What is its current speed of movement?	Movable indicator, current speed, possibly max speed
firstResponderOrganizationMovableMetric	Can the entity be moved or not? What is its current speed of movement?	Movable indicator, current speed, possibly max speed

Weaponry

The *weaponry* subcategory is a *physical metricType* that designates the types (hand-carried or militarySystem-mounted) and quantities of weapons the entity has. Its *MType* value is 270 and its variable indicates weaponry types and quantities; may also include original or desired values. Table 8.9 shows a sample of the *weaponry* Metrics.

Table 8.9 *weaponry* metrics sample

ClassName	Defining questions	Variables in vector
govtTypeUnitWeaponry	What are the types (hand-carried and militarySystem-mounted) and quantities of weapons the entity has?	Weaponry types and quantities; may also include original or desired values
externalForceOrganizationAdvocatingConflictWeaponry	What are the types (hand-carried and militarySystem-mounted) and quantities of weapons the entity has?	Weaponry types and quantities; may also include original or desired values
politicalFactionWeaponry	What are the types (hand-carried and militarySystem-mounted) and quantities of weapons the entity has?	Weaponry types and quantities; may also include original or desired values
lawEnforcementOrganizationWeaponry	What are the types (hand-carried and militarySystem-mounted) and quantities of weapons the entity has?	Weaponry types and quantities; may also include original or desired values

Damage

The *damage* subcategory is a *physical metricType* that designates the damage directly caused by the Action. (Remember that an Action may imply a desired result, but not guarantee that result will occur. The actual result is registered in a Metric.) Its *MType* value is 280 and its variable indicates severity (numeric or categorical), quantity (numeric), and coverage (over all components) by damaged entity. Table 8.10 shows a sample of the *damage* Metrics.

Table 8.10 *damage* metrics sample

ClassName	Defining questions	Variables in vector
economicDamageActionMoP	What is the amount of damage to the economy or financial infrastructure?	Severity, quantity, and coverage
civilianHousingCapacityDamagedMoP	What is the amount of civilian housing that has been damaged?	Severity, quantity, and coverage
religiousBuildingsCapacityDamagedMoP	What is the amount of religious buildings that has been damaged?	Severity, quantity, and coverage
manufacturingInfraCapacityDamagedMoP	What amount of manufacturing structures has been damaged?	Severity, quantity, and coverage

8.4 Flow State Variables

The *flow* MType category is a *metricType* that describes flows. This category has five subcategories: *capacityFlowrate*, *munitionConsumption*, *supplyConsumption*, *powerConsumption*, and *moneyConsumption*. Figure 8.5 shows the *physical* part of the State Variable taxonomy. The "metric" class box represents all of the Metric classes that are children of the subcategories. The toilet syphon diagram represents the capacity flowrate MType subcategory.

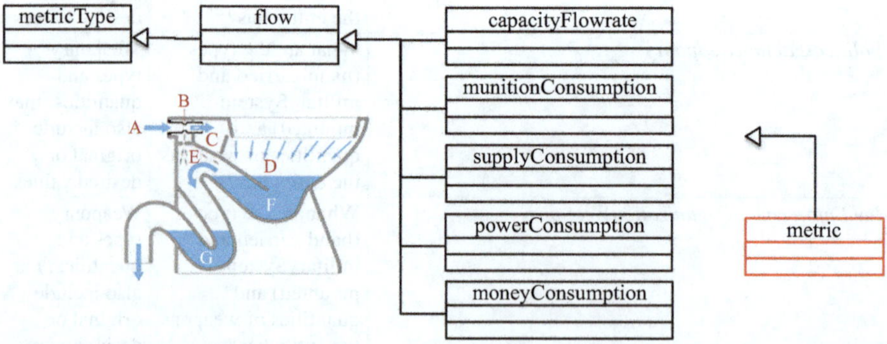

Fig. 8.5 *flow* state variables taxonomy

Capacity Flowrate

The *capacityFlowrate* subcategory is a *flow metricType* that designates the carrying capacity of the entity in terms of flowrates and in terms of quantities. Its *MType* value is 290, and its variable indicates number or volume per day; weight, height, and volume capacity and may include original or desired values. Table 8.11 shows a sample of the *capacityFlowrate* Metrics.

Table 8.11 *capacityFlowrate* metrics sample

ClassName	Defining questions	Variables in vector
civilianHousingCapacityRating	What is the carrying capacity of civilian housing?	Flowrate and absolute quantity
refugeeCampsCapacityRating	What is the carrying capacity of refugee camps and temporary shelters?	Flowrate and absolute quantity
religiousBuildingsCapacityRating	What is the carrying capacity of religious buildings?	Flowrate and absolute quantity

Munition Consumption

The *munitionConsumption* subcategory is a *flow metricType* that designates the munitions consumed by the action. Its *MType* value is 291 and its variable indicates the amount. Table 8.12 shows a sample of the *munitionConsumption* Metrics.

Table 8.12 *munitionConsumption* metrics sample

ClassName	Defining questions	Variables in vector
strikeFinancialInfrastructureMunitionsConsumptionMoP	What is the quantity of munitions consumed, by type, in the action?	
strikeCivilianHousingMunitionsConsumptionMoP	What is the quantity of munitions consumed, by type, in the action?	
strikeReligiousFacilityMunitionsConsumptionMoP	What is the quantity of munitions consumed, by type, in the action?	
strikeManufacturingInfrastructureMunitionsConsumptionMoP	What is the quantity of munitions consumed, by type, in the action?	

Supply Consumption

The *supplyConsumption* subcategory is a *flow metricType* that designates the supplies consumed by the action. Its *MType* value is 292, and its variable indicates the amount of food, petroleum, oil and lubricants (POL), water, etc. Table 8.13 shows a sample of the *supplyConsumption* Metrics.

Table 8.13 *supplyConsumption* metrics sample

ClassName	Defining questions	Variables in vector
strikeFinancialInfrastructureSupplyConsumptionMoP	What is the quantity of supplies consumed, by type, in the action?	Food, POL, water, etc.
strikeCivilianHousingSupplyConsumptionMoP	What is the quantity of supplies consumed, by type, in the action?	Food, POL, water, etc.
strikeReligiousFacilitySupplyConsumptionMoP	What is the quantity of supplies consumed, by type, in the action?	Food, POL, water, etc.
strikeManufacturingInfrastructureSupplyConsumptionMoP	What is the quantity of supplies consumed, by type, in the action?	Food, POL, water, etc.

Power Consumption

The *powerConsumption* subcategory is a *flow metricType* that designates the electric power consumed by the action. Its *MType* value is 293 and its variable indicates the amount. Table 8.14 shows a sample of the *powerConsumption* Metrics.

Table 8.14 *powerConsumption* metrics sample

ClassName	Defining questions	Variables in vector
strikeFinancialInfrastructurePowerConsumptionMoP	What is the amount of electric power consumed in the action?	
strikeCivilianHousingPowerConsumptionMoP	What is the amount of electric power consumed in the action?	
strikeReligiousFacilityPowerConsumptionMoP	What is the amount of electric power consumed in the action?	
strikeManufacturingInfrastructurePowerConsumptionMoP	What is the amount of electric power consumed in the action?	

Money Consumption

The *moneyConsumption* subcategory is a *flow metricType* that designates the money consumed by the action. (Normally, we say money is spent, not consumed; however, when it is spent, it is no longer a resource available to the spender. Using "consumption" allows for a common word for usage among the four resources.) Its *MType* value is 294 and its variable indicates the amount. Table 8.15 shows a sample of the *moneyConsumption* Metrics.

Table 8.15 *moneyConsumption* metrics sample

ClassName	Defining questions	Variables in vector
strikeFinancialInfrastructureMoneyConsumptionMoP	What is the amount of money consumed in the action?	
strikeCivilianHousingMoneyConsumptionMoP	What is the amount of money consumed in the action?	
strikeReligiousFacilityMoneyConsumptionMoP	What is the amount of money consumed in the action?	
strikeManufacturingInfrastructureMoneyConsumptionMoP	What is the amount of money consumed in the action?	

8.5 Relational State Variables

The *relational* MType category is a *metricType* that describes relations. This category has seven subcategories: *affiliation*, *hierarchy*, *owner*, *initiator*, *recipient*, *activity*, and *availability*. Figure 8.6 shows the *relational* part of the State Variable taxonomy. The "metric" class box represents all of the Metric classes that are children of the subcategories. The "likes" icon (used in counting likes) in the figure represents the affiliation Mtype subcategory.

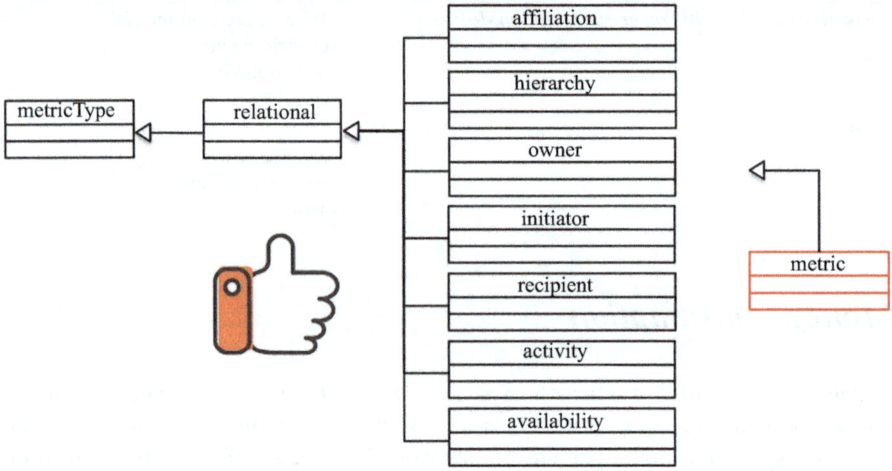

Fig. 8.6 *Relational* state variables taxonomy

Affiliation

The *affiliation* subcategory is a *relational metricType* that designates the level of affiliation which the entity holds for each side. Its *MType* value is 310, and its variable indicates the name of the side with which the Actor is affiliated and an intensity of affiliation in the 0–100 range. The sum over all sides can exceed 100. Table 8.16 shows a sample of the *affiliation* Metrics.

Table 8.16 *affiliation* metrics sample

ClassName	Defining questions	Variables in vector
workerPopulationAffiliation	What is the level of affiliation which the entity holds for each side	Side name, intensity 0–100, sum over all sides can exceed 100
govtDecisionAuthorityAffiliation	What is the level of affiliation which the entity holds for each side	Side name, intensity 0–100, sum over all sides can exceed 100
firstResponderOrganizationAffiliation	What is the level of affiliation which the entity holds for each side	Side name, intensity 0–100, sum over all sides can exceed 100

Hierarchy

The *hierarchy* subcategory is a *relational metricType* that designates the Actor's authority level, name of superior, and type of distribution of authority. (Collectively, these define a hierarchy when combined with the other Actors in the hierarchy.) Its *MType* value is 320 and its variables give the values. Table 8.17 shows a sample of the *hierarchy* Metrics.

Table 8.17 *hierarchy* metrics sample

ClassName	Defining Questions	Variables in vector
workerPopulationHierarchy	What is the Actor's authority level, name of superior, and type of distribution of authority (define hierarchy)?	
govtDecisionAuthorityHierarchy	What is the Actor's authority level, name of superior, and type of distribution of authority (define hierarchy)?	
firstResponderOrganizationHierarchy	What is the Actor's authority level, name of superior, and type of distribution of authority (define hierarchy)?	

Owner

The *owner* subcategory is a *relational metricType* that designates the Object's owner. Its *MType* value is 330 and its variable gives the value. Table 8.18 shows a sample of the *owner* Metrics.

Table 8.18 *owner* metrics sample

ClassName	Defining Questions	Variables in vector
governanceOwner	What is the object's owner?	
changeInGovtLeadershipOwner	What is the object's owner?	
govtDecisionMakingOwner	What is the object's owner?	

Initiator

The *initiator* subcategory is a *relational metricType* that designates the Actor that initiated the Action. Its *MType* value is 331 and its variable gives the value. Table 8.19 shows a sample of the *initiator* Metrics.

Table 8.19 *initiator* metrics sample

ClassName	Defining questions	Variables in vector
mediateNegotiateOrPersuadeConflictingPartiesInitiator	What is the Actor that initiated the Action?	
provideAdvisorsToGovtOfficialsInitiator	What is the Actor that initiated the Action?	
educateGovtPersonnelInitiator	What is the Actor that initiated the Action?	

Recipient

The *recipient* subcategory is a *relational metricType* that designates the recipient or recipients (direct or indirect object) of an action. Its *MType* value is 332 and its variable gives the value. Table 8.20 shows a sample of the *recipient* Metrics.

Table 8.20 *recipient* metrics sample

ClassName	Defining questions	Variables in vector
placeMineMoP	What is the target of the mine (mines, unexploded ordinance, IEDs) placement Action?	anyone in the area, particular convoy, etc.
provideAdvisorsToGovtOfficialsObjectOfAction	What is the identity of the entity or entities that are the recipients (direct or indirect object) of the action?	
educateGovtPersonnelObjectOfAction	What is the identity of the entity or entities that are the recipients (direct or indirect object) of the action?	

Activity

The *activity* subcategory is a *relational metricType* that designates the activity of the entity. Its *MType* value is 340 and its variable gives the value of coverage of efforts, intensity of efforts, and quantity of work. Table 8.21 shows a sample of the *activity* Metrics.

Table 8.21 *activity* metrics sample

ClassName	Defining questions	Variables in vector
firstResponderOrgActivityRating	How active is the first responder organization?	Coverage of efforts, intensity of efforts, quantity of work
govtTypeOrganizationActivityRating	How active is the government organization (includes administration and legislative organizations)?	Coverage of efforts, intensity of efforts, quantity of work
politicalPopulationActivityRating	How active is the politically active population?	Coverage of efforts, intensity of efforts, quantity of work
sideActivityRatingMOPE	How close to winning or losing is the side - activity level?	Defeated, discouraged, destroyed, disintegrated, isolated, or opposite; coverage of efforts, intensity of efforts, quantity of work

Availability

The *availability* subcategory is a *relational metricType* that designates the availability of the entity. Its *MType* value is 340, and its variable gives the numeric or categorical level of availability and may include original or desired levels. Table 8.22 shows a sample of the *availability* Metrics.

Table 8.22 *availability* metrics sample

ClassName	Defining questions	Variables in vector
acceptableJobAvailabilityRating	What is the level of acceptable job availability?	
foodSupplyRating	What is the rating on the available food supply?	
potableWaterSupplyRating	What is the rating on the availability of potable water?	
informationAvailabilityRating	What is the availability level of information to the public?	

8.6 HSCB State Variables

The *HSCB* MType category is a *metricType* that describes HSCB attributes. This category has ten subcategories: *decisionMaking*, *influence*, *fairnessCorruption*, *effectiveness*, *efficiency*, *operatingHealth*, *levelRating*, *progress*, *professionalism*, and *transparency*. Figure 8.7 shows the *HSCB* part of the State Variable taxonomy. The "metric" class box represents all of the Metric classes that are children of the subcategories. The photograph showing money being passed represents a measure of the fairness/corruption MType subcategory.

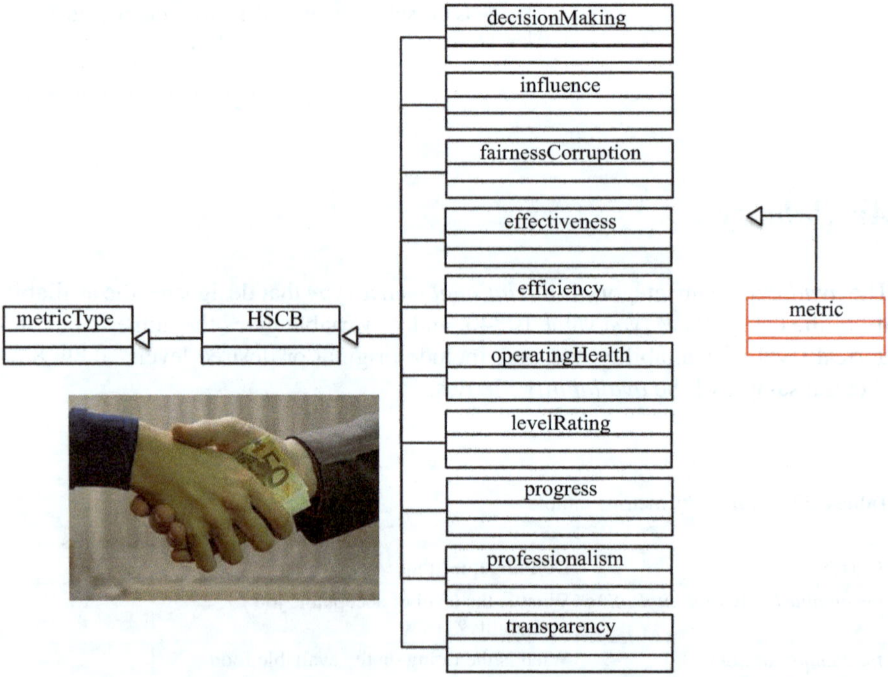

Fig. 8.7 *HSCB* state variables taxonomy

Decision Making

The *decisionMaking* subcategory is an *HSCB metricType* that describes the decision-making process and the quality of the decision-making of the entity. Its *MType* value is 410 and its variable gives the decision-making process and quality of decision-making. Table 8.23 shows a sample of the *decisionMaking* Metrics.

Table 8.23 *decisionMaking* metrics sample

ClassName	Defining questions	Variables in vector
govtDecisionAuthorityRating	What is the decision-making process and the quality of the decision-making?	Decision-making process, quality of decision-making
keyLeaderPoliticalDecisionMakingRating	What is the process and quality of the key political leader's decision-making?	Decision-making process, quality of decision-making
keyLeaderMilitaryDecisionMakingRating	What is the process and quality of the key military leader's decision-making?	Decision-making process, quality of decision-making

Influence

The *influence* subcategory is an *HSCB metricType* that designates the influence of the entity. Its *MType* value is 420 and its variable gives the numeric or categorical level. Table 8.24 shows a sample of the *influence* Metrics.

Table 8.24 *influence* metrics sample

ClassName	Defining questions	Variables in vector
keyLeaderPoliticalProPeaceInfluenceRating	What is the influence of the pro-peace key political leader?	
keyLeaderPoliticalAgitatorInfluenceRating	What is the influence of the anti-peace key political leader?	
keyLeaderPoliticalInfluenceRating	What is the influence of the key political leader?	
externalAgitatorForcesInfluenceRating	What is the influence of external forces advocating conflict?	

Fairness/Corruption

The *fairnessCorruption* subcategory is an *HSCB metricType* that designates the level of fairness/corruption of the entity. Its *MType* value is 430 and its variable gives the numeric or categorical level. Table 8.25 shows a sample of the *fairness-Corruption* Metrics.

Table 8.25 *fairnessCorruption* metrics sample

ClassName	Defining questions	Variables in vector
corruptionInCultureRating	What are the perceived and actual levels of corruption that is prevalent in the culture?	Corruption, institutional bias
corruptionInSocialServicesRating	What is the existing level of corruption in the social services?	Corruption, institutional bias
basicNaturalResourceFairnessRating	What is the Host Nation basic natural resources fairness rating?	Corruption, institutional bias

Effectiveness

The *effectiveness* subcategory is an *HSCB metricType* that designates the effectiveness of the entity. Its *MType* value is 440 and its variable gives the numeric or categorical level. Table 8.26 shows a sample of the *effectiveness* Metrics.

Table 8.26 *effectiveness* metrics sample

ClassName	Defining questions	Variables in vector
centralAuthorityEffectivenessRating	How effectively does it govern?	Political capacity (tax collection capability, etc.)
socialServicesAdequacyRating	How effective is the social service?	Child services, or elderly care, etc.; coverage (distribution throughout society) and effectiveness
judicialOrganizationEffectivenessRating	What is the effectiveness of the court type organization, government or alternative?	
govtLawEnforcementEffectivenessRating	What is the effectiveness of the government law enforcement organization?	

Efficiency

The *efficiency* subcategory is an *HSCB metricType* that designates the economic efficiency of the entity. Its *MType* value is 450 and its variable gives the numeric or categorical level. Table 8.27 shows a sample of the *efficiency* Metrics.

Table 8.27 *efficiency* metrics sample

ClassName	Defining questions	Variables in vector
agricultureBusinessEfficiency	What is the economic efficiency of the entity?	Numeric or categorical level
criminalOrganizationEfficiency	What is the economic efficiency of the entity?	Numeric or categorical level
energyBusinessEfficiency	What is the economic efficiency of the entity?	Numeric or categorical level

Operating Health

The *operatingHealth* subcategory is an *HSCB metricType* that designates the operating health of the entity. Its *MType* value is 460 and its variable gives the numeric or categorical level. Table 8.28 shows a sample of the *operatingHealth* Metrics.

Table 8.28 *operatingHealth* metrics sample

ClassName	Defining questions	Variables in vector
agBusinessRating	What is the health of the agricultural business?	
energyBusinessRating	What is the health of the energy business?	
financialServicesIndustryRating	What is the health of the financial business?	

Level Rating

The *levelRating* subcategory is an *HSCB metricType* that designates the rating level for the entity. Its *MType* value is 461 and its variable gives its value. Table 8.29 shows a sample of the *levelRating* Metrics.

Table 8.29 *levelRating* metrics sample

ClassName	Defining questions	Variables in vector
governanceRating	How well does it govern; what are the checks on its power?	
govtLeaderChangeRating	How does leadership change: elections, hereditary succession, coup, revolution?	
govtDecisionMakingRating	How and how well does the government make decisions?	Distribution of power, number of political parties, checks and balances, elections, dispute resolution
relationshipWithIntervenorsRating	What is the level of the relationship between the parties?	Ranging from adversarial to very close

Progress

The *progress* subcategory is an *HSCB metricType* that designates the progress directly resulting from the action. Its *MType* value is 462 and its variable gives the value. Table 8.30 shows a sample of the *progress* Metrics.

Table 8.30 *progress* metrics sample

ClassName	Defining questions	Variables in vector
govtReformChangeMoP	What is the progress in reforming all parts and levels of government?	
electionsConductedChangeMoP	What is the progress in planning, executing, monitoring, and publicizing elections?	
constitutionReformChangeMoP	What is the progress in constitutional reform?	

Professionalism

The *professionalism* subcategory is an *HSCB metricType* that designates the professionalism, morale and leadership qualities of the entity. Its *MType* value is 470 and its variable gives the numeric or categorical level of equipment, manpower, doctrine, training level, resources, leadership, organizational culture, morale, history, civil-military relations. Table 8.31 shows a sample of the *professionalism* Metrics.

Table 8.31 *professionalism* metrics sample

ClassName	Defining questions	Variables in vector
lawEnforcementOrganizationRating	What is the professionalism, morale and leadership qualities rating of the law enforcement organization?	Numeric or categorical level of equipment, manpower, doctrine, training level, resources, leadership, organizational culture, morale, history, civil-military relations
armedForcesProfessionalismRating	What is the professionalism, morale and leadership qualities rating of the Host Nation military?	Numeric or categorical level of equipment, manpower, doctrine, training level, resources, leadership, organizational culture, morale, history, civil-military relations
govtDecisionAuthorityProfessionalism	What is the professionalism, morale and leadership qualities of the entity?	Numeric or categorical level of equipment, manpower, doctrine, training level, resources, leadership, organizational culture, morale, history, civil-military relations

Transparency

The *transparency* subcategory is an *HSCB metricType* that designates the transparency of operations of the entity. Its *MType* value is 480 and its variable gives the numeric or categorical level. Table 8.32 shows a sample of the *transparency* Metrics.

Table 8.32 *transparency* metrics sample

ClassName	Defining questions	Variables in vector
publicRecordsTransparencyRating	What is the transparency rating for government information?	Records exist, availability, public reporting (push), visibility of actions, etc.
governanceTransparency	What is the transparency of operations of the entity?	Numeric or categorical level
changeInGovtLeadershipTransparency	What is the transparency of operations of the entity?	Numeric or categorical level
govtDecisionMakingTransparency	What is the transparency of operations of the entity?	Numeric or categorical level

8.7 Patterns of State Variables

Each type of Metric (identified by its *MType* value), such as *transparency*, might be assignable to every Actor, Object, and Action class. However, many possible assignments would be meaningless because of the real-world referent of the particular class. For example, unless *transparency* referred to a physical quality (such as possessed by glass), attaching it to an infrastructure class would not yield meaningful information. It turns out that the rules for such attachments are best defined at the subcategory level for Actor, Object, and Action classes.

Note: in building an ontology, categories and subcategories are expected to exhibit similarities. Thus, it is not surprising that there should be similarities of properties (now called MType subcategories) among things that are supposed to be otherwise similar.

State Variables of Actors

In Fig. 8.8, each of the Actor categories and subcategories (described in Chap. 4) is listed on one axis and each of the metric MTypes (within their categories) is listed on the other axis. The cells show which MTypes are attached to each Actor subcategory (blackened cells). (Because all of the Actor classes in a given subcategory have the same MType attachment structure, it is not necessary to list them.)

- This figure shows that all Actor classes have metrics attached for each of the three *metricVectorKey* state variables.
- In the *physical* Metric category, all Actor classes have *location*, *members*, and *movable* state variables. Only those Actor classes in the *armedForce*, *politicalOrgArmed*, *militarySystem*, *simpleCompositeActor*, and *hierarchicalActor* subcategories have *weaponry* state variables. No Actor classes have *quantity*, *disasterIndicators*, or *damage* state variables.
- No Actor classes have *flow* state variables.
- In the *relational* Metric category, all Actor classes, except for those in the *nonhumanActor* subcategory, have *affiliation* and *hierarchy* state variables and only the *nonhumanActors* have an *owner* state variable. All Actor classes have an *activity* state variable. No Actor classes have *initiator*, *recipient*, or *availability* state variables.
- The *HSCB* category of state variables is more complicated. All Actor classes, except the *nonhumanActor* category, have the *decisionMaking* and *influence* state variables. Only those in the *nonhumanActor* category and the *economicOrganization* subcategory have an *operatingHealth* state variable. The Actor classes of the *individualActor*, *significantGroupActor*, and *compositeActor* categories all have *fairnessCorruption*, *effectiveness*, and *professionalism* state variables. Only the Actor classes of the *economicOrganization* subcategory have an affinity state variable. Only the *politicalOrgUnarmed* and *politicalOrArmed* subcategory Actor classes and the *compositeActor* category Actor classes have a *transparency* state variable. No Actor classes have *levelRating* or *progress* state variables.

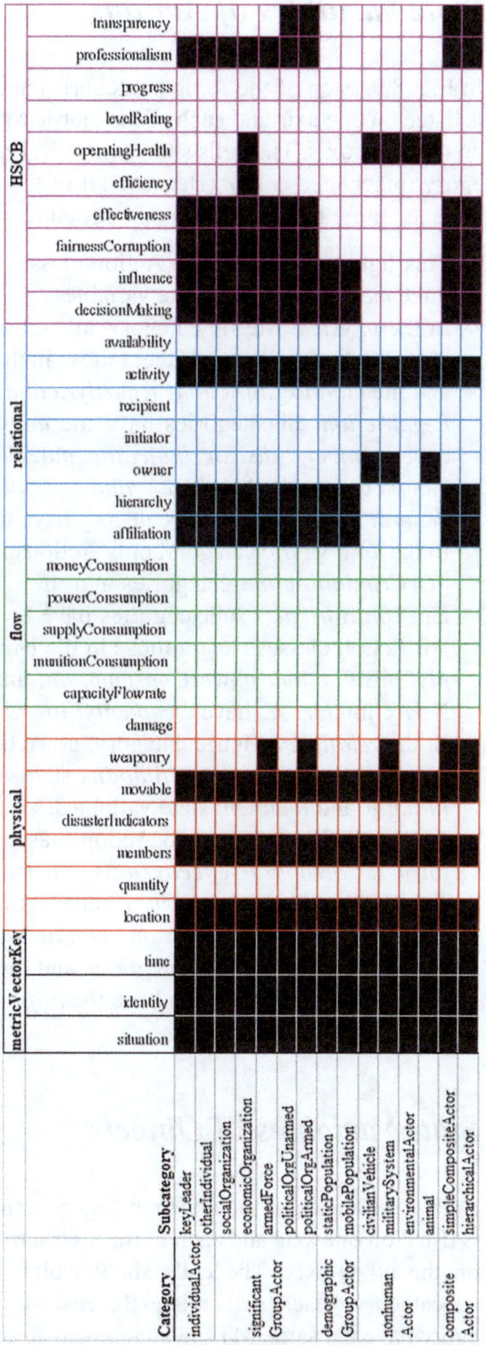

Fig. 8.8 State variables of actors

State Variables of Actions

In Fig. 8.9, each of the Action categories and subcategories (described in Chap. 5) is listed on one axis and each of the metric MTypes (within their categories) is listed on the other axis. The cells show which MTypes are attached to each Action subcategory (blackened cells). (Because all of the Action classes in a given subcategory have the same MType attachment structure, it is not necessary to list them.)

- This figure shows that all Action classes have metrics attached for each of the three *metricVectorKey* state variables.
- In the *physical* Metric category, all Action classes have the *location* state variable. All Action classes, except those in the *strikeAction* and *C2Action* categories and the *monitoringAction*, *securityAction*, *generalConflictAction*, and *policyOrLegalAction* subcategories have the *quantity* state variable. No Action classes have *members*, *disasterIndicator*, *movable*, or *weaponry* state variables. Only Action classes in the *strikeAction* and *policingOrCriminalAction* categories and *generalConflictAction* subcategory have the *damage* state variable.
- In the *flow* Metric category, only Action classes in the *conflictAction* and *policingOrCriminalAction* categories and the *strikeTargetAction*, *strikeMethodAction*, and *controlAction* subcategories have the *munitionConsumption* state variable. All Action classes except those in the *envirnStrikeAction* subcategory have *supplyConsumption*, *powerConsumption*, and *MoneyConsumption* state variables. No Action classes have a *capacityFlowrate* state variable.
- In the *relational* Metric category, no Action classes have the *affiliation*, *hierarchy*, *owner*, *activity*, or *availability* state variables and all Action classes have the *initiator* and *recipient* state variables.
- In the *HSCB* category, no Action classes have the *decisionMaking*, *influence*, *fairnessCorruption*, *effectiveness*, *efficiency*, *operatingHealth*, *levelRating*, *professionalism*, or *transparency* state variables. Only the Action classes in the *C2Action*, *informationAction*, *conflictAction*, *policingOrCriminalAction*, and *civilGovernmentAction* categories and the *conflictOrganizationAction* and *socialAidAction* subcategories have the *progress* state variable.

State Variables of Objects

In Fig. 8.10, each of the Object categories and subcategories (described in Chap. 6) is listed on one axis and each of the metric MTypes (within their categories) is listed on the other axis. The cells show which MTypes are attached to each Object subcategory (blackened cells). (Because all of the Object classes in a given subcategory have the same MType attachment structure, it is not necessary to list them.)

- This figure shows that all Object classes have metrics attached for each of the three *metricVectorKey* state variables.

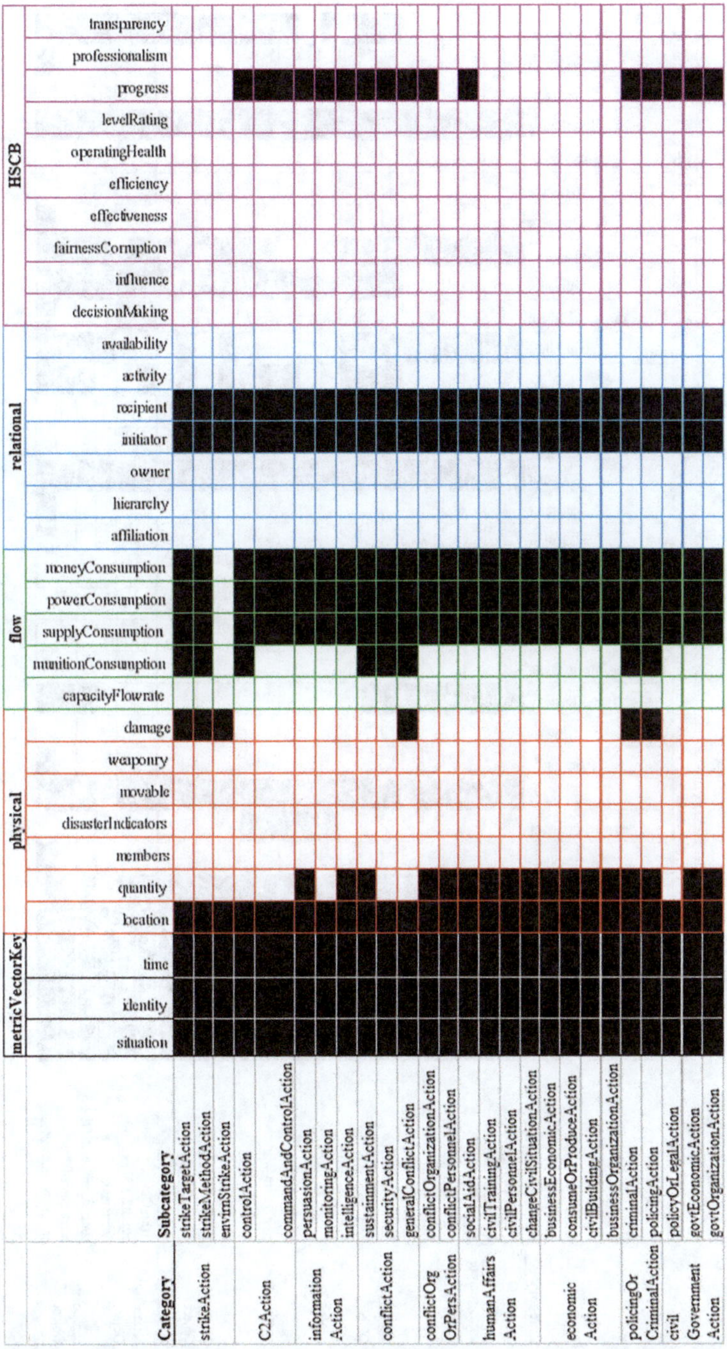

Fig. 8.9 State variables of actions

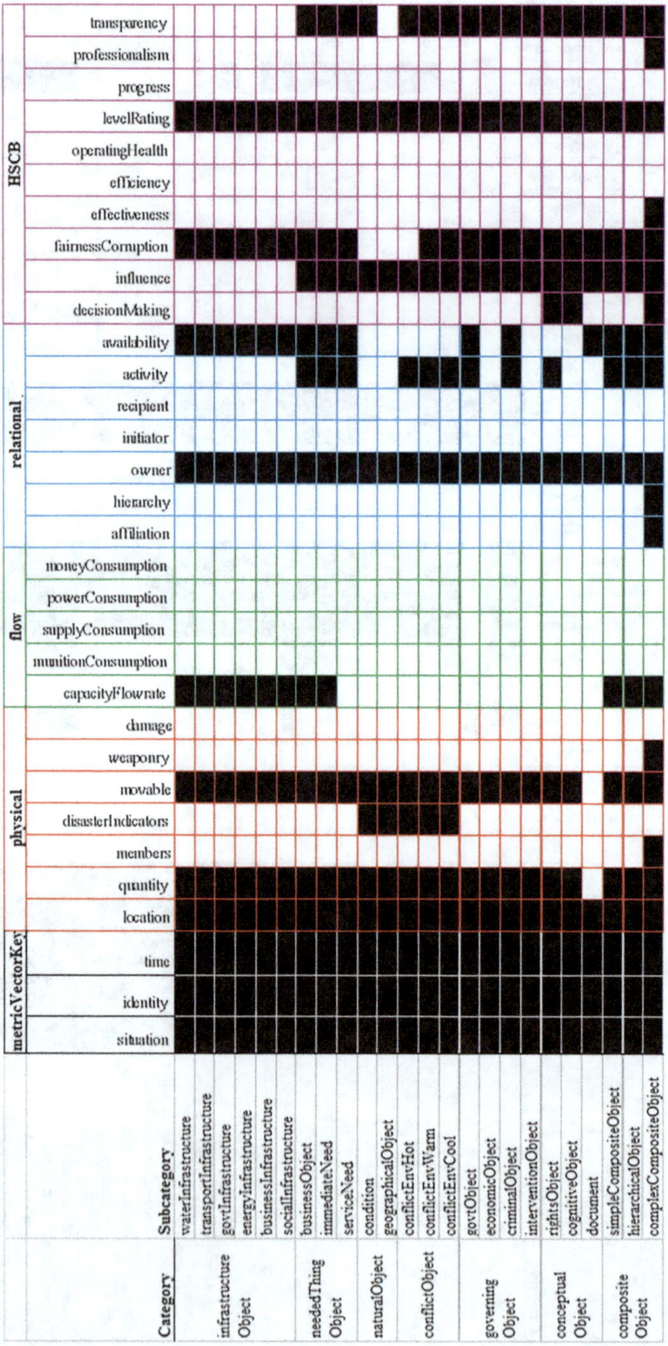

Fig. 8.10 State variables of objects

- In the *physical* Metric category, all Object classes have the *location* state variable. All Object classes, except those in the *document* subcategory, have the *quantity* and *movable* state variables. Only Object classes in the *complexCompositeObject* subcategory have the *members* or *weaponry* state variables. Only Object classes in the *naturalObject* and *conflictObject* categories have the *disasterIndicators* state variable. No Object classes have a *damage* state variable.
- In the *flow* Metric category, only Object classes in the *infrastructureObject* and *compositeObject* categories and the *businessObject* and *immediateNeed* subcategories have the *capacityFlowrate* state variable. No Object classes have the *munitionConsumption*, *supplyConsumption*, *powerConsumption*, or *MoneyConsumption* state variables.
- In the *relational* Metric category, only Object classes in the *complexCompositeObject* subcategory have the *affiliation* or *hierarchy* state variables. All Object classes have the *owner* state variable. Only Object classes in the *neededThingObject*, *conflictObject* and *compositeObject* categories and *govtObject*, *criminalObject*, and *rightsObject* subcategories have the *activity* state variable. Only Object classes in the *infrastructureObject*, *neededThingObject* and *compositeObject* categories and the *govtObject*, *economicObject* and *document* subcategories have the *availability* state variable. No Object classes have the *initiator* or *recipient* state variables.
- In the *HSCB* category, no Object classes have the *efficiency*, *operatingHealth*, or *progress* state variables. All Object classes have the *levelRating* state variable. All Object classes except those in the *infrastructureObject* category have the *influence* state variable. All Object classes except those in the *naturalObject* category and the *conflictEnvHot* subcategory have the *fairnessCorruption* state variable. All Object classes except those in the *infratructureObject* category and the *geographicalObject* subcategory have the *transparency* state variable. Only the Object classes in the *rightsObject*, *cognitiveObject* and *complexCompositeObject* subcategories have the *decisionMaking* state variable. Only the Object classes in the *complexCompositeObject* subcategory have the *effectiveness* or *professionalism* state variables.

8.8 State Variable Relationships

The only relationships used in this chapter are the *is-a*, the implied *hasMetric* and the *hasPart* relationships, explained as follows:

is-a: Each instance of A is an instance of B and each property of B is a property of A. Its inverse relationship is *superClassOf*.

hasMetric: A has Metric B (also shown as *described by*). Its inverse relationship is *metricOf*.

hasPart: An entity has another entity as a part. Its inverse relationship is *isPartOf*. The version that is used is one in which the relationship is an aggregation.

8.9 State Variables Review

State variables or Metrics are the things that define the property types (MTypes) of the Actor, Action and Object classes. The Metric instances hold the property values of the related Actor, Action and Object instances. The Metrics correspond (very roughly) to adjectives and adverbs, modifying the nouns (Actors and Objects) and verbs (Actions), respectively. However, the Metrics provide more specific information than do standard adjectives and adverbs by providing current state variable values for the Actors, Objects and Actions.

The MType categories and subcategories organize the Metrics into meaningful groups, with the first category, *metricVectorKey*, providing the link to join all of the values for a particular observation and connect to the correct *situation*, instance *identity*, and *time*.

Each of the Actor, Action, and Object classes has a defined set of MTypes. The set of MTypes is uniform for all of the classes within the particular subcategory of the class. (There are a few classes for which a particular type of Metric is defined, but not really applicable. In these cases, the values are simply left as null.)

Metrics may also be categorized by the PMESII paradigm. The PMESII Ontology and the MType Metric Ontology provide meta-data showing different associations of the Metrics. The PMESII categorization of Metrics is discussed in Chap. 9.

Chapter 9
MCO DIME Ontology and PMESII Metric Ontology

The current standard paradigm for understanding conflict involves consideration of the instruments of power, that is, the means for effecting change in a situation and the state variables that describe the situation at any given time. The simplest listing of the instruments of power is Diplomatic, Informational, Military, and Economic (DIME). The simplest listing of the state variable categories is Political, Military, Economic, Social, Information, and Infrastructure (PMESII). Each of the DIME and PMESII categories can be decomposed down to observable and measurable elements. This chapter discusses how the DIME/PMESII paradigm is implemented in the Modern Conflict Ontology (MCO).

9.1 DIME+ Ontology

Figure 3.8 showed how the DIME Ontology fits into the overall MCO structure (with Fig. 9.1 illustrating its icon as used in the Context Diagram, Fig. 3.5).

Fig. 9.1 DIME icon

D. S. Hartley III, *An Ontology of Modern Conflict*, Understanding Complex Systems, https://doi.org/10.1007/978-3-030-53214-7_9

The DIME concept rests on the observation that a nation has many instruments of power with which it can attempt to achieve its international goals. These instruments or levers of power can be collected into a small set of categories to make remembering them easier. The most popular set is DIME – Diplomatic, Information (or Informational), Military, and Economic. Variations to DIME have been proposed, such as DIMEFIL, which adds Financial, Intelligence, and Law Enforcement to the DIME package. We add an Other category, resulting in DIME+.

Obviously, the components of DIME are Actions. After describing the decomposition of DIME+ into categories and subcategories, we will describe how the Actions are related to this decomposition.

There are five DIME+ categories: diplomatic, informational, military, economic, and others. Figure 9.2 shows the top level of the DIME Ontology. The "action" class box represents all of the Action classes that are part of the subcategories. Note that there are two relationship types in this figure, so we have an ontology, not a taxonomy. The *is-a* relationship is represented by the triangle connections on the left and the diamond connection on the right represents the *isPartOf* relationship. This latter relationship does permit a single Action to be part of multiple DIME subcategories; however, in the current version of the MCO, each Action class is part of only one DIME subcategory. The presence of only one connection from the "action" indicates this. The treaty signing picture in the figure represents all of the possible DIME Actions.

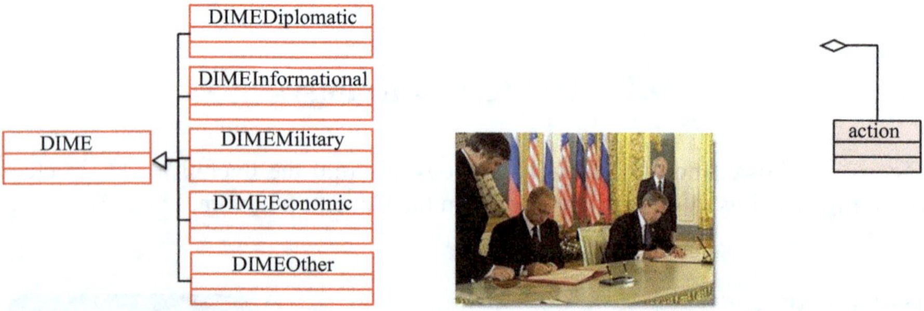

Fig. 9.2 DIME+ ontology overview

The DIMEDiplomatic Category

This category is defined as a *DIME* action that is diplomatic inter- or intragovernmental. **Diplomatic power** rests in negotiations and agreements. To the extent that a nation regards itself as bound to honor its agreements, diplomacy can result in changes in the actions of one or more nations. However, there are advantages to be had short of binding agreements. Offering to provide or withhold favors among countries can also result in changes of actions.

There are three subcategories of the *DIMEDiplomatic* category: diplomatic external, diplomatic internal government policy, and diplomatic internal government organizations. Figure 9.3 shows the diplomatic level of the ontology. The "action" class box represents all of the Action classes that are part of the subcategories. The seal for the Diplomatic Security Service in the figure represents one of the more physically active parts of the category.

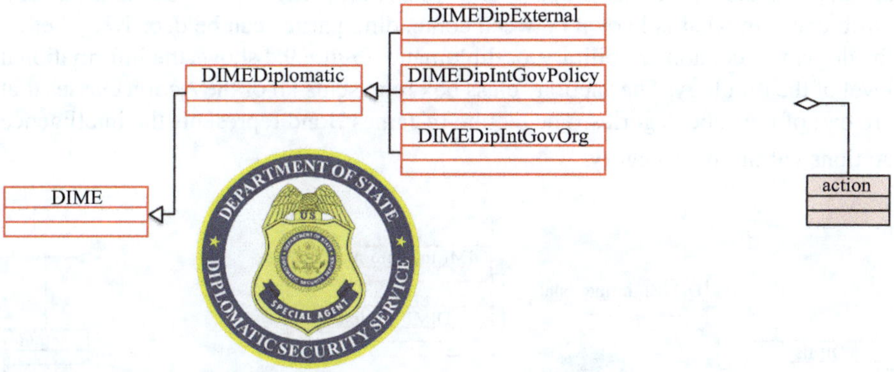

Fig. 9.3 DIME diplomatic ontology

The DIMEDipExternal Subcategory

This subcategory is defined as a *DIMEDiplomatic* action that originates externally from the recipient. An example is *provideAdvisorsToLeaders*, defined as a *persuasionAction* that provides advisors to government or other organizations – other than judicial organizations.

The DIMEDipIntGovPolicy Subcategory

This subcategory is defined as a *DIMEDiplomatic* action that involves <u>internal governmental policy</u> actions, which may or may not originate externally. An example is *reduceLikelihoodOfPopulationMovements*, defined as a *persuasionAction* that reduces likelihood of population movements.

The DIMEDipIntGovOrg Subcategory

This subcategory is defined as a *DIMEDiplomatic* action that involves <u>internal governmental organization</u> actions, which may or may not originate externally. An example is *decreaseNGOOrganizations*, defined as a *socialAidAction* that decreases the number of NGOs, IOs, or IGOs.

The DIMEInformational Category

This category is defined as a *DIME* action that is informational or intelligence-related. **Information power** lies in gaining information from others and in controlling the information desired by others. With cyberwarfare, remotely controlled sabotage is also possible (Lubold and Volz 2018; Taylor and Germano 2018). Differences in what is known between contending parties can be decisive, whether the domain is economic, military, or diplomatic. Figure 9.4 shows the informational level of the ontology. The "action" class box represents all of the Action classes that are part of the subcategories. The picture of James Bond represents the intelligence Actions within this category.

Fig. 9.4 DIME informational ontology

The DIMEInfoInformational Subcategory

This subcategory is defined as a *DIMEInformational* action that involves informational actions, including education. An example is *strikeCommunicationsProcess*, defined as a *strikeTargetAction* that is aimed at military communications equipment or systems.

The DIMEInfoIntel Subcategory

This subcategory is defined as a *DIMEInformational* action that involves intelligence, cyber, or electronic warfare actions. An example is *assessOperational-Progress*, defined as a *commandAndControlAction* that assesses progress toward accomplishing tasks, creating conditions and achieving objectives.

The DIMEMilitary Category

This category is defined as a *DIME* action that is military or law enforcement related. **Military power** is an obvious component. However, the fact that it has limits to its effectiveness and on its use should serve warning that the other levers of power also have limits (both individually and collectively). Figure 9.5 shows the military level of the ontology. The "action" class box represents all of the Action classes that are part of the subcategories. The photograph of armed troops in the snow in this figure represents one part of this category.

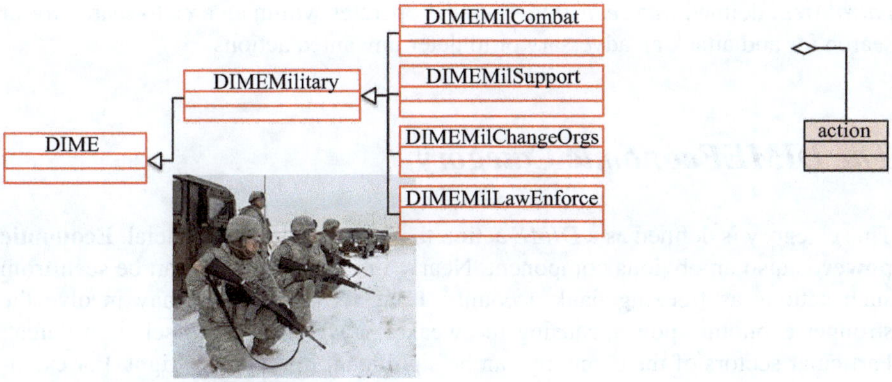

Fig. 9.5 DIME military ontology

The DIMEMilCombat Subcategory

This subcategory is defined as a *DIMEMilitary* action that is directly combat related. An example is *strikeMilitaryVehicle*, defined as a *strikeTargetAction* that is aimed at a military system.

The DIMEMilSupport Subcategory

This subcategory is defined as a *DIMEMilitary* action that supports combat or peace operations. An example is *protectAttachedCivilian*, defined as a *securityAction* that protects civilians and contractors authorized to accompany the force.

The DIMEMilChangeOrgs Subcategory

This subcategory is defined as a *DIMEMilitary* action that changes the number of a type of organization or its size. An example is *increaseParamilitaryOrganizations*, defined as a *conflictOrganizationAction* that increases the number of paramilitary organizations.

The DIMEMilLawEnforce Subcategory

This subcategory is defined as a *DIMEMilitary* action that is predominantly law enforcement related but may be performed by the military, also. An example is *patrolArea*, defined as a *securityAction* that operates within an area to search for or search for and attack an adversary or to deter unwanted actions.

The DIMEEconomic Category

This category is defined as a *DIME* action that is economic or financial. **Economic power** is also an obvious component. Nearly immediate effects can be seen from such actions as freezing bank accounts. Long-term strategies may involve the stronger economic power causing the weaker power to spend itself into defeat. Particular sectors of the economy can be significant in their own right. For example, the need to import oil and fuels (energy) costs the importing nation directly and may have less direct costs in the form of obligations and possible coercion. Energy independence allows a nation to avoid these problems and becoming a net exporter adds to the nation's influence (Olson 2018; Wall Street Journal Editorial Staff 2018). Figure 9.6 shows the economic level of the ontology. The "action"

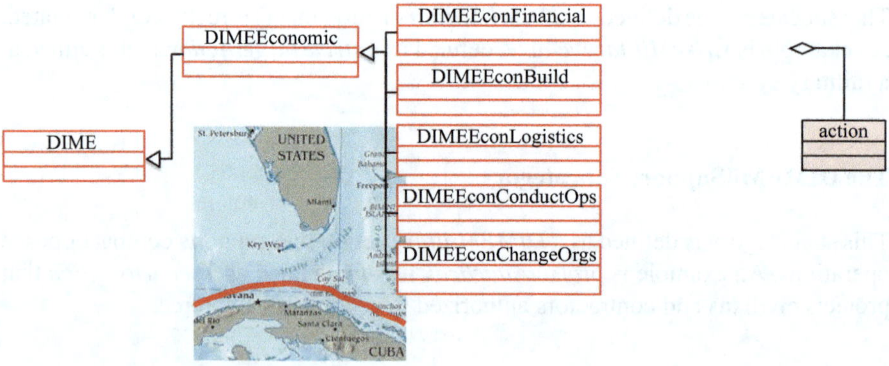

Fig. 9.6 DIME economic ontology

class box represents all of the Action classes that are part of the subcategories. The red line between the United States and Cuba represents an economic embargo imposed by the United States on Cuban goods, an economic Action.

The DIMEEconFinancial Subcategory

This subcategory is defined as a *DIMEEconomic* action that is financial in nature. An example is *disburseFunds*, defined as a *businessEconomicAction* that spends funds for various purposes.

The DIMEEconBuild Subcategory

This subcategory is defined as a *DIMEEconomic* action that involves building or rebuilding. An example is *rebuildCivilianHousing*, defined as a *civilBuildingAction* that builds or rebuilds civilian housing.

The DIMEEconLogistics Subcategory

This subcategory is defined as a *DIMEEconomic* action that involves suppling or moving material things. An example is *distributeEquipmentOrMaterial*, defined as a *sustainmentAction* that distributes equipment or material.

The DIMEEconConductOps Subcategory

This subcategory is defined as a *DIMEEconomic* action that consists of conducting an economic operation. An example is *conductLaborStrikes*, defined as a *businessEconomicAction* that conducts labor strikes against business or government organizations.

The DIMEEconChangeOrgs Subcategory

This subcategory is defined as a *DIMEEconomic* action that changes the number of a type of organization or its size. An example is *decreaseContractorPersonnel*, defined as a *civilPersonnelAction* that decreases number of contractor personnel.

The DIMEOther Category

This category is defined as a *DIME* action that is related to criminal, terror, natural, or self-related Actions. **Other power** includes criminal and environmental (e.g., natural disasters) forces. Figure 9.7 shows the other level of the ontology. The "action" class box represents all of the Action classes that are part of the subcategories. The photograph of the tornado and lightning represents a natural Action.

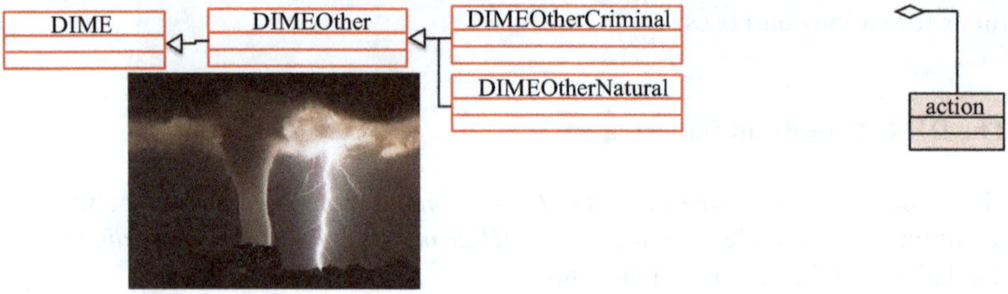

Fig. 9.7 DIME other ontology

The DIMEOtherCriminal Subcategory

This subcategory is defined as a *DIMEOther* action that is criminal or terroristic. An example is *conductPiracyOperation*, defined as a *generalConflictAction* that is part of or supports a piracy operation.

The DIMEOtherNatural Subcategory

This subcategory is defined as a *DIMEOther* action that is natural or self-related. An example is *moveSelf*, defined as a *controlAction* in which the Actor changes location or location distribution, for example, maneuvers forces to achieve a position of advantage over the enemy.

Relating Action Classes to DIME+

The Action classes have been defined in Chap. 5, where they are grouped by structural categories and subcategories. However, they also can be assigned to the DIME+ categories and subcategories. This latter grouping is more subjective and hence more debatable. Rather than list the Action classes again, the assignments are reported in Fig. 9.8 as the numbers of classes in each structural subcategory that have been assigned to each DIME+ category.

Category	Subcategory	Diplomatic	Information	Military	Economic	Other
strikeAction	strikeTargetAction		7	25	1	6
	strikeMethodAction		2	13		4
	envirnStrikeAction					7
C2Action	controlAction			8		5
	commandAndControlAction		5	4		
informationAction	persuasionAction	8	10		1	
	monitoringAction	4				
	intelligenceAction	1	12			
conflictAction	sustainmentAction			1	12	
	securityAction	6	2	15		
	generalConflicAction	6	1	24	1	2
conflictOrgOrPersAction	conflicOrganizationAction	2	3	13	2	2
	conflicPersonnelAction	4	3	15	2	2
humanAffairsAction	socialAidAction	7	1		12	
	civilTrainingAction		8	1		
	civilPersonnelAction	17	4	2	9	2
	changeCivilSituationAction	1	6	1		
economicAction	businessEconomicAction				8	
	consumeOrProduceAction				13	
	civilBuildingAction		3		26	
	businessOrganizationAction		2		30	2
policingOrCriminalAction	criminalAction					9
	policingAction			6		
civilGovernmentAction	policyOrLegalAction	26				
	govtEconomicAction	3			13	
	govtOrganizationAction	15	2	3	2	

Fig. 9.8 Action categories broken into DIME categories

This raw DIME profile is based on one entry for each Action class. The colors help identify the weightings. A similar DIME profile can be created for the Actions of a particular party (side), with each use of an Action class providing its own entry. You can use the DIME profile of other parties' actions to help infer their agendas and capabilities (or capacities). Once you have established approximate capability levels, you can then look at the percent of their capability they are exerting.

The raw DIME data consist of numbers of actions in each category. These are "inputs," not "outputs," of the party. That is, they are measures of effort, not measures of effect. If you apply cost estimates per type of action (money or man-hours or man-hours by type of expertise), you can get an estimate of total expenditures and expenditures by category.

9.2 PMESII Ontology

Figure 3.7 showed how the PMESII Ontology or PMESII Metric Ontology fits into the overall MCO structure (with Fig. 9.9 illustrating its icon as used in the Context Diagram, Fig. 3.5).

The traditional thinking about conventional conflict focused on military actions and military metrics. Clearly, more is needed to organize thinking about unconventional conflict. Figure 1.1, derived from *Joint Operations* (Chairman of the Joint Chiefs of Staff 2011), shows a figure illustrating this point. The five domains, political, military, economic, social, information (or informational, as shown in the figure), and infrastructure (**PMESII**), are shown as parts of the operational environment. Each has its own link and node structure; however, all of the domains are intercon-

Fig. 9.9 PMESII icon

nected. The figure emphasizes that the operational center of gravity (COG) may differ from the strategic COG. In the military, the COG of an organization is the source of strength of the organization. The concept was invented by von Clausewitz to describe the location of the cohesive strength of a military force – the point at which an attack should be aimed (Clausewitz 1993).

Variations to PMESII have been proposed, such as PMESII-PT, which adds physical environment and time to the package. However, the central PMESII construct has proved to be durable and useful and that term will generally be used to refer to all variants. In the MCO, we have added the kinetic and environmental domains, resulting in PMESII+.

There are eight PMESII+ categories: political, military, economic, social, information, infrastructure, kinetic, and environmental. There is also a "null" category to connect to Metric classes with no legitimate PMESII association. Figure 9.10 shows the top level of the PMESII Ontology. The "metric" class box represents all of the Action classes that are part of the subcategories. Note that there are two relationship types in this figure, so we have an ontology, not a taxonomy. The *is-a* relationship is represented by the triangle connections on the left and the diamond connection on the right represents the *isPartOf* relationship. The latter relationship permits multiple connections between the "metric" and the PMESII subcategories. In the DIME

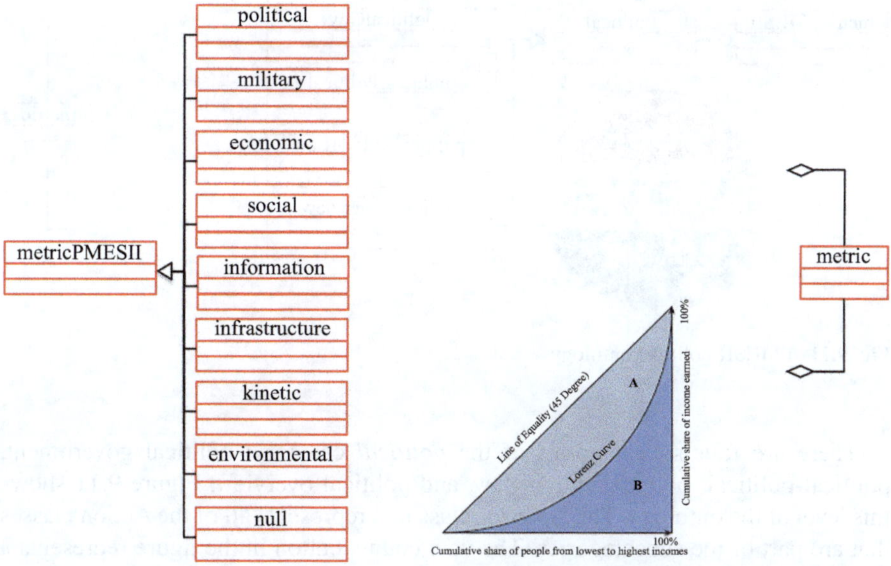

Fig. 9.10 PMESII+ ontology overview

case, this possibility has not been implemented; however, in the PMESII case, multiple connections are implemented. This is the reason that two connections are shown in the figure. The convention is that this indicates one or more connections, not precisely two. The large triangle in the figure represents the calculation of the GINI coefficient that is a measure of inequities.

Note that the Metrics that are categorized in the PMESII Ontology are the same Metrics in the MType Ontology described in Chap. 8. This means that each one is made up of the state variables described there. Following the descriptions of the PMESII categories and subcategories, there is a section that relates the two ontologies.

To reiterate, each Metric class may be part of more than one subcategory.

The political Category

This category is defined as a *metricPMESII* that is a political Metric. The primary components of the political domain are governance (policies, personnel, organizations, freedom, etc.), the rule of law (judiciary, law enforcement, crime, etc.), and politics (leadership, factions, etc.), as well as some miscellaneous items (intervenor status, stability/peace and legitimacy ratings, etc.). Measuring the status of some of these items is difficult, but obviously necessary.

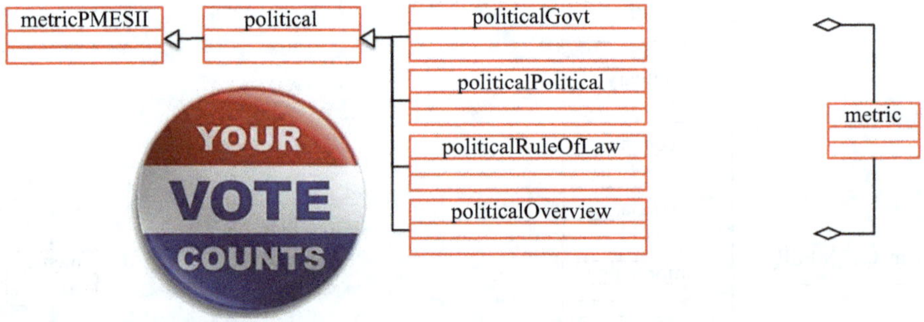

Fig. 9.11 PMESII political ontology

There are four subcategories of the *political* category: political government, political-political, political rule of law, and political overview. Figure 9.11 shows this level of the ontology. The “metric” class box represents all of the Action classes that are part of the subcategories. The vote counts button in the figure represents a measure of political activity.

The politicalGovt Subcategory

This subcategory is defined as a *political* Metric that involves the government. An example is *firstResponderOrgActivityRating*, with the defining question "How active is the first responder organization?"

The politicalPolitical Subcategory

This subcategory is defined as a *political* Metric that involves politics. An example is *keyLeaderPoliticalAgitatorInfluenceRating*, with the defining question "What is the influence of the anti-peace key political leader?"

The politicalRuleOfLaw Subcategory

This subcategory is defined as a *political* Metric that involves the rule of law. An example is *lawEnforcementOrganizationRating*, with the defining question "What is the professionalism, morale and leadership qualities rating of the law enforcement organization?"

The politicalOverview Subcategory

This subcategory is defined as a *political* Metric that involves overarching things. An example is *increaseKeyIntervenorDiplomatMoP*, with the defining question "How much is the increase in key intervenor diplomatic personnel?"

The military Category

This category is defined as a *metricPMESII* that is a military Metric. The primary components of the military domain are conflict, government (the relation with, intelligence services, organizational sizes, etc.), and security, as well as other items (insurgents, terrorists, paramilitary forces, capacities, etc.). Measuring the status of these items is easier than measuring the political status. (The US military traditionally measures itself in terms of capacity, capability, and readiness.)

There are four subcategories of the *military* category: military conflict, military government, military security, and military other. Figure 9.12 shows this level of the ontology. The "metric" class box represents all of the Action classes that are part of the subcategories. The photograph of a bombed out building in the figure represents a measure of military damages.

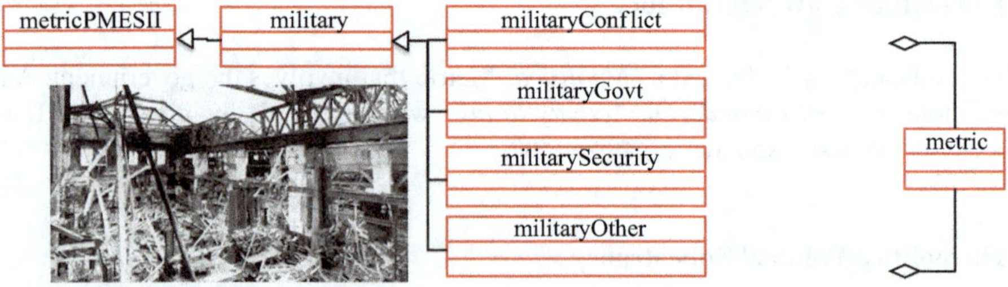

Fig. 9.12 PMESII military ontology

The militaryConflict Subcategory

This subcategory is defined as a *military* Metric that involves conflict. An example is *conventionalWarActionMoP*, with the defining question "What is the progress of the conventional war Action?"

The militaryGovt Subcategory

This subcategory is defined as a *military* Metric that involves the government. An example is *cooperationBetweenHNMilitaryAndIntervenorsRating*, with the defining question "What is the level of cooperation between the Host Nation military and the intervenors?"

The militarySecurity Subcategory

This subcategory is defined as a *military* Metric that involves security. An example is *refugeeCampSecurityActionMoP*, with the defining question "What is the progress in security provided for refugee camps?"

The militaryOther Subcategory

This subcategory is defined as a *military* Metric that involves other military things. An example is *clearMineActionMoP*, with the defining question "What is the progress of the mine clearance Action?"

The economic Category

This category is defined as a *metricPMESII* that is an economic Metric. The primary components of the economic domain are agriculture, crime, energy, finance, governmental economic actions, jobs, and "other" things. Measuring the status of economic items is a relatively well-defined process compared to some other PMESII items.

There are seven subcategories of the *economic* category: economic agriculture, economic crime, economic energy, economic finance, economic government, economic jobs, and economic other. Figure 9.13 shows this level of the ontology. The "metric" class box represents all of the Action classes that are part of the subcategories. The graph in the figure represents two measures of economic activity.

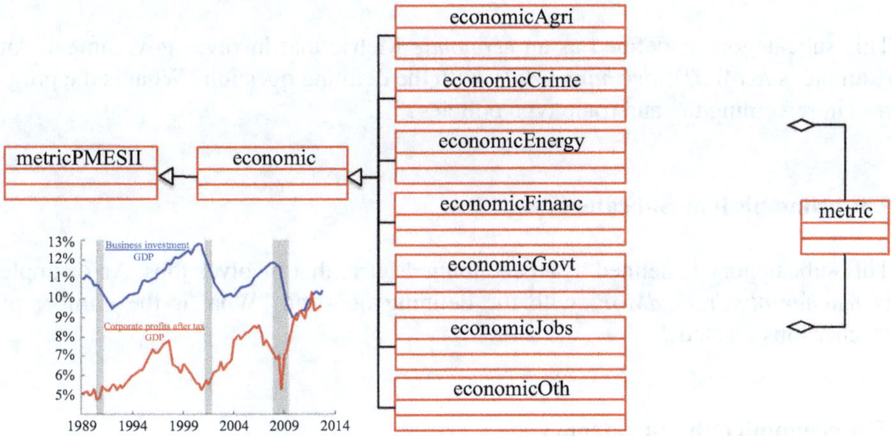

Fig. 9.13 PMESII economic ontology

The economicAgri Subcategory

This subcategory is defined as an *economic* Metric that involves agriculture. An example is *agInvestmentMoP*, with the defining question "What is the monetary investment in agricultural support programs and other agricultural transactions?"

The economicCrime Subcategory

This subcategory is defined as an *economic* Metric that involves crime. An example is *drugUseRating*, with the defining question "What is the drug use level?"

The economicEnergy Subcategory

This subcategory is defined as an *economic* Metric that involves energy. An example is *electricDistributionCapacityRating*, with the defining question "What is the carrying capacity of the electricity distribution infrastructure?"

The economicFinanc Subcategory

This subcategory is defined as an *economic* Metric that involves finance. An example is *currencyChangeMoP*, with the defining question "What is the progress in reforming the currency?"

The economicGovt Subcategory

This subcategory is defined as an *economic* Metric that involves government. An example is *taxAndTradeChangeMoP*, with the defining question "What is the progress in reforming tax and trade type policies?"

The economicJobs Subcategory

This subcategory is defined as an *economic* Metric that involves jobs. An example is *teacherJobsCreatedMoP*, with the defining question "What is the number of teacher jobs created?"

The economicOth Subcategory

This subcategory is defined as an *economic* Metric that involves other economic things. An example is *economicDamageActionMoP*, with the defining question "What is the amount of damage to the economy or financial infrastructure?"

The social Category

This category is defined as a *metricPMESII* that is a social Metric. The primary components of the social domain are basic needs (food, water, shelter, etc.), education, health, movement (freedom of-, restrictions on-, forced-, etc.), safety, and other items. Some of the items are relatively easy to measure, while some of the "other items" that include religious and associational metrics are quite difficult to measure.

There are six subcategories of the *social* category: social basic needs, social education, social health, social movement, social safety, and social other. Figure 9.14 shows this level of the ontology. The "metric" class box represents all of the Action classes that are part of the subcategories. The photograph of a stack of diplomas represents a measure of education.

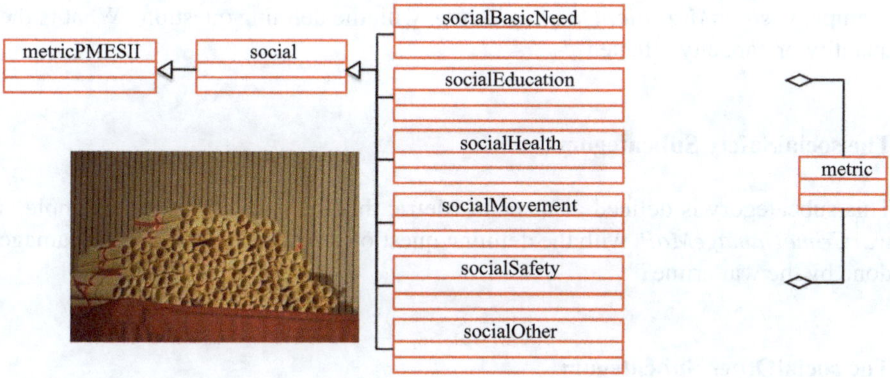

Fig. 9.14 PMESII social ontology

The socialBasicNeed Subcategory

This subcategory is defined as a *social* Metric that involves basic needs: food, water, shelter, and medical attention. An example is *trashDisposalRating*, with the defining question "What is the effectiveness of trash disposal?"

The socialEducation Subcategory

This subcategory is defined as a *social* Metric that involves education. An example is *teacherJobsCreatedMoP*, with the defining question "What is the number of teacher jobs created?" (Notice that this is an example of a Metric contributing to more than one PMESII subcategory – here in *socialEducation* and above in *economicJobs*.)

The socialHealth Subcategory

This subcategory is defined as a *social* Metric that involves health. An example is *reduceDrugDemandActionMoP*, with the defining question "What is the reduction of drug demand?"

The socialMovement Subcategory

This subcategory is defined as a social Metric that involves movement of people. An example is *stressMigrationQuantityRating*, with the defining question "What is the quantity or capacity affected?"

The socialSafety Subcategory

This subcategory is defined as a *social* Metric that involves safety. An example is *warCrimeDamageMoP*, with the defining question "What is the severity of damage done by the war crime?"

The socialOther Subcategory

This subcategory is defined as a *social* Metric that involves other social things. An example is *corruptionInCultureRating*, with the defining question "What are the perceived and actual levels of corruption that is prevalent in the culture?"

The information Category

This category is defined as a *metricPMESII* that is an information Metric. The primary components of the information domain are general information items (primarily decision-making ratings), media (capacity, freedom, etc.), opinions (of various actors about legitimacy, satisfaction, etc.), and information operations (including gathering and disseminating information in a neutral sense and "information ops" in the military sense).

There are four subcategories of the *information* category: information general, information media, information opinions, and information operations. Figure 9.15 shows this level of the ontology. The "metric" class box represents all of the Action classes that are part of the subcategories. The chart in the figure represents information measures.

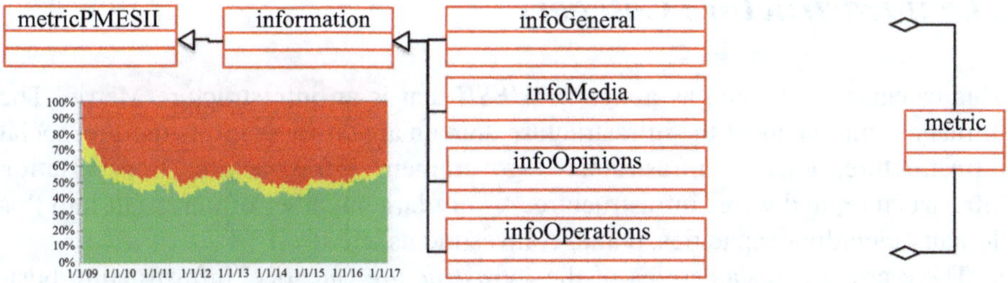

Fig. 9.15 PMESII information ontology

The infoGeneral Subcategory

This subcategory is defined as an *information* Metric that involves general information. An example is *informationAvailabilityRating*, with the defining question "What is the availability level of information to the public?"

The infoMedia Subcategory

This subcategory is defined as an *information* Metric that involves the media. An example is *mediaInfraCapacityRating*, with the defining question "What is the carrying capacity of the media infrastructure?"

The infoOpinions Subcategory

This subcategory is defined as an *information* Metric that involves opinions. An example is *keyLeaderOpinionRating*, with the defining question "For this issue and leader, what is the opinion level?"

The infoOperations Subcategory

This subcategory is defined as an *information* Metric that involves information operations. An example is *positivePRActionMoP*, with the defining question "What is the direct result of the positive public relations Action?"

The infrastructure Category

This category is defined as a *metricPMESII* that is an infrastructure Metric. The primary components of the infrastructure domain are business infrastructure, social infrastructure, energy infrastructure, government infrastructure, transportation infrastructure, and water infrastructure. A very large number of items fall into this domain, including capacities, damage, investments, etc.

There are six subcategories of the *infrastructure* category: infrastructure business, infrastructure social, infrastructure energy, infrastructure government, infrastructure transport, and infrastructure water. Figure 9.16 shows this level of the ontology. The "metric" class box represents all of the Action classes that are part of the subcategories. The photograph of highway bridges in the figure represents a measure of transportation infrastructure.

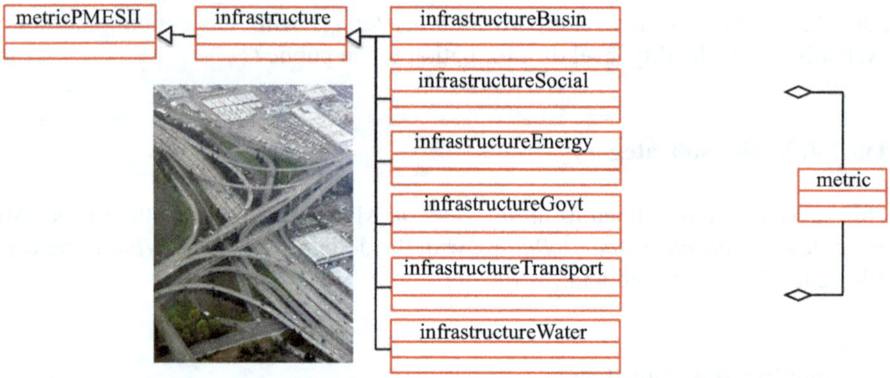

Fig. 9.16 PMESII infrastructure ontology

The infrastructureBusin Subcategory

This subcategory is defined as an *infrastructure* Metric for business infrastructure. An example is *manufacturingInfraCapacityRating*, with the defining question "What is the carrying capacity of manufacturing structures?"

The infrastructureSocial Subcategory

This subcategory is defined as an *infrastructure* Metric for social infrastructure. An example is *educationInfraCapacityDamagedMoP*, with the defining question "What amount of the education infrastructure has been damaged?"

The infrastructureEnergy Subcategory

This subcategory is defined as an *infrastructure* Metric for energy infrastructure. An example is *electricProductionInvestmentMoP*, with the defining question "What is the monetary investment in rebuilding the electricity production infrastructure?"

The infrastructureGovt Subcategory

This subcategory is defined as an *infrastructure* Metric for government infrastructure. An example is *militaryStructuresJobsCreatedMoP*, with the defining question "How many jobs have been created in rebuilding the military infrastructure?"

The infrastructureTransport Subcategory

This subcategory is defined as an *infrastructure* Metric for transportation infrastructure. An example is *roadCapacityRebuiltMoP*, with the defining question "What capacity of the road infrastructure has been rebuilt?"

The infrastructureWater Subcategory

This subcategory is defined as an *infrastructure* Metric for water infrastructure. An example is *damsCapacityRating*, with the defining question "What is the carrying capacity of the dam infrastructure?"

The kinetic Category

This category is defined as a *metricPMESII* that is a kinetic Metric. The primary components of the kinetic domain are logistics, damage and attrition, and command, control, communications, computers, and intelligence (C4I).

There are four subcategories of the *kinetic* category: kinetic logistic, kinetic damage and attrition, kinetic general, and kinetic C4I. C4I is included here because it has kinetic effects. It could easily have been included as a subcategory of Information. This is an example of the fact that creating ontologies involves art as well as science. Figure 9.17 shows this level of the ontology. The "metric" class box represents all of the Action classes that are part of the subcategories. The picture in the figure represents various methods of logistics.

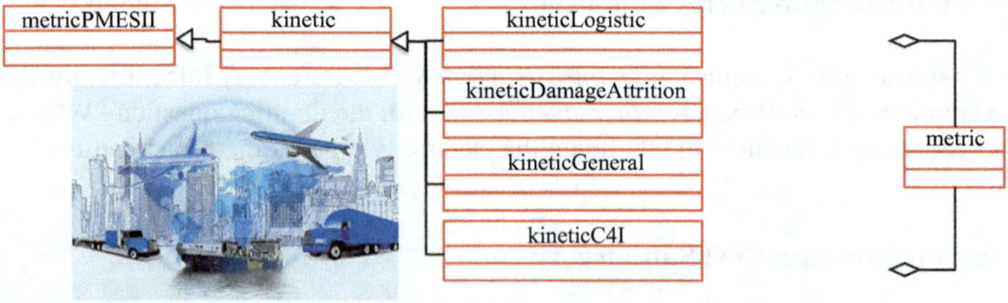

Fig. 9.17 PMESII kinetic ontology

The kineticLogistic Subcategory

This subcategory is defined as a *kinetic* Metric that involves logistics. An example is *logisticsAcquiredMoP*, with the defining question "What quantity has been acquired?"

The kineticDamageAttrition Subcategory

This subcategory is defined as a *kinetic* Metric that involves damage and attrition. An example is *manmadeDisasterDamageMoP*, with the defining question "What is the damage from man-made disasters (nuclear powerplant, hazardous materials, etc.)?"

The kineticGeneral Subcategory

This subcategory is defined as a *kinetic* Metric that involves general kinetic things. An example is *pollutionDisasterIndicators*, with the defining question "Is the entity a disaster or a condition; is it man-made or natural?"

The kineticC4I Subcategory

This subcategory is defined as a *kinetic* Metric that involves command, control, communications, computers, and intelligence. An example is *HNC4IRating*, with the defining question "What is the rating level of the Host Nation C4I?"

The environmental Category

This category is defined as a *metricPMESII* that is an environmental Metric. The primary components of the environmental domain are atmospheric, earth, water, and fire effects.

There are two subcategories of the *environmental* category: environmental atmospheric and environmental earth, water, and others. Figure 9.18 shows this level of the ontology. The "metric" class box represents all of the Action classes that are part of the subcategories. The picture in the figure of a house destroyed by a hurricane represents a measure of environmental damage.

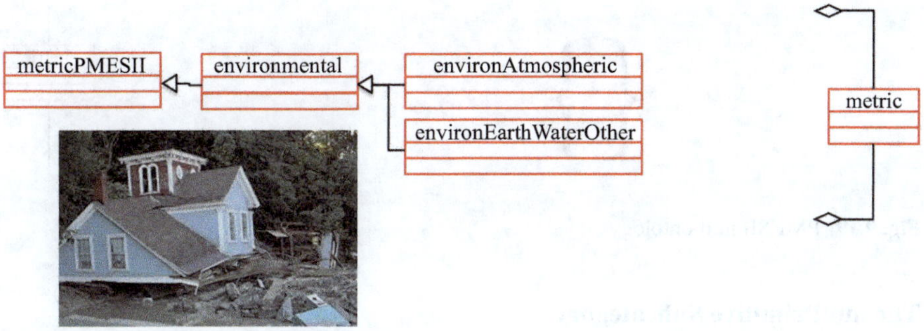

Fig. 9.18 PMESII environmental ontology

The environAtmospheric Subcategory

This subcategory is defined as an *environmental* Metric that covers atmospheric effects. An example is *airMovementDamageMoP*, with the defining question "What is the damage from air movement (storms, blizzards, hurricanes, thunderstorms, tornados, etc.)?"

The environEarthWaterOther Subcategory

This subcategory is defined as an *environmental* Metric that covers earth, water, and other effects. An example is *naturalFeaturesRating*, with the defining question "What is the attractiveness rating of the natural feature?"

The null Category

This category is defined as a *metricPMESII* that is for null-impact Metrics.

There is one subcategory of the *null* category: null primitive. Figure 9.19 shows this level of the ontology. The "metric" class box represents all of the Action classes that are part of the subcategories. The empty parentheses in the figure (representing the null set) indicate that the classes in this category do not contribute to a PMESII value.

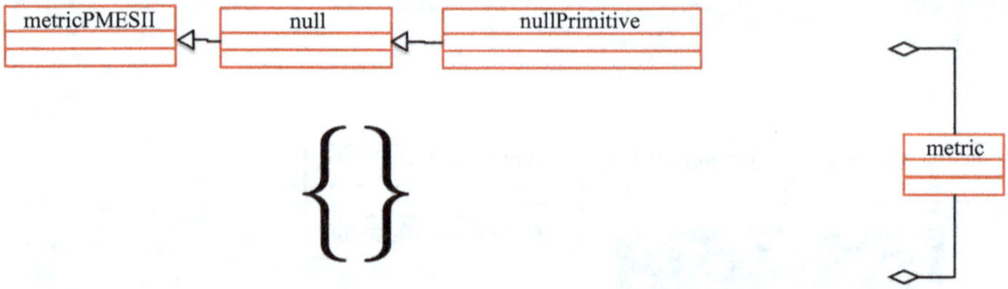

Fig. 9.19 PMESII null ontology

The nullPrimitive Subcategory

This subcategory is defined as a *null* Metric that includes primitive metrics. These are *situation*, *identity*, *time*, and *location*, which are separated from the others in terms of PMESII because they carry no PMESII information. An example is location, with the defining question "What is the current location description?"

Relating Metrics to PMESII+

The Actor classes were defined in Chap. 4; the Action classes were defined in Chap. 5; and the Object classes were defined in Chap. 6. They were each assigned sets of Metrics, which were grouped by structural categories and subcategories in Chap. 8. However, the Metrics also can be assigned to the PMESII+ categories and subcategories. As noted above, the latter assignment is more subjective and, hence more debatable. For example, each class, by its definition, may have some potential PMESII relationship; but each MType of Metric class may have some potential PMESII relationship; and the combination might imply some other PMESII relationship. Because there are almost 6000 Metrics, there is the human problem of consistency in assignment. However, this large number has the advantage of covering up small errors when the results are viewed in aggregations.

Rather than list the classes again, the assignments are reported in Fig. 9.20 as number of Actor classes in each structural subcategory that have been assigned to each PMESII+ category. Similarly, the Action class assignments are reported in Fig. 9.21, and the Object class assignments are reported in Fig. 9.22.

Category	Subcategory	political	military	economic	social	information	infrastructure	kinietic	environmental
individualActor	keyLeader	126	28	32	128	42			
	otherIndividual	64	22	14	55	31			
significantGroupActor	socialOrganization	12			55	9			
	economicOrganization	32		105	38	35	8		
	armedForce	17	48		18	12			
	politicalOrgUnarmed	63			22	15			
	politicalOrgArmed	36	19		13	8			
demographicGroupActor	StaticPopulation	17		4	35	11			1
	mobilePopulation	11	3		23	9			
nonhumanActor	civilianVehicle			4	4		17	3	
	militarySystem		19		4		4	3	
	environmentalActor				4			3	17
	animal		1	8	2				
compositeActor	simpleCompositeActor	6	15		4	4			
	hierarchicalActor	24	12	7	17	10			

Fig. 9.20 Actor categories broken into PMESII categories

The seven domains of PMESII+, described above, are concerned with describing a situation. The numbers in the figures are the numbers of metrics assigned to each category. There is no assigned method of converting the metrics' values into a level of impact on the category. The user may create such a method. The numbers do provide an idea of the amount of information that may be available to assess the value of each category. The colors are provided to help identify where the concentrations are.

As shown in Fig. 9.20, the Actor classes supply few Metrics in the last four PMESII+ categories. However, the Action classes supply numerous Metrics there (Fig. 9.21). The Action classes are most heavily concentrated in the *economic* and *social* PMESII+ categories. The Object classes also supply numerous entries to the last four PMESII+ categories and are fairly evenly balanced (Fig. 9.22).

Category	Subcategory	political	military	economic	social	information	infrastructure	kinietic	environmental
strikeAction	strikeTargetAction	4	49	136	51	3	63	47	
	strikeMethodAction	7	37	65	28	6	20	23	2
	envirnStrikeAction				8			7	11
C2Action	controlAction		27	39	15	3		13	
	commandAndControlAction		18	27	9	1		10	
informationAction	persuasionAction	15	6	61	34	26			
	monitoringAction	8		13	4				
	intelligenceAction	5	16	43	14	20		7	
conflictAction	sustainmentAction	2	29	43	13		8	22	
	securityAction	4	77	69	27	2		2	1
	generalConflicAction	14	90	107	40	2		36	
conflictOrgOrPersAction	conflicOrganizationAction	10	56	66	22				
	conflicPersonnelAction	10	44	88	28			4	
humanAffairsAction	socialAidAction	2		61	75		1	8	1
	civilTrainingAction	12		37	15	2			
	civilPersonnelAction	20		118	67	4	4	4	
	changeCivilSituationAction	8	2	27	19	1			
economicAction	businessEconomicAction			39	10				
	consumeOrProduceAction			57	19		10		1
	civilBuildingAction	4	1	154	39	2	107		
	businessOrganizationAction			166	38	2			
policingOrCriminalAction	criminalAction	5	10	45	17			9	
	policingAction	11	6	32	8			6	
civilGovernmentAction	policyOrLegalAction	42	1	97	30	1			
	govtEconomicAction	7		101	18		6	1	1
	govtOrganizationAction	50	3	66	41	1			

Fig. 9.21 Action categories broken into PMESII categories

Category	Subcategory	political	military	economic	social	information	infrastructure	kinetic	environmental
infrastructureObject	waterInfrastructure				6		22		
	transportInfrastructure			7	7		42		
	govtInfrastructure	2	11		9		28	19	
	energyInfrastructure			4	4		24		
	businessInfrastructure			7	6	5	33		
	socialInfrastructure	1	1	2	14	1	43		
neededThingObject	businessobject			94	13	13	15	1	
	immediateNeed				34	4	4	2	
	serviceNeed			1	51	15	2	1	
naturalObject	condition				30	13		14	41
	geographicalObject			1	5	5		5	15
conflictObject	conflictEnvHot	1	45		11	10	7	9	
	conflictEnvWarm	25	27	6	33	21		12	
	conflictEnvCool	20	4	1	29	16		8	
governingObject	govtObject	104	8	4	32	22	3		2
	economicObject	8		20	5	5			3
	criminalObject	14	5	77	25	15			
	interventionObject	12	7		3	3			
conceptualObject	rightsObject	38			30	36			
	cognitiveObject	8			37	119			
	document	5	9		3	3		1	
compositeObject	simpleCompositeObject	1		16	12	4			7
	hierarchicalObject	2		40	12	12	50		7
	complexCompositeObject	14	9	8		4	14	4	

Fig. 9.22 Object categories broken into PMESII categories

Figure 9.23 sums the Metrics over the Actor, Action, and Object classes to provide a profile of the total PMESII+ categories. The results show that the *economic* and *social* categories receive the most weight and the *kinetic* and *environmental* categories receive the least weight.

These PMESII profiles are based on one entry for each Metric. Similar PMESII profiles can be created for the Metrics of a particular party, with each use of a Metric providing its own entry. You can use the Metric profiles of other parties to help infer their span of effects.

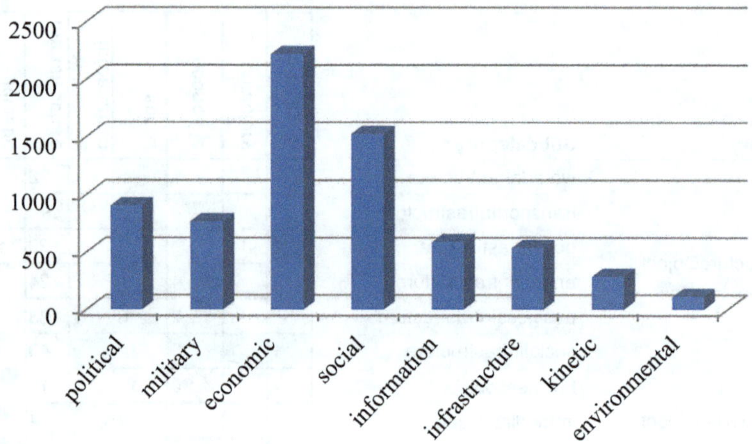

Fig. 9.23 Distribution of PMESII Metrics

9.3 DIME and PMESII Relationships

The only relationships used in this chapter are the *is-a*, and the *hasPart* relationships, explained as follows:

is-a: Each instance of A is an instance of B and each property of B is a property of A. Its inverse relationship is *superClassOf*.

hasPart: entity has another entity as a part. Its inverse relationship is *isPartOf*. The version that is used is one in which the relationship is an aggregation.

9.4 DIME/PMESII Recapitulation

While the DIME/PMESII paradigm provides a good philosophical paradigm for modern conflict, we have found that as a practical matter it is difficult to assign particular classes to the DIME+ and PMESII+ subcategories. The assignments do provide useful information when viewed in the aggregate, where individual errors or assignment uncertainties can cancel each other.

DIME Action Recapitulation

DIME is the realm of actions. There are two ways of organizing actions, the structural Action Ontology described in Chap. 5, and the DIME Ontology described in this chapter. The two Action Ontologies provide two ways of organizing, thinking about, visualizing, etc. the same thing.

The Action Ontology contains 474 Action Classes, which are categorized by type of Action, e.g., strike, C2, information, etc. The DIME Ontology contains the same 474 Action Classes, categorized using the DIME levers-of-power paradigm:

- Diplomatic actions – "message sending," "agreements," etc.
- Information actions – collecting, analyzing and disseminating information
- Military actions – applying military power (including law enforcement)
- Economic actions – applying economic power
- Other actions – criminal, natural forces, etc.

Within the DIME paradigm, the action originator can be of any type, e.g., a military force can engage in diplomatic or economic actions. The chart in Fig. 9.8 shows the number of Actions in each of the structural Action subcategories that are assigned to each DIME category.

PMESII Metric Recapitulation

PMESII is in the realm of state variables or metrics. Metrics describe the results of actions on Actors and Objects and success/failure of Actions. There are also two ways of organizing metrics, the MType Metric Ontology described in Chap. 8, and the PMESII Ontology described in this chapter. The two ontologies provide two ways of organizing, thinking about, visualizing, etc. the same thing.

The MType Metric Ontology contains ~6000 Metric Classes. Each element class (Actor, Object, and Action) has a number of Metric Classes based on type of element class. The instantiation of the Metric class gives state variable information about the instantiation of the element class. Metric Classes are categorized by type of Metric, e.g., physical, flow, relationship, etc. The PMESII Ontology contains the same ~6000 Metric Classes, categorized using the PMESII paradigm describing the state of the world:

- Political metrics – the state of the political world
- Military metrics – the state of the military world
- Economic metrics – the state of the economic world
- Social metrics – the state of the social world
- Information metrics – the state of the information world
- Infrastructure metrics – the state of the infrastructure world
- Kinetic metrics – the state of kinetic things, e.g., logistics, damage, etc.
- Environment metrics – the state of the natural environment, e.g., sea state, storms, etc.

In the MType Metric Ontology, 779 element classes (114 Actor, 474 Action, and 191 Object classes) are assigned to ~6000 Metric classes, which are categorized by type. Three charts are required to visualize the assignments (Figs. 8.8, 8.9, and 8.10). Note the assignments are identical at the subcategory levels of each element type, so the cells are just blacked in.

In the PMESII Ontology, the 779 element classes are assigned to the same ~6000 Metric classes; however, here the Metrics are assigned to PMESII categories. Three charts are required to visualize the assignments (Figs. 9.20, 9.21, and 9.22). Here the number of Metric classes is of interest for each cell, with color coding to help show where the concentrations are. A fourth chart (Fig. 9.23) is provided to show the distribution across all element classes.

Metrics really live at the instance level. A metric value is given for an instance of a class, e.g., this bridge can only operate at half capacity. To make this work, Metric classes are assigned to element classes, e.g., a bridge class can have a capacity metric. For example, the road infrastructure (including bridges and tunnels) between two points may be operating at half capacity because the capacity metric for one bridge is at half of its capacity – or the road may be operating a nearly full capacity because the bridge can be bypassed – or the road capacity may be ½ for large trucks, but full capacity for automobiles. The charts emphasize the difficulties of working with such a large number of Metric classes.

DIME/PMESII Assignments

The assignment of Actions to DIME categories is somewhat problematic, e.g., is Action *A25* really diplomatic or military or economic. DIME works best at the national level – at lower levels the assignment question can get in the way of thinking about what to do.

Similarly, the assignment of Metrics to PMESII categories is very problematic, i.e., is Metric *M3045* really political or military or economic. PMESII works OK at all levels; however, combining them by "rolling up" requires modeling assumptions and inferring connections in other ways requires additional modeling assumptions.

If you want to organize by DIME/PMESII, consider Fig. 9.24. Actions are categorized by DIME; Actions change the state of the world (possibly null change); Actions seldom change just one thing or just one aspect of the thing; and the state of the world is categorized by PMESII. This involves modeling problems. Which instances (things) are affected by the selected action instance? How are their Metric instances changed by the action? What roll-ups of the Metrics are desired and how should they be computed?

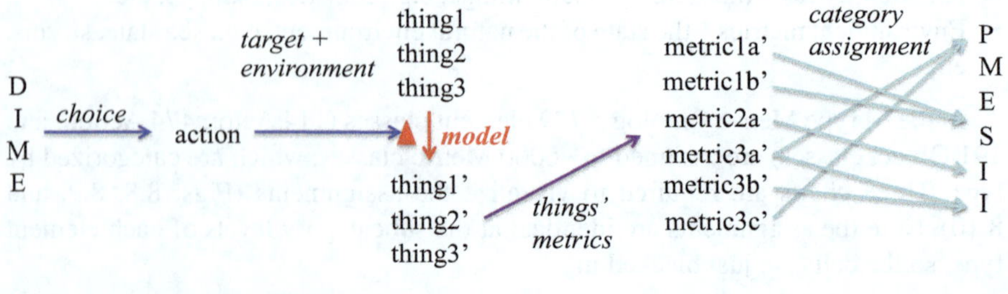

Fig. 9.24 DIME to PMESII

Summary

I've been working with concepts similar to DIME/PMESII since before the terms were invented. The concepts are good – better than what we had before. However, in my experience, they are better as concepts than as working tools. For example, Actions are what DIME is about; but they are better organized by their types than by DIME categories and subcategories. The real point about DIME is that it isn't just military actions that are useful. Once you get past that, you are really just arguing about how many angels can dance on the head of a pin.

State Variables or Metrics are the same. The important point of PMESII is that you can't just measure damage and know what is going on. State Variables are really worse than actions because there are so many of them. Each element (Actor class, Object class, and Action class) has attached state variables or metrics. For the 779 element classes, there are about 6000 metric classes. "But there is more...." Most metric classes are actually sub-vectors. If you want to discuss the capacity of a bridge, you might want to talk about its weight capacity and its throughput capacity (number of lanes for traffic) and its current values versus its ideal values (undamaged or expected values when the bridge is complete), etc. A full simulation would not have 6000 slots, but three to ten times that number. Then you have to remember that these numbers refer to classes, not instances. Each instance of a bridge has its own instances of the metrics. With any metric ontology, we have an implicit conception of rolling up the details, for instance metrics for the entire Infrastructure. The methodology chosen for such a roll-up is an explicit modeling decision. If we were to say that the infrastructure is 47% destroyed, what does that mean? What should it mean? How does a user understand what it means versus what the modeling decision really says? Then there are cross-terms: the economy depends on the infrastructure. How? This requires another explicit modeling decision.

This chapter concludes the discussion of the concrete parts of the MCO. The next chapters discuss the more abstract parts of the MCO.

Chapter 10
MCO Concept Ontologies

A *concept* is a thing that defines relations among elements. The Modern Conflict Ontology (MCO) sub-ontologies that represent two such relations are described here: semantic concepts and stocks-and-flows. The first involves similarities in the meanings of words or phrases. The second involves standard active relationships among the Actor, Action, and Object classes. These relate to common usage relationships among entities. The relationships in both of these ontologies are obvious to humans, but have to be specified to a computer. Specifying these relationships has the additional benefit of providing cues to humans building a restricted ontology or knowledge base (instantiated ontology). In creating such a restricted ontology, the idea is to omit as many elements as possible; however, the relationships in these two ontologies provide a reminder to reconsider elements that might be otherwise discarded.

10.1 The Semantic Concept Ontology

A *semanticConcept* is a *concept* that is a semantic or "meaning" concept. A semantic concept is expressed as a word or phrase to which a number of Actor, Action, and/or Object classes are similar in meaning. These similarities are obvious to a human reader, but not to a computer. The Semantic Concept Ontology makes these connections for the computer.

Figure 3.9 showed how the Semantic Concept Ontology fits into the overall MCO structure (with Fig. 10.1 illustrating its icon as used in the context diagram, Fig. 3.5).

D. S. Hartley III, *An Ontology of Modern Conflict*, Understanding Complex Systems, https://doi.org/10.1007/978-3-030-53214-7_10

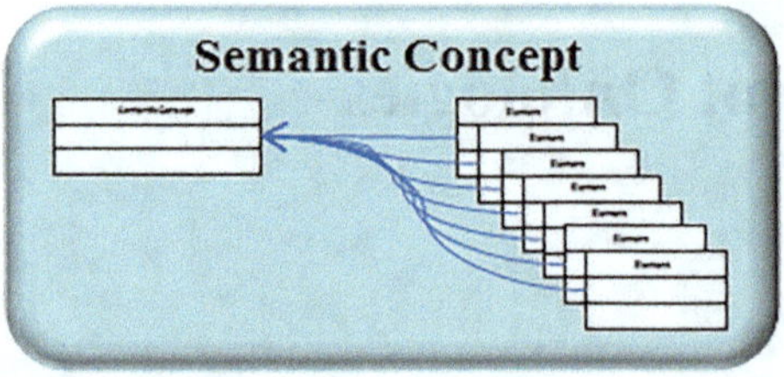

Fig. 10.1 Semantic concept icon

Overview of Semantic Concepts

There are seven categories of *semanticConcepts*: business concepts, government concepts, needs concepts, social concepts, military concepts, other concepts, and small concepts. Figure 10.2 shows the top level of the Semantic Concept Ontology. The picture of a person with an idea in the figure represents an arbitrary concept.

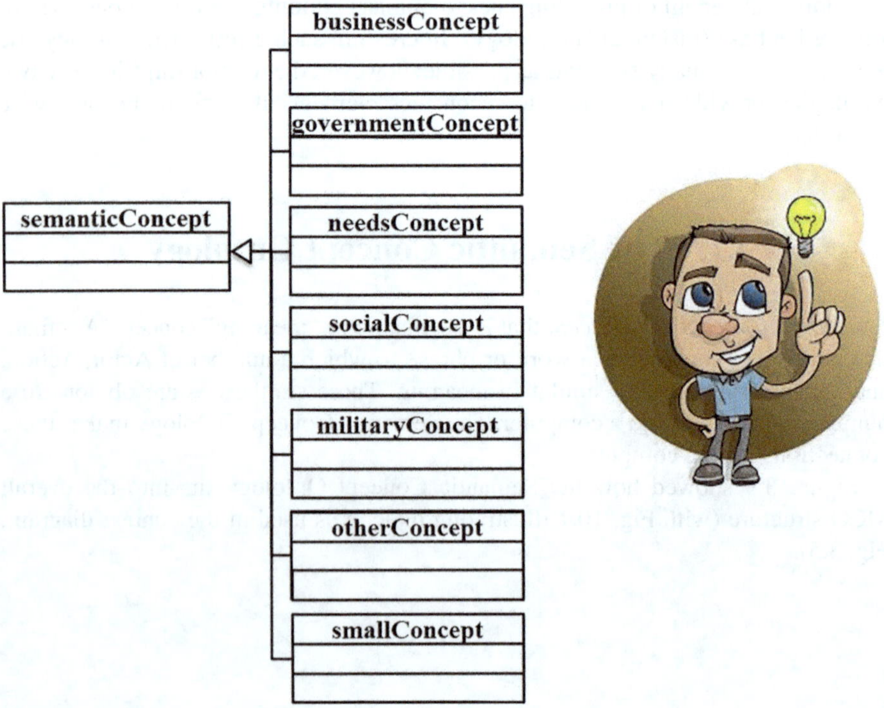

Fig. 10.2 Semantic concept ontology overview

Changes to semanticConcepts from the UCO

Two new categories have been added, *militaryConcept* and *smallConcept*. This resulted in moving some of the *conceptNames*. Each *conceptName* has the suffix "Issue" appended to the old name to emphasize its identity and "issue" has replaced *conceptName* as the descriptor. For the most part, the tables of the semantic concepts contained in the book about the UCO describe the content adequately (Hartley 2018). (However, the overview figure, Figure 8.3, of the ontology incorrectly switched the contents of the *needsConcept* and *socialConcept* categories.)

There are changes to class names, as has been discussed before. New classes have also been added, as appropriate. In this process, a few classes that should have been included in an issue, but weren't, have been added. The tables below include only the additions to the UCO. The tables also have been changed slightly. In the UCO, the headers indicating the type of class were "A," "E," and "D," for Actor, Environmental Element, and Action ("D" for DIME so that there weren't two "A" headers). In the MCO, the headers are "A," "O," and "D," for Actor, Object, and Action. A "T" (for "true") in a column indicates that the element is of that type.

An Insight into semanticConcepts

The operative descriptor for the Semantic Concept Ontology is "similarity." The "issues" are the subcategories of the ontology, each naming a concept. All of the element classes (Actor, Object, and Action) that are "similar" to the concept are connected by an *isSimilarTo* relationship to the concept. An element class can be similar to many different concepts. This is the reason that the *isSimilarTo* relationship has a two-headed arrow in the figure below.

Business Concepts

A *businessConcept* is a *semanticConcept* concerning business. This category is comprised of eight issues: commerce, economy/finance, transportation, energy, media, other business, produce/consume, and employment. Figure 10.3 shows the business concept part of the Semantic Concept Ontology. The "element" class box represents all of the classes that are similar to the issues. The picture of a person holding a checklist in the figure represents an arbitrary business concept.

The *commerceIssue* is a *businessConcept* that concerns commerce (Table 10.1).

The *economyFinanceIssue* is a *businessConcept* that concerns economy and finance (Table 10.2).

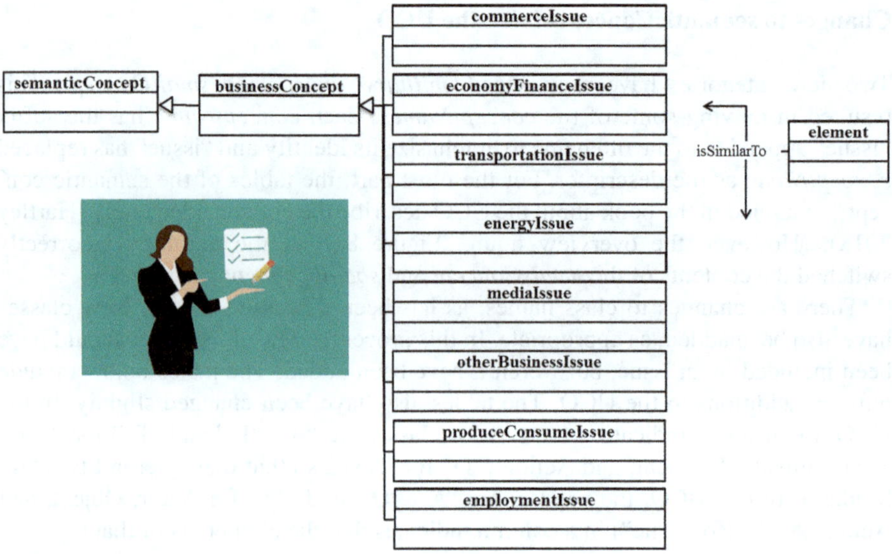

Fig. 10.3 Business concept ontology

Table 10.1 Commerce issue additions

ID	OE element	A	O	D
284	*privatizeBusiness*			T
310	*market*		T	
2111	*nationalizeBusiness*			T

Table 10.2 Economy/finance issue additions

ID	OE element	A	O	D
276	*seekInvestmentCapital*			T
1991	*rebuildFinancialInfrastructure*			T
2029	*obtainFunds*			T
2112	*inhibitInvestmentCapital*			T
2113	*disburseFunds*			T
2172	*conductEconomicWarOperation*			T

The *transportationIssue* is a *businessConcept* that concerns transportation (Table 10.3)

The *energyIssue* is a *businessConcept* that concerns energy. No additions were made.

The *mediaIssue* is a *businessConcept* that concerns the media (Table 10.4).

The *otherBusinessIssue* is a *businessConcept* that concerns other businesses (besides, commerce, finance, transportation, energy, and the media) (Table 10.5).

Table 10.3 Transportation issue additions

ID	OE element	A	O	D
575	*militaryVehicle*		T	
588	*rebuildReplaceMilitaryVehicle*			T
592	*strikeMilitaryVehicle*			T
1856	*increaseTransportationBusinesses*			T
1857	*decreaseTransportationBusinesses*			T
2080	*landVehicle*	T		
2081	*landSystem*	T		
2082	*watercraft*	T		
2083	*aircraft*	T		
2084	*spacecraft*	T		
2085	*seaSystem*	T		
2086	*airSystem*	T		
2087	*spaceSystem*	T		
2167	*operateCivilTransportInfrastructure*			T
2195	*increaseMilitarySystems*			T
2196	*decreaseMilitarySystems*			T
2197	*increaseCivilianVehicles*			T
2198	*decreaseCivilianVehicles*			T

Table 10.4 Media issue additions

ID	OE element	A	O	D
476	*collectData*			T
480	*controlInformation*			T

Table 10.5 Other business issue additions

ID	OE element	A	O	D
284	*privatizeBusiness*			T
296	*keyLaborLeader*	T		
309	*keyBusinessIndividual*	T		
311	*commercialSectorUnit*		T	
496	*shopAndCommercialInfrastructure*		T	
500	*commercialEquipment*		T	
513	*rebuildShopsOrCommercialInfrastructure*			T
517	*strikeShopsOrCommercialInfrastructure*			T
767	*naturalResource*		T	
2111	*nationalizeBusiness*			T
2123	*businessInfrastructureUnit*		T	

The *produceConsumeIssue* is a *businessConcept* that concerns production and consumption. For instance, *foodSupply* is an Object that is produced or consumed. The MCO does not have an added Action for producing food, as this Action already exists. However, *distributeFood* is an Action that relates to consuming food that has been added. (From the distributor's point of view, the food supply on hand is reduced by distribution, which is semantically related to consumption.) (Table 10.6)

Table 10.6 Produce/consume issue additions

ID	OE element	A	O	D
345	*foodSupply*		T	
355	*distributeFood*			T
365	*distributeDurableGoodsRelief*			T
699	*acquireEquipmentOrMaterialOrInfrastructure*			T
701	*warehouseEquipmentOrMaterial*			T
703	*distributeEquipmentOrMaterial*			T
705	*movePeopleEquipmentOrMaterialOnTheGround*			T
707	*movePeopleEquipmentOrMaterialThroughTheAir*			T
709	*movePeopleEquipmentOrMaterialOverTheWater*			T
711	*movePeopleEquipmentOrMaterialUnderTheWater*			T
1923	*goodsAndEquipment*		T	
2145	*supplyForce*			T
2191	*strikeSuppliesOrSupplyLine*			T

The *employmentIssue* is a *businessConcept* that concerns employment (Table 10.7).

Table 10.7 Employment issue additions

ID	OE element	A	O	D
295	*workerOrganization*	T		
427	*generalPopulation*	T		
740	*strikeCivilians*			T
1844	*increaseWorkerOrganizations*			T
1845	*decreaseWorkerOrganizations*			T
1872	*increaseGeneralPopulation*			T
1873	*decreaseGeneralPopulation*			T
2026	*increaseFirstRespondersPersonnel*			T
2064	*workerPopulation*	T		

Government Concepts

A *governmentConcept* is a *semanticConcept* concerning government. This category is comprised of six issues: government services, governing, rights/freedom, stability/unrest, factions, and crime/corruption. Figure 10.4 shows the government con-

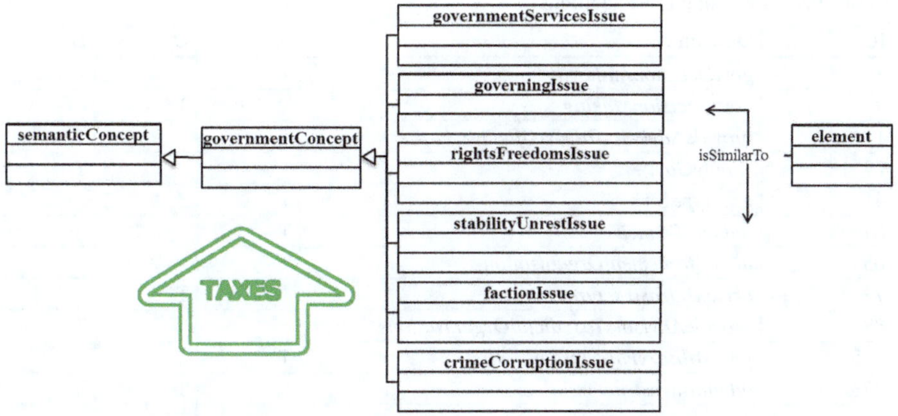

Fig. 10.4 Government concept ontology

cept part of the Semantic Concept Ontology. The "element" class box represents all of the classes that are similar to the issues. The picture of rising taxes in the figure represents a typical government concept.

The *governmentServicesIssue* is a *governmentConcept* that concerns government services (Table 10.8).

Table 10.8 Government services issue additions

ID	OE element	A	O	D
1792	*provideProgramOrPersonnelSupport*			T
2019	*increaseFirstResponderOrganizations*			T
2020	*decreaseFirstResponderOrganizations*			T
2052	*increaseSocialServicesOrganizations*			T
2053	*decreaseSocialServicesOrganizations*			T
2124	*socialInfrastructureUnit*		T	
2143	*provideEmergencyManagement*			T
2168	*operateSocialInfrastructure*			T

The *governingIssue* is a *governmentConcept* that concerns governing (Table 10.9).

The *rightsFreedomIssue* is a *governmentConcept* that concerns rights and freedoms. The *stabilityUnrestIssue* is a *governmentConcept* that concerns stability, instability, and unrest. The *factionIssue* is a *governmentConcept* that concerns factions. No additions have been made to these three subcategories.

The *crimeCorruptionIssue* is a *governmentConcept* that concerns crime and corruption (Table 10.10).

Table 10.9 Governing issue additions

ID	OE element	A	O	D
3	*govtDecisionAuthority*	T		
9	*govtDecisionMaking*		T	
13	*provideAdvisorsToGovtOfficials*			T
27	*supplyGovt*			T
41	*govtTypeUnit*	T		
64	*judicialBranch*	T		
65	*lawEnforcementOrganization*	T		
72	*prisonInfrastructure*		T	
88	*provideAdvisorsToJudicialOrganizations*			T
145	*govtMilitaryForceOrganization*	T		
573	*adminInfrastructure*		T	
574	*militaryInfrastructure*		T	
576	*rebuildGovtInfratructure*			T
580	*rebuildMilitaryInfrastructure*			T
584	*strikeGovtInfrastructure*			T
586	*strikeMilitaryInfrastructure*			T
1814	*buildPrisonInfrastructure*			T
1815	*strikePrisonInfrastructure*			T
1819	*increasePoliticalPopulation*			T
1820	*decreasePoliticalPopulation*			T
1927	*keyLawEnforcementLeader*	T		
1930	*govtMilitaryPerson*	T		
1980	*increaseMilitaryForcesPersonnel*			T
2009	*increaseLawEnforcementOrganizations*			T
2010	*decreaseLawEnforcementOrganizations*			T
2011	*increaseBureaucracyOrganizations*			T
2012	*decreaseBureaucracyOrganizations*			T
2013	*increaseGovtOrganizations*			T
2014	*decreaseGovtOrganizations*			T
2015	*increaseJudicialOrganizations*			T
2016	*decreaseJudicialOrganizations*			T
2040	*increaseMilitaryOrganizations*			T
2041	*decreaseMilitaryOrganizations*			T
2091	*legislativeBranchPresence*		T	
2092	*executiveBranchPresence*		T	
2115	*wholeOfGovt*	T		
2122	*govtInfrastructureUnit*		T	
2141	*protectMilitaryInfrastructure*			T
2169	*operateNonMilGovtInfrastructure*			T
2173	*conductInformationWarOperation*			T
2188	*disruptElection*			T
2189	*strikeGovtInstitutionOrKeyOfficial*			T

Table 10.10 Crime/corruption issue additions

ID	OE element	A	O	D
1778	*reduceDrugDemand*			T
2104	*conductHumanTrafficking*			T
2105	*conductAntiWhiteCollarCrimeOp*			T
2157	*commitWarCrime*			T
2177	*conductCounterdrugOperation*			T

Needs Concepts

A *needsConcept* is a *semanticConcept* concerning needs. This category is comprised of five issues: displaced persons, water, housing/shelter/camps, waste/sewage/pollution, and health. Figure 10.5 shows the needs concept part of the Semantic Concept Ontology. The "element" class box represents all of the classes that are similar to the issues. The picture of water for washing in the figure represents a water issue concept.

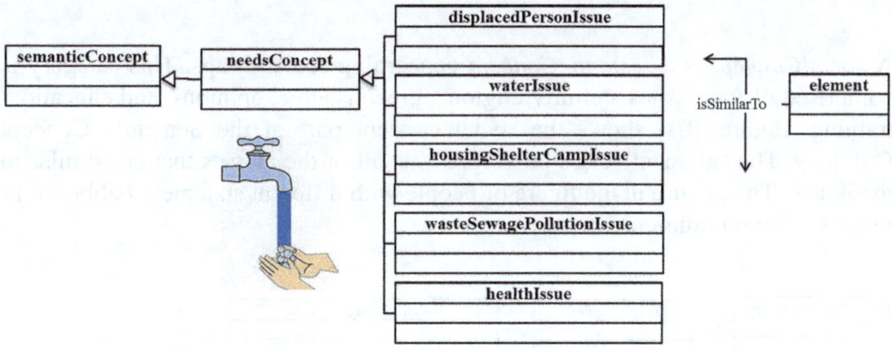

Fig. 10.5 Needs concept ontology

The *displacedPersonIssue* is a *needsConcept* that concerns displaced persons (Table 10.11).

The *waterIssue* is a *needsConcept* that concerns need for water (Table 10.12).

The *housingShelterIssue* is a *needsConcept* that concerns housing. No additions have been made.

The *wasteSewagePollutionIssue* is a *needsConcept* that concerns waste. No additions have been made.

The *healthIssue* is a *needsConcept* that concerns health (Table 10.13).

Table 10.11 Displaced person issue additions

ID	OE element	A	O	D
415	*reduceLikelihoodOfPopulationMovements*			T
2107	*increaseLikelihoodOfPopulationMovements*			T

Table 10.12 Water issue additions

ID	OE element	A	O	D
2120	*waterInfrastructureUnit*		T	
2166	*operateWaterInfrastructure*			T

Table 10.13 Health issue additions

ID	OE element	A	O	D
2073	*healthTrauma*	T		
2079	*healthConditionAfflicts*			T
2137	*provideForceHealthProtection*			T

Social Concepts

A *socialConcept* is a *semanticConcept* concerning social things. This category is comprised of four issues: culture/religion, agree/disagree, opinions, and education/training. Figure 10.6 shows the social concept part of the Semantic Concept Ontology. The "element" class box represents all of the classes that are similar to the issues. The picture in the figure of people with different statement bubbles represents different opinions.

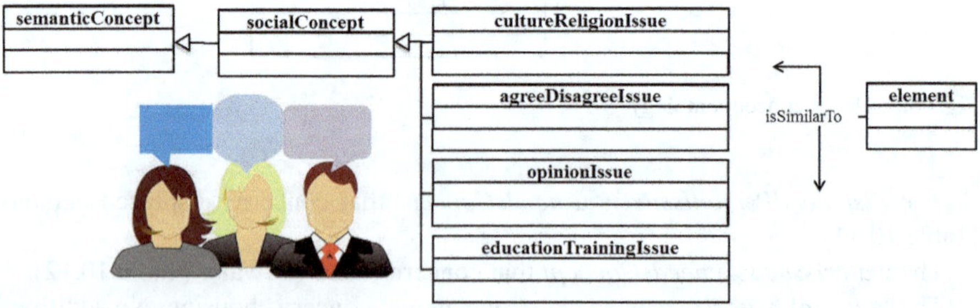

Fig. 10.6 Social concept ontology

The *cultureReligionIssue* is a *socialConcept* that concerns culture and religion (Table 10.14).

The *agreeDisagreeIssue* is a *socialConcept* that concerns agreements and disagreements. No additions have been made.

Table 10.14 Culture/religion issue additions

ID	OE element	A	O	D
443	*rebuildReligiousFacility*			T
445	*strikeReligiousFacility*			T

Table 10.15 Opinion issue addition

ID	OE element	A	O	D
10	*relationshipWithIntervenors*		T	

Table 10.16 Education/training issue addition

ID	OE element	A	O	D
165	*conductMilitaryExercise*			T

The *opinionIssue* is a *socialConcept* that concerns opinions (Table 10.15).

The *educationTrainingIssue* is a *socialConcept* that concerns education and training (Table 10.16).

Military Concepts

A *militaryConcept* is a *semanticConcept* concerning military things. This category is comprised of four issues: attack, defense, operations, and weapons. Figure 10.7 shows the military concept part of the Semantic Concept Ontology. The "element" class box represents all of the classes that are similar to the issues. The picture of a fort in the figure represents defense issues.

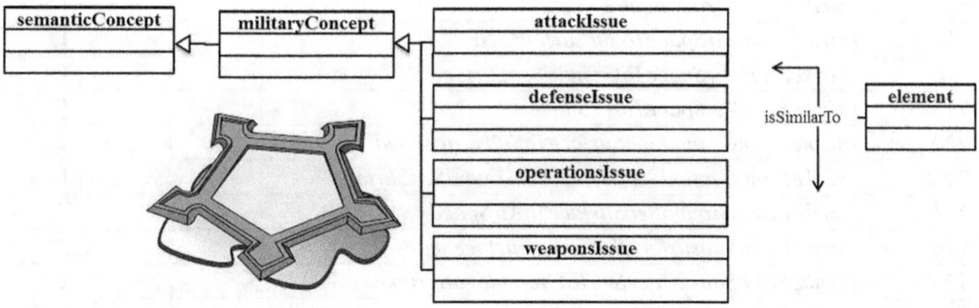

Fig. 10.7 Military concept ontology

The *attackIssue* is a *militaryConcept* that concerns attacks (Table 10.17).

The *defenseIssue* is a *militaryConcept* that concerns defense (Table 10.18).

The *operationsIssue* is a *militaryConcept* that concerns operations (Table 10.19).

The *weaponsIssue* is a *militaryConcept* that concerns weapons (Table 10.20).

Table 10.17 Attack issue additions

ID	OE element	A	O	D
121	*deathAndInjuryOfCiviliansFromConflict*		T	
122	*deathAndInjuryOfCombatantsFromConflict*		T	
123	*propertyDestructionFromConflict*		T	
331	*strikeFinancialInfrastructure*			T
377	*strikeCivilianHousing*			T
445	*strikeReligiousFacility*			T
505	*strikeManufacturingInfrastructure*			T
511	*strikeAgricultureInfrastructureLivestock*			T
517	*strikeShopsOrCommercialInfrastructure*			T
524	*strikeInformationAndMediaInfrastructure*			T
533	*strikeSchoolOrEducationalInfrastructure*			T
540	*strikeHealthcareInfrastructure*			T
564	*strikeElectricityProductionPlant*			T
566	*strikeElectricityDistributionInfrastructure*			T
568	*strikeExtractiveEnergyProductionInfrastructure*			T
570	*strikeExtractiveEnergyTransportationInfrastructure*			T
584	*strikeGovtInfrastructure*			T
586	*strikeMilitaryInfrastructure*			T
592	*strikeMilitaryVehicle*			T
647	*strikeRoadInfrastructure*			T
649	*strikeRailroadInfrastructure*			T
651	*strikeBridgeOrTunnelInfrastructure*			T
653	*strikeSeaportInfrastructure*			T
655	*strikeAirportInfrastructure*			T
657	*strikeVehicle*			T
659	*strikeWaterwaysInfrastructure*			T
677	*strikeWaterDistributionInfrastructure*			T
679	*strikeWaterOrSewageTreatmentFacilities*			T
681	*strikeDamInfrastructure*			T
714	*strikeFromAirSpaceToSubsurfaceSea*			T
716	*strikeFromAirSpaceToGroundSeaSurface*			T
718	*strikeFromAirSpaceToAirSpace*			T
720	*strikeFromGroundSeaSurfaceToSubsurfaceSea*			T
722	*strikeFromGroundSeaSurfaceToGroundSeaSurface*			T
724	*strikeFromGroundSeaSurfaceToAirSpace*			T
726	*strikeFromSubsurfaceSeaToSubsurfaceSea*			T
728	*strikeFromSubsurfaceSeaToGroundSeaSurface*			T
730	*strikeFromSubsurfaceSeaToAirSpace*			T
732	*strikeWithHighYieldExplosives*			T
734	*strikeWithChemicals*			T
736	*strikeWithBiologicals*			T
738	*strikeWithRadiologicals*			T

(continued)

Table 10.17 (continued)

ID	OE element	A	O	D
740	*strikeCivilians*			T
1805	*terrorism*		T	
1812	*strikeCommunicationsProcess*			T
1815	*strikePrisonInfrastructure*			T
1905	*strikeMiningInfrastructure*			T
1918	*strikeCulturalInfrastructure*			T
2034	*strikeRefugeeCampOrTemporaryShelter*			T
2106	*placeMinesIEDs*			T
2158	*conductCyberAttack*			T
2161	*strikeWithNuclearWeapon*			T
2185	*conductElectronicWarAttack*			T
2189	*strikeGovtInstitutionOrKeyOfficial*			T
2190	*strikeCombatPersonnel*			T
2191	*strikeSuppliesOrSupplyLine*			T
2192	*performSuicideBombingOrInsiderStrike*			T
2195	*increaseMilitarySystems*			T
2196	*decreaseMilitarySystems*			T
2197	*increaseCivilianVehicles*			T
2198	*decreaseCivilianVehicles*			T

Table 10.18 Defense issue additions

ID	OE element	A	O	D
138	*conductBorderControlBoundarySecurityFreedomOfMovement*			T
170	*establishConfidenceBuildingOrSecurityMeasure*			T
174	*provideSecurityAssistance*			T
176	*conductSecurityCoordination*			T
178	*provideForceSecurity*			T
180	*provideSecurityForHumanitarianAssistanceActivities*			T
182	*provideSecurityForPeaceOperationActivities*			T
184	*provideSecurityForStabilityActivities*			T
211	*clearMinesIEDs*			T
419	*provideRefugeeCampSecurity*			T
1973	*securityInRefugeeCamp*		T	
1974	*forceAndOperationsSecurity*		T	
2024	*civilDefensePlan*		T	
2025	*executeCivilDefensePlan*			T
2137	*provideForceHealthProtection*			T
2138	*provideAirSpaceMissileDefense*			T
2139	*protectAttachedCivilian*			T
2140	*provideElectronicAndCyberDefense*			T
2141	*protectMilitaryInfrastructure*			T
2142	*provideCBRNDefense*			T
2154	*defendSelf*			T
2156	*disengageSelfFromCombat*			T
2186	*patrolArea*			T

Table 10.19 Operations issue additions

ID	OE element	A	O	D
125	*foreignConflict*		T	
126	*conductPeacekeepingOperation*			T
128	*conductConventionalWarOperation*			T
130	*conductIrregularWarOperation*			T
167	*conductStabilityOperation*			T
186	*conductPersonnelRecoveryOperation*			T
199	*conductTerrorismOperation*			T
209	*provideConsequenceManagementSupport*			T
213	*conductPiracyOperation*			T
215	*conductHumanitarianAssistanceOperation*			T
217	*conductNonCombatantEvacuationOperation*			T
482	*conductBenignPublicInformationOperation*			T
484	*conductNegativeInformationOperation*			T
705	*movePeopleEquipmentOrMaterialOnTheGround*			T
707	*movePeopleEquipmentOrMaterialThroughTheAir*			T
709	*movePeopleEquipmentOrMaterialOverTheWater*			T
711	*movePeopleEquipmentOrMaterialUnderTheWater*			T
1800	*borderControl*		T	
1801	*militaryOperation*		T	
1807	*piracy*		T	
2102	*conductAntiOrCounterterrorismOp*			T
2103	*conductAntiPiracyOperation*			T
2125	*organizeForce*			T
2126	*commandSubordinate*			T
2127	*issuePlan*			T
2128	*communicate*			T
2129	*assessOperationalProgress*			T
2135	*removeObstacle*			T
2136	*impedeMovement*			T
2146	*provideGeneralForceSupport*			T
2147	*fundForce*			T
2148	*shapeSituation*			T
2149	*deployIntoTheater*			T
2150	*redeploy*			T
2151	*searchForAdversary*			T
2155	*surrenderSelf*			T
2159	*controlFires*			T
2160	*engageInCombat*			T
2164	*militaryPlan*		T	
2170	*holdInPlace*			T
2171	*conductCBRNOperation*			T
2172	*conductEconomicWarOperation*			T

(continued)

Table 10.19 (continued)

ID	OE element	A	O	D
2173	*conductInformationWarOperation*			T
2174	*conductShowOfForceOperation*			T
2175	*conductSanctionMIOQuarantineOperation*			T
2176	*conductFONOverflightOperation*			T
2177	*conductCounterdrugOperation*			T
2178	*conductCounterinsurgencyOperation*			T
2179	*conductInsurgencyOperation*			T
2180	*conductDisasterReliefOperation*			T
2181	*conductPeaceEnforcementOperation*			T
2182	*conductNationAssistanceOperation*			T
2183	*conductOtherPeaceOperation*			T
2184	*conductRelocationOperation*			T
2187	*coordinateActivity*			T

Table 10.20 Weapons issue additions

ID	OE element	A	O	D
1806	*mineAndIED* (includes land- and naval-mines and IEDs)		T	
2081	*landSystem*	T		
2085	*seaSystem*	T		
2086	*airSystem*	T		
2087	*spaceSystem*	T		
2094	*munition*		T	
2095	*weapon*		T	
2099	*milSystemsGroup*	T		
2195	*increaseMilitarySystems*			T
2196	*decreaseMilitarySystems*			T

Other Concepts

An *otherConcept* is a *semanticConcept* concerning other things. This category is comprised of four issues: environment, computer/management information systems (MIS)/C4I, vehicles, and organizations. Figure 10.8 shows the other concept part of the Semantic Concept Ontology. The "element" class box represents all of the classes that are similar to the issues. The picture of the computer network in the figure represents computer issues.

The *environmentIssue* is an *otherConcept* that concerns environment (Table 10.21).

The *computerMISC4IIssue* is an *otherConcept* that concerns computer-related things (Table 10.22).

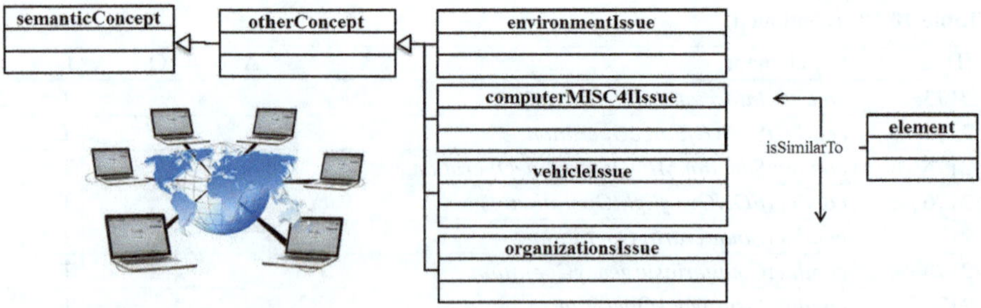

Fig. 10.8 Other concept ontology

Table 10.21 Environment issue additions

ID	OE element	A	O	D
2068	*earthMovement*	T		
2069	*storm*	T		
2070	*waterMovement*	T		
2071	*manmadeDisaster*	T		
2072	*fireOrWildfire*	T		
2074	*earthShifts*			T
2075	*airBlows*			T
2076	*waterMoves*			T
2077	*fireBurns*			T
2078	*systemFails*			T
2089	*geographicalSubdivision*		T	
2093	*obscureVision*			T
2097	*genericAnimal*	T		
2098	*animalGroup*	T		
2101	*humanPhysicalFeature*		T	
2114	*boundary*		T	
2119	*compositeGeographicalUnit*		T	
2199	*increaseAnimals*			T
2200	*decreaseAnimals*			T

The *vehicleIssue* is an *otherConcept* that concerns vehicles (Table 10.23).

The *organizationsIssue* is an *otherConcept* that concerns organizations other than government and business organizations (Table 10.24).

Table 10.22 Computer/MIS/C4I issue additions

ID	OE element	A	O	D
146	*intelligenceServiceOrganization*	T		
743	*establishSensorProcess*			T
1811	*strikeSensorProcess*			T
1936	*keyIntelligenceServiceLeader*	T		
2023	*increaseIntelligenceServicesOrganizations*			T
2110	*disseminateInformation*			T
2130	*controlIntelligenceProcess*			T
2131	*processAndAnalyse*			T
2132	*employIRC*			T
2133	*counterDeception*			T
2134	*supportDeception*			T
2144	*identifyNeutralizeInsiderThreatOrSuicideBombing*			T
2165	*dataInformation*		T	

Table 10.23 Vehicle issue additions

ID	OE element	A	O	D
575	*militaryVehicle*		T	
588	*rebuildReplaceMilitaryVehicle*			T
592	*strikeMilitaryVehicle*			T
618	*vehicleNonMilitary*		T	
639	*rebuildReplaceVehicle*			T
657	*strikeVehicle*			T
2080	*landVehicle*	T		
2081	*landSystem*	T		
2082	*watercraft*	T		
2083	*aircraft*	T		
2084	*spacecraft*	T		
2085	*seaSystem*	T		
2086	*airSystem*	T		
2087	*spaceSystem*	T		
2099	*milSystemsGroup*	T		
2195	*increaseMilitarySystems*			T
2196	*decreaseMilitarySystems*			T
2197	*increaseCivilianVehicles*			T
2198	*decreaseCivilianVehicles*			T

Table 10.24 Organizations issue additions

ID	OE element	A	O	D
104	*interventionOrganization*	T		
142	*keyMilitaryIndividual*	T		
144	*interventionForceOrganization*	T		
147	*regimeSponsoredNonMilitaryArmedForceOrganization*	T		
148	*demobilizedArmedForce*	T		
190	*keyNonGovtArmedOfficial*	T		
194	*paramilitaryForceOrganization*	T		
195	*privateSecurityForceOrganization*	T		
196	*insurgentOrganization*	T		
320	*compositeInfrastructureUnit*		T	
343	*NGOOrganization*	T		
344	*keyNGOIndividual*	T		
1862	*increaseNGOOrganizations*			T
1863	*decreaseNGOOrganizations*			T
1926	*nonGovtArmedIndividual*	T		
1929	*interventionForcePerson*	T		
1954	*interventionOrganizationEnvironment*		T	
2017	*increaseIntervenorOrganizations*			T
2018	*decreaseIntervenorOrganizations*			T
2021	*increaseIntervenorForceOrganizations*			T
2022	*decreaseIntervenorForceOrganizations*			T
2042	*increaseRegimeSponsoredNonMilitaryArmedForceOrganizations*			T
2043	*decreaseRegimeSponsoredNonMilitaryArmedForceOrganizations*			T
2044	*increaseParamilitaryOrganizations*			T
2045	*decreaseParamilitaryOrganizations*			T
2046	*increasePrivateSecurityOrganizations*			T
2047	*decreasePrivateSecurityOrganizations*			T
2048	*increaseInsurgentOrganizations*			T
2049	*decreaseInsurgentOrganizations*			T
2050	*increaseTerroristOrganizations*			T
2051	*decreaseTerroristOrganizations*			T
2090	*demobilizedArmedForcePresence*		T	
2100	*armedForceUnit*	T		
2117	*organizationUnit*	T		
2118	*populationUnit*	T		
2193	*geopoliticalUnit*		T	
2194	*facilityUnit*		T	

Small Concepts

A *smallConcept* is a *semanticConcept* with few connections. This category is comprised of three issues: explosive devices, non-nation-states, and self/things. Figure 10.9 shows the small concept part of the Semantic Concept Ontology. The "element" class box represents all of the classes that are similar to the issues. The photograph of the improvised explosive devices (IEDs) in the figure represents explosive device issues.

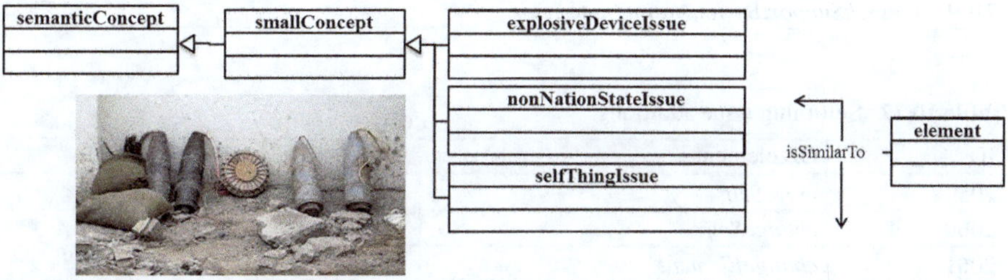

Fig. 10.9 Small concept ontology

The *explosiveDeviceIssue* is a *smallConcept* that concerns explosive devices. This subcategory provides an example that illustrates the need for the double-headed arrow for the *isSimilarTo* relationship in the figure. The *mineAndIED* class is similar to the classes in the *explosiveDeviceIssue* subcategory and is also similar to the classes in the *weaponsIssue* subcategory of the *militaryConcept* category, shown above (Table 10.25).

Table 10.25 Explosive device issue additions

ID	OE element	A	O	D
211	*clearMinesIEDs*			T
1806	*mineAndIED*		T	
2106	*placeMinesIEDs*			T

The *nonNationStateIssue* is a *smallConcept* that concerns non-nation-states (Table 10.26).

The *selfThingIssue* is a *smallConcept* that concerns control of self or things (Table 10.27).

Table 10.26 Non-nation-state issue additions

ID	OE element	A	O	D
191	*nonNationStateActorFunding*		T	
192	*nonNationStateActorRecruiting*		T	
193	*nonNationStateActorSupport*		T	
201	*IDOrInterdictFundingOfNonNationStateActor*			T
203	*IDInderdictOrInterruptRecruitmentByNonNationStateActor*			T
205	*IDInstitutionalOrLocalSupportForNonNationStateActor*			T
2039	*seekRecruitsByNonNationStateActor*			T
2108	*seekFundsByNonNationStateActor*			T
2109	*seekSupportByNonNationStateActor*			T

Table 10.27 Self/thing issue additions

ID	OE element	A	O	D
2059	*moveSelf*			T
2060	*changeSelf*			T
2061	*changeElement*			T
2062	*controlElement*			T
2096	*changeOwnActivityLevel*			T
2152	*concealSelf*			T
2153	*concealEntity*			T
2154	*defendSelf*			T
2155	*surrenderSelf*			T
2156	*disengageSelfFromCombat*			T

10.2 The Stocks-and-Flows Ontology

A stock-and-flow relationship is a generalization of the concept of increasing or decreasing (flow) the amount of something (stock). This concept expresses a commonality of usage. Thus, the basic relationship is between an Action class and an Object class. The generalization adds other relationships, including relationships involving Actor classes. A *stocksAndFlows* class is a *concept* that is a stocks-and-flows idea. These relationships are obvious to a human reader, but not to a computer. The Stocks-and-Flows Ontology makes these connections for the computer.

Figure 3.8 showed how the Stocks-and-Flows Ontology fits into the overall MCO structure (with Fig. 10.10 illustrating its icon as used in the context diagram, Fig. 3.5).

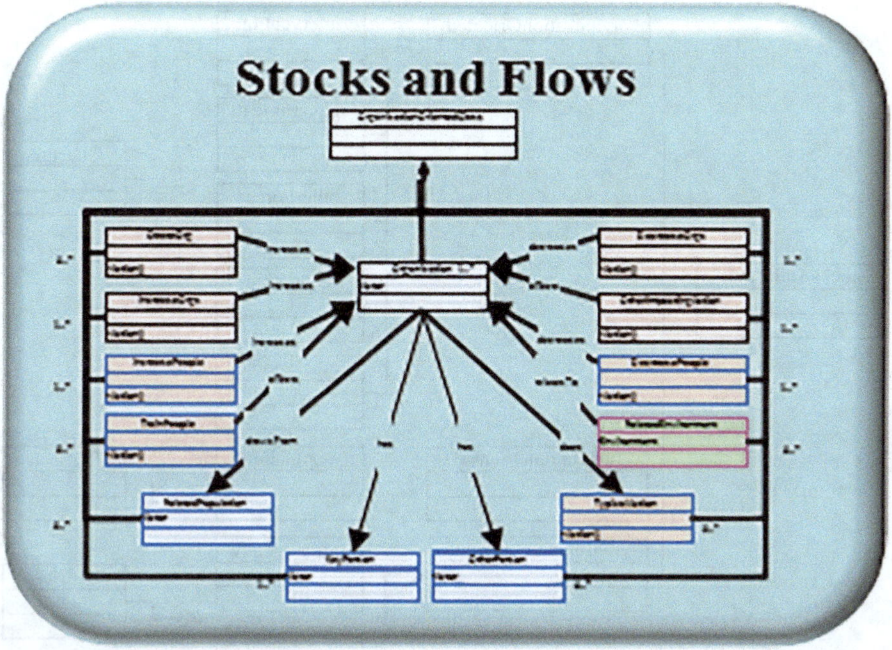

Fig. 10.10 Stocks-and-Flows icon

Overview of Stocks-and-Flows Concepts

There are three categories of *stocksAndFlows*: organization-, population-, and environment-oriented concepts. Figure 10.11 shows part of the Stocks-and-Flows (SaF) Ontology. It collapses the SaF classes (that belong in the fourth column) into three classes. (Because "Stocks-and-Flows" and "*stocksAndFlows*" are tedious to repeat, we will often use the abbreviation "SAF" as a prefix.)

The structure starts with the familiar top-level class, then category classes, and then subcategory classes (third column). The subclasses of each subcategory are too numerous to show in this figure and will be described below.

SaF Roles

Each of the subclasses (SaF classes) of the subcategories has its own set of "roles." However, each of the SaF classes of a particular category (column 2) has the same set of roles. Thus, all of the SaF classes of the *intervention* subcategory and all of the SaF classes of the *government* subcategory, through the SaF classes of the *otherPeople* subcategory, are represented by the *OOClass* box because of this similarity. The fact that the open triangle of the *OOClass* arrowhead does not touch any subcategory box is meant to represent that each SaF class represented by *OOClass* is a subclass of exactly one of the subcategories.

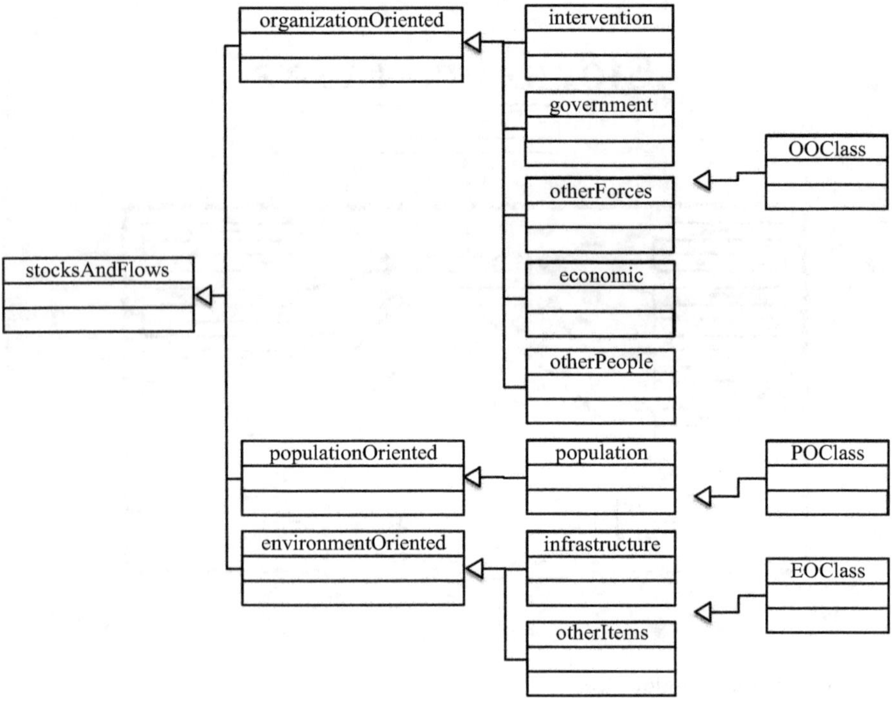

Fig. 10.11 Stocks-and-Flows ontology, part 1

Part 2 of the SaF Ontology diagram is shown in Fig. 10.12. Any element, whether Actor class, Action class, or Object class, may have a role in one or more of the classes represented by the three boxes, *OOClass*, *POClass*, and *EOClass*. This relationship is represented by the arrow leading from the "element" box to the class boxes, with a relationship label indicating its role. We use the generic "SAF Role" relationship to refer to these relationships.

If we consider the relationship in the other direction, we can talk about the particular SAF Role-type relationship. (This is the way the SaF classes are actually constructed.) For example, a SaF class that is an Organization-Oriented class has a SAF Role named *orgType* (first entry in Table 10.28). Thus, this SaF class has a *hasOrgPart* relationship with some element (in particular, some organization), which, as explained in the table, means that the element is the organization entity of the SaF class. (As we will see below, the organization is the central entity in an Organization-Oriented class.)

The roles are listed in Table 10.28, grouped by category. The SAF Role column names the role and the third column provides a verbal description. The final column of the table gives the cardinality of the role – that is, whether there is only one element that has this role or some number between 1 and *n* (represented by the "*") or between 0 and *n* (meaning there might be no element).

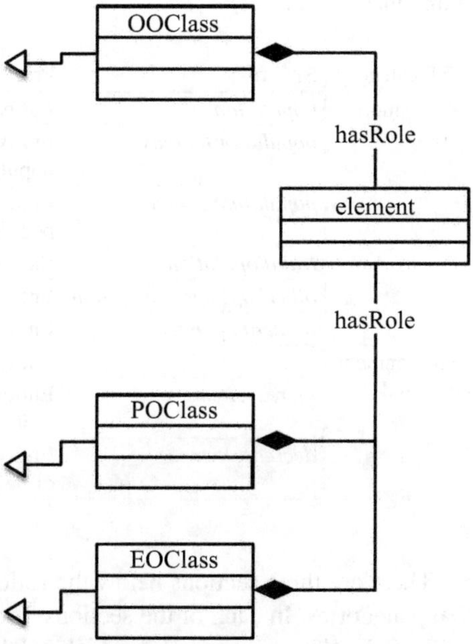

Fig. 10.12 Stocks-and-Flows ontology, part 2

Table 10.28 SaF class roles

SaF category	SAF role	Property description	# in class
Organization oriented	*orgType*	Entity has an organization entity as a part	1..*
	createOrg	Entity has an action that creates it	0..*
	orgIncrease	Entity has an action that increases its number of organizations	1..*
	orgDecrease	Entity has an action that decreases its number of organizations	1..*
	peopleIncrease	Entity has an action that increases its number of people	1..*
	peopleDecrease	Entity has an action that decreases its number of people	1..*
	trainPeople	Entity has an action that trains its people	0..*
	otherImpactingAction	Entity has an action that impacts its organization entity	1..*
	relatedPopulation	Entity has a related population	0..*
	keyPerson	Entity has a key person	1..*
	relatedActor	Entity has a related actor	0..*
	nonhumanIncrease	Entity has an action that increases its number of nonhuman actors	0..*
	nonhumanDecrease	Entity has an action that decreases its number of nonhuman actors	0..*
	relatedObject	Entity has an object related to its organization entity	0..*
	typicalAction	Entity has a typical action	0..*

(continued)

Table 10.28 (continued)

SaF category	SAF role	Property description	# in class
Population oriented	*population*	Entity has a population entity as a part	1
	populationIncrease	Entity has an action that increases its population	1..*
	populationDecrease	Entity has an action that decreases its population	1..*
	trainPopulation	Entity has an action that trains its population	0..*
	otherPopImpactingAction	Entity has an action that impacts its population	0..*
	relatedPopObject	Entity has an object related to its population	0..*
Environment oriented	*item*	Entity has an object as a part	1..*
	increaseItem	Entity has an action that increases its number of items	1..*
	decreaseItem	Entity has an action that decreases its number of items	1..*

There are three sections below the following two subsections, one for each of the SaF categories. In each of the sections, the first part of the discussion gives a picture of the SaF roles of Table 10.28, which will help explain the SaF classes of the category.

Changes to SaF Ontology from the UCO

With the change of terminology form Environmental Element classes to Object classes, it would be proper to change terminology for the *environmentOriented* concept to *objectOriented* concept; however, the terminology of "object-oriented" things has already been applied to too many other things. For this reason, we will leave this terminology as is and just remember that it refers to our Objects. For the most part, the tables of the SaF Concepts contained in the book about the UCO describe the content adequately (Hartley 2018).

Two new roles have been created, *nonhumanIncrease* and *nonhumanDecrease*, in the *organizationOriented* classes. This accounts for the use of dogs to detect buried people in disasters, to detect drugs by the police, and to detect hidden explosives by the military. We will not repeat the role assignment tables for these tables, as their changes are obvious.

Three additional SaF classes have been added, the *legitimacySAF* class, relating changes to legitimacy; the *populationGeographicSAF*, describing the relationships of a geographic population; and the *wholeOfGovernmentSAF* class, describing the relationships in the whole of government. These are environment oriented, population oriented, and organization oriented, respectively. Tables that show the role assignments for these SaF classes are included, both because they did not appear in the UCO and to illustrate the different types of role assignments.

In addition, there have been new elements added to the ontology and additional assignments of elements to the SaF classes; however, the additions to the role assignments for the affected classes are not included here because of space considerations.

An Insight into SaF

The roles are used in assigning element classes (Actor, Action, and Object classes) to the SaF classes. The simplest category, *environmentOriented*, provides the simplest example of roles. Each subclass will have one or more Objects that serve in the *item* role. This is the "stock," which has an associated quantity of some sort. Each SaF class will also have one or more Actions that serve in the *increaseItem* role and one or more Actions that serve in the *decreaseItem* role. These are the "flows" that increase or decrease the quantity associated with the "stock."

There are 99 SaF classes, divided among the 8 subcategories of the 3 categories of *stocksAndFlows*. Each of these SaF classes is connected to various Actor, Action, and Object classes based on the relationships among the various classes – which determine the numbers and types of roles for the categories. That is, the contents of Table 10.28 are derived from the actual relationships and describe them, rather than prescribing them.

As mentioned above, the explicit statement of these relationships provides a computer-interpretable connection among the various elements, whereas the simple statement of the elements' definitions would be insufficient. On the other hand, although these relationships are clear to a human reader, humans are fallible and can easily forget the existence of one or more of them. Their explicit descriptions can serve as reminders. For example, in building a model, a particular variable (representing an element) might be thought to be unimportant. However, if it is related through a SaF class to an important variable, the model builder has an opportunity to reconsider its importance.

Environment-Oriented Stocks-and-Flows

This category is a *stocksAndFlows* concept that is environment- (Object-) oriented. Figure 10.13 shows the role structure of its SaF classes, which were described in Table 10.28. The picture of the water pail filling and emptying in the figure illustrates the concept.

Figure 10.13 has an interesting property: the relationships that are named (increases and decreases) are implied relationships, and the relationships that are actually specified in the ontology are not named (the relationships that connect the classes to the environmentOrientedClass). In this structural diagram, the box labeled environmentOrientedClass (not italicized, as it is not the name of an actual class) represents a SaF class. The connection between it and the item box (shaded green to convey that it will be an Object class) is the *hasItemPart* relationship. The connection between the box labeled environmentOrientedClass and the increaseItem box (shaded pink to convey that it will be an Action class) is the *hasItemIncreaser* relationship. The arrow labeled "increases" from the increaseItem box to the item box is the reason for choosing the actual Action and Object classes to be components of this particular SaF class. That is, an instantiation of that Action class acts to increase the contents of the instantiation of that Object class.

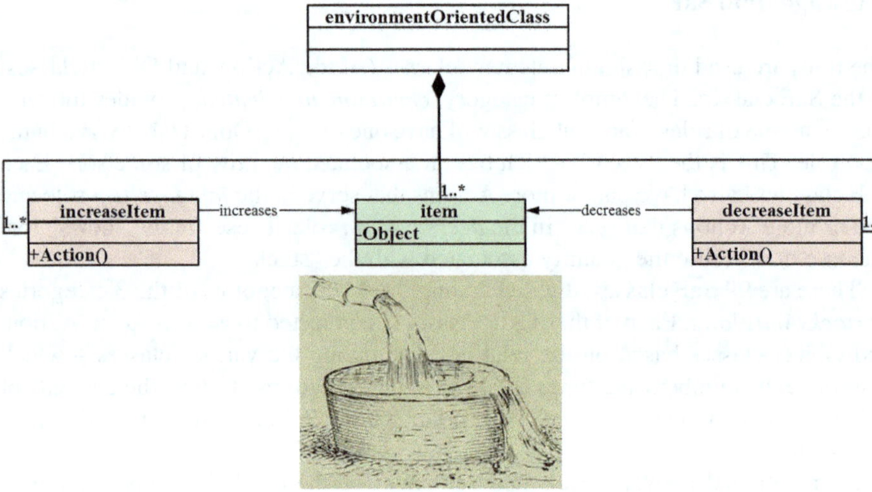

Fig. 10.13 SaF environmentOrientedClass structure

An example of an environmentOrientedClass will serve to reify the description above. The *legitimacySAF* class is new in the MCO and thus needs to be described. Its role assignments are shown in Table 10.29. As you can see, there are two classes with item roles, the *domesticLegitimacyOfActor* class and the *internationalLegitimacyOfActor* class. The legitimacy of instances of these classes can be increased by instances of the *increaseLegitimacyOfActor* class and decreased by instances of the *decreaseLegitimacyOfActor* class.

Table 10.29 *legitimacySAF* class role assignments

SAF Role	ID	OE element
item	106	*domesticLegitimacyOfActor*
item	107	*internationalLegitimacyOfActor*
increaseItem	1794	*increaseLegitimacyOfActor*
decreaseItem	1793	*decreaseLegitimacyOfActor*

This category has two subcategories, as shown in Fig. 10.11, the *infrastructureSAFSubcat* and the *otherItemsSAFSubcat*. (The figure leaves out the "SAFSubcat" portion of the names to save space.)

The *infrastructureSAFSubcat* is an *environmentOriented* SaF that is an infrastructure environment. There are 33 SaF classes in this subcategory, as shown in Table 10.30.

The *otherItemsSAFSubcat* is an *environmentOriented* SaF that is some other environment. There are 17 SaF classes in this subcategory, as shown in Table 10.31.

Table 10.30 SaF classes of the *infrastructureSAFSubcat*

SAFClass	Definition
commerceSectorSAF	An infrastructureSAFSubcat that is infrastructure of the commerce sector
marketsSAF	An infrastructureSAFSubcat that is infrastructure of the markets
transportAirportsSAF	An infrastructureSAFSubcat that is infrastructure of the airports
transportBridgesAndTunnelsSAF	An infrastructureSAFSubcat that is infrastructure of the bridges and tunnels
transportRailroadsSAF	An infrastructureSAFSubcat that is infrastructure of the railroad
transportRoadsSAF	An infrastructureSAFSubcat that is infrastructure of the roads
transportSeaportsSAF	An infrastructureSAFSubcat that is infrastructure of the seaports
transportVehiclesSAF	An infrastructureSAFSubcat that is infrastructure of the vehicles
transportWaterwaysSAF	An infrastructureSAFSubcat that is infrastructure of the waterways
energyExtractiveProductionInfrastructureSAF	An infrastructureSAFSubcat that is infrastructure of extractive energy production
energyExtractiveTransportationInfrastructureSAF	An infrastructureSAFSubcat that is infrastructure of extractive energy transportation
energyElectricityProductionInfrastructureSAF	An infrastructureSAFSubcat that is infrastructure of electricity production
energyElectricityDistributionInfrastructureSAF	An infrastructureSAFSubcat that is infrastructure of electricity distribution
waterDamsSAF	An infrastructureSAFSubcat that is infrastructure of the dams
waterWaterAndSewageSystemsInfrastructureSAF	An infrastructureSAFSubcat that is infrastructure of the water and sewage systems
waterDistributionInfrastructureSAF	An infrastructureSAFSubcat that is infrastructure of water distribution
governmentGovtInfrastructureSAF	An infrastructureSAFSubcat that is infrastructure of the government infrastructure
governmentPrisonInfrastructureSAF	An infrastructureSAFSubcat that is infrastructure of the prison infrastructure

(continued)

Table 10.30 (continued)

SAFClass	Definition
governmentMilitaryInfrastructureSAF	An infrastructureSAFSubcat that is infrastructure of the military
governmentMilitaryVehiclesSAF	An infrastructureSAFSubcat that is infrastructure of military vehicles
businessAgricultureInfrastructureSAF	An infrastructureSAFSubcat that is infrastructure of agriculture
businessFinancialInfrastructureSAF	An infrastructureSAFSubcat that is infrastructure of the business financial system
businessManufacturingInfrastructureSAF	An infrastructureSAFSubcat that is infrastructure of the manufacturing system
businessMediaInfrastructureSAF	An infrastructureSAFSubcat that is infrastructure of the media
businessGovtMISComputerSystemsSAF	An infrastructureSAFSubcat that is infrastructure of the MIS and computer systems
businessMiningInfrastructureSAF	An infrastructureSAFSubcat that is infrastructure of mining
businessShopsAndCommercialInfrastructureSAF	An infrastructureSAFSubcat that is infrastructure of the shops and commercial systems
businessCulturalInfrastructureSAF	An infrastructureSAFSubcat that is infrastructure of the cultural businesses
socialCivilianHousingSAF	An infrastructureSAFSubcat that is infrastructure of the civilian housing
socialTemporaryShelterSAF	An infrastructureSAFSubcat that is infrastructure of the temporary shelters
socialEducationInfrastructureSAF	An infrastructureSAFSubcat that is infrastructure of the education system
socialHealthcareInfrastructureSAF	An infrastructureSAFSubcat that is infrastructure of the healthcare system
socialReligiousInfrastructureSAF	An infrastructureSAFSubcat that is infrastructure of the religious organizations

Table 10.31 SaF classes of the *otherItemsSAFSubcat*

SAFClass	Definition
crimeCorruptionSAF	An otherItemsSAFSubcat that concerns corruption
crimeDrugSAF	An otherItemsSAFSubcat that concerns drugs
crimeGeneralSAF	An otherItemsSAFSubcat that concerns general crime
crimeNonStateActorRecruitingFundingSupportSAF	An otherItemsSAFSubcat that concerns non-nation-state actor recruiting, funding, and seeking support
supplyGoodsAndEquipmentSAF	An otherItemsSAFSubcat that concerns goods and equipment
supplyEducationSAF	An otherItemsSAFSubcat that concerns education supplies
supplyHealthcareSAF	An otherItemsSAFSubcat that concerns healthcare supplies
supplyPotableWaterSAF	An otherItemsSAFSubcat that concerns potable water supplies
supplyFoodSAF	An otherItemsSAFSubcat that concerns food supplies
supplyNaturalResourcesSAF	An otherItemsSAFSubcat that concerns natural resources supplies
wasteSAF	An otherItemsSAFSubcat that concerns waste
fundsSAF	An otherItemsSAFSubcat that concerns funds
C4ISAF	An otherItemsSAFSubcat that concerns command, control, communications, computers, and intelligence
senseOfCommunitySAF	An otherItemsSAFSubcat that concerns a sense of community
governmentPoliciesSAF	An otherItemsSAFSubcat that concerns government policies
civilDisturbanceSAF	An otherItemsSAFSubcat that concerns civil disturbance
legitimacySAF	An otherItemsSAFSubcat that concerns legitimacy

Population-Oriented Stocks-and-Flows

This category is a *stocksAndFlows* concept that is population-oriented. Figure 10.14 shows the role structure of its SaF classes, described in Table 10.28.

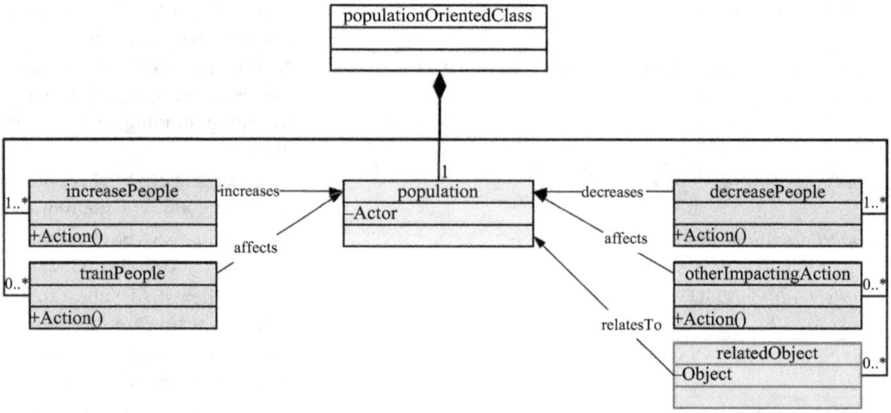

Fig. 10.14 SaF populationOrientedClass structure

Note that as with Fig. 10.13, showing the environmentOrientedClass structure, the labeled relationships are the implied relationships and the unlabeled relationships are the specified relationships of the ontology. In this figure, the reason for this convention is clearer: there are too many overlapping specified relationships to allow labeling.

This category has only one subcategory, as shown in Fig. 10.11, the *populationSAFSubcat*.

The *populationSAFSubcat* is a *populationOriented* SaF that is a population. There are 10 SaF classes in this subcategory, as shown in Table 10.32.

Table 10.32 SaF classes of the *populationSAFSubcat*

SAFClass	Definition
HNDemobArmedForcesSAF	A populationSAFSubcat that is a demobilized armed forces population
DPMigrantsSAF	A populationSAFSubcat that is a migrants population
DPIDPsSAF	A populationSAFSubcat that is an internally displaced persons population
DPRefugeesSAF	A populationSAFSubcat that is a refugees population
DPExpatriatesSAF	A populationSAFSubcat that is an expatriates population
populationGeneralSAF	A populationSAFSubcat that is a general population
populationCulturalSAF	A populationSAFSubcat that is a cultural population
populationPoliticalSAF	A populationSAFSubcat that is a political population
populationReligiousSAF	A populationSAFSubcat that is a religious population
populationGeographicSAF	A populationSAFSubcat that is a geographically specified population

The *populationGeographicSAF* class is new. Its role assignments are shown in Table 10.33. Note that there are two populations that are represented here and a large number of "other" Actions that impact the populations.

Table 10.33 populationGeographicSAF class role assignments

SAF role	ID	OE element
population	765	*geographicalPopulation*
population	2193	*geopoliticalUnit*
populationIncrease	409	*freedomOfMovement*
populationDecrease	409	*freedomOfMovement*
trainPopulation		
otherPopImpactingAction	404	*stressMigration*
otherPopImpactingAction	410	*forcedPopulationMovement*
otherPopImpactingAction	413	*resettlePeople*
otherPopImpactingAction	1864	*increaseIDPs*
otherPopImpactingAction	1865	*decreaseIDPs*
otherPopImpactingAction	1866	*increaseMigrants*
otherPopImpactingAction	1867	*decreaseMigrants*
otherPopImpactingAction	1868	*increaseRefugees*
otherPopImpactingAction	1869	*decreaseRefugees*
otherPopImpactingAction	1870	*increaseExpatriates*
otherPopImpactingAction	1871	*decreaseExpatriates*
otherPopImpactingAction	1872	*increaseGeneralPopulation*
otherPopImpactingAction	1873	*decreaseGeneralPopulation*
relatedPopObject	2089	*geographicalSubdivision*
relatedPopObject	2119	*compositeGeographicalUnit*

Organization-Oriented Stocks-and-Flows

This category is a *stocksAndFlows* concept that is organization-oriented. Figure 10.15 shows the role structure of its SaF classes, described in Table 10.28.

Note that as with Figs. 10.13 and 10.14, which show the other two class structures, the labeled relationships are the implied relationships and the unlabeled relationships are the specified relationships of the ontology. In this figure, the reason for this convention is abundantly clear: there are far too many overlapping specified relationships to allow labeling.

This category has five subcategories, as shown in Fig. 10.11, the *interventionSAFSubcat*, the *governmentSAFSubcat*, the *otherForcesSAFSubcat*, the *economicSAFSubcat*, and the *otherPeopleSAFSubcat*.

The *interventionSAFSubcat* is an *organizationOriented* SaF that is an intervention organization. There are two SaF classes in this subcategory, as shown in Table 10.34.

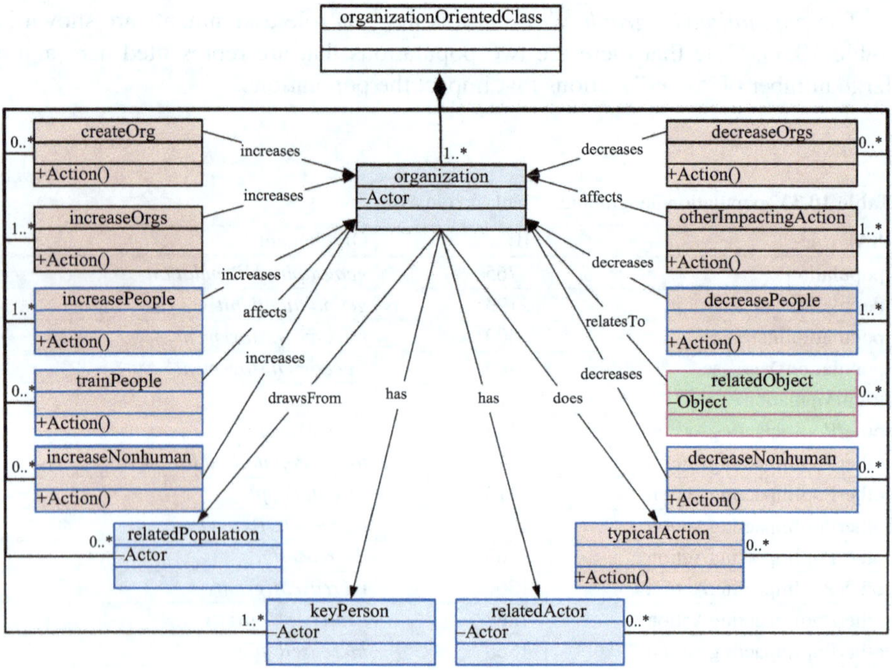

Fig. 10.15 SaF organizationOrientedClass structure

Table 10.34 SaF classes of the *interventionSAFSubcat*

SAFClass	Definition
interventionForceSAF	An interventionSAFSubcat that is an intervention military force
interventionOrgSAF	An interventionSAFSubcat that is an intervention civilian government organization

The *governmentSAFSubcat* is an *organizationOriented* SaF that is a government organization. There are 12 SaF classes in this subcategory, as shown in Table 10.35.

The *wholeOfGovernmentSAF* class is new. Its role assignments are shown in Table 10.36. This SaF class has a large number of components that have the *orgType* role. This class has similarities to and differences from the hierarchical composite Actor class *wholeOfGovt*. It addresses the same concept; however, this class emphasizes the relationships among other types of classes, whereas the composite class emphasizes the hierarchical relationships among Actor classes.

The *otherForcesSAFSubcat* is an *organizationOriented* SaF that is some other force organization. There are six SaF classes in this subcategory, as shown in Table 10.37.

Table 10.35 SaF classes of the *governmentSAFSubcat*

SAFClass	Definition
wholeOfGovernmentSAF	A governmentSAFSubcat that is a whole-of-government organization
executiveSAF	A governmentSAFSubcat that is a government executive branch organization
legislativeSAF	A governmentSAFSubcat that is a government legislative branch organization
judiciarySAF	A governmentSAFSubcat that is a judicial branch organization
bureaucracySAF	A governmentSAFSubcat that is a bureaucracy organization
militarySAF	A governmentSAFSubcat that is a military organization
intelligenceServiceSAF	A governmentSAFSubcat that is an intelligence organization
lawEnforcementSAF	A governmentSAFSubcat that is a law enforcement organization
firstResponderSAF	A governmentSAFSubcat that is a first responder organization
educationSAF	A governmentSAFSubcat that is an education organization
healthcareSAF	A governmentSAFSubcat that is a healthcare organization
socialServiceSAF	A governmentSAFSubcat that is a social service organization

Table 10.36 *wholeOfGovernmentSAF* class role assignments

SAF role	ID	OE element
orgType	5	*firstResponderOrganization*
orgType	8	*socialServicesOrganization*
orgType	41	*govtTypeUnit*
orgType	64	*judicialBranch*
orgType	65	*lawEnforcementOrganization*
orgType	145	*govtMilitaryForceOrganization*
orgType	146	*intelligenceServiceOrganization*
orgType	1887	*executiveBranch*
orgType	1888	*legislativeBranch*
orgType	1937	*healthcareOrganization*
orgType	1938	*educationOrganization*
orgType	1940	*govtBureaucracyOrganization*
orgType	2115	*wholeOfGovt*
createOrg	23	*createGovt*
createOrg	78	*rebuildOrMonitorLawEnforcementAndFirstResponder Organizations*
createOrg	153	*createOrReformOrMonitorMilitary*
createOrg	159	*createOrReformOrMonitorIntelligenceServices*
createOrg	421	*provideSocialProtectionProgram*
createOrg	1790	*assistInCreatingSocialServices*
createOrg	2004	*changeEducationPolicy*
createOrg	2005	*changeHealthcarePolicy*
orgIncrease	1988	*increaseHealthcareOrganizations*
orgIncrease	2007	*increaseEducationOrganizations*
orgIncrease	2009	*increaseLawEnforcementOrganizations*

(continued)

Table 10.36 (continued)

SAF role	ID	OE element
orgIncrease	2011	*increaseBureaucracyOrganizations*
orgIncrease	2013	*increaseGovtOrganizations*
orgIncrease	2015	*increaseJudicialOrganizations*
orgIncrease	2019	*increaseFirstResponderOrganizations*
orgIncrease	2023	*increaseIntelligenceServicesOrganizations*
orgIncrease	2040	*increaseMilitaryOrganizations*
orgIncrease	2052	*increaseSocialServicesOrganizations*
orgDecrease	1829	*decreaseIntelligenceServicesOrganizations*
orgDecrease	1989	*decreaseHealthcareOrganizations*
orgDecrease	2008	*decreaseEducationOrganizations*
orgDecrease	2010	*decreaseLawEnforcementOrganizations*
orgDecrease	2012	*decreaseBureaucracyOrganizations*
orgDecrease	2014	*decreaseGovtOrganizations*
orgDecrease	2016	*decreaseJudicialOrganizations*
orgDecrease	2020	*decreaseFirstResponderOrganizations*
orgDecrease	2041	*decreaseMilitaryOrganizations*
orgDecrease	2053	*decreaseSocialServicesOrganizations*
peopleIncrease	1882	*increaseGovtPersonnel*
peopleIncrease	1978	*increaseHealthcarePersonnel*
peopleIncrease	1980	*increaseMilitaryForcesPersonnel*
peopleIncrease	1981	*increaseIntelligenceServicesPersonnel*
peopleIncrease	1984	*increaseEducators*
peopleIncrease	2026	*increaseFirstRespondersPersonnel*
peopleIncrease	2027	*increaseLawEnforcementPersonnel*
peopleDecrease	1816	*decreaseFirstRespondersPersonnel*
peopleDecrease	1828	*decreaseMilitaryForcesPersonnel*
peopleDecrease	1878	*decreaseLawEnforcementPersonnel*
peopleDecrease	1880	*decreaseIntelligenceServicePersonnel*
peopleDecrease	1883	*decreaseGovtPersonnel*
peopleDecrease	1920	*decreaseEducators*
peopleDecrease	1979	*decreaseHealthcarePersonnel*
trainPeople	15	*educateGovtPersonnel*
trainPeople	17	*trainFirstResponders*
trainPeople	21	*trainNewPoliticalLeaders*
trainPeople	80	*trainLawEnforcementPersonnel*
trainPeople	155	*trainMilitaryForces*
trainPeople	161	*trainIntelligenceService*
trainPeople	386	*trainEducators*
trainPeople	1977	*educateStudents*
otherImpactingAction		
relatedPopulation		

(continued)

Table 10.36 (continued)

SAF role	ID	OE element
keyPerson	142	*keyMilitaryIndividual*
keyPerson	381	*keyEducationIndividual*
keyPerson	1927	*keyLawEnforcementLeader*
keyPerson	1928	*keyBureaucrat*
keyPerson	1931	*keyJudicialLeader*
keyPerson	1932	*keyLegislator*
keyPerson	1933	*keyGovtExecutive*
keyPerson	1934	*keyHealthcareLeader*
keyPerson	1935	*keyFirstResponderLeader*
keyPerson	1936	*keyIntelligenceServiceLeader*
relatedActor	1877	*lawEnforcementPerson*
relatedActor	1879	*intelligenceServicePerson*
relatedActor	1881	*govtPerson*
relatedActor	1919	*educator*
relatedActor	1924	*healthcarePerson*
relatedActor	1925	*firstResponderPerson*
relatedActor	1930	*govtMilitaryPerson*
relatedActor	2080	*landVehicle*
relatedActor	2082	*watercraft*
relatedActor	2083	*aircraft*
relatedActor	2084	*spacecraft*
relatedActor	2097	*genericAnimal*
nonhumanIncrease	2197	*increaseCivilianVehicles*
nonhumanIncrease	2199	*increaseAnimals*
nonhumanDecrease	2198	*decreaseCivilianVehicles*
nonhumanDecrease	2200	*decreaseAnimals*
relatedObject		
typicalAction		

Table 10.37 SaF classes of the *otherForcesSAFSubcat*

SAFClass	Definition
armedRegimeSponsoredNonMilitarySAF	An otherForcesSAFSubcat that is a regime-sponsored non-military armed force organization
armedParamilitarySAF	An otherForcesSAFSubcat that is a paramilitary organization
armedPrivateSecuritySAF	An otherForcesSAFSubcat that is a private security organization
armedInsurgentSAF	An otherForcesSAFSubcat that is an insurgent organization
armedTerroristSAF	An otherForcesSAFSubcat that is a terrorist organization
conflictAdvocateSAF	An otherForcesSAFSubcat that is an external organization advocating conflict

The *economicSAFSubcat* is an *organizationOriented* SaF that is an economic organization. There are 15 SaF classes in this subcategory, as shown in Table 10.38.

The *otherPeopleSAFSubcat* is an *organizationOriented* SaF that represents organizations not included in other subcategories. There are four SaF classes in this subcategory, as shown in Table 10.39.

Table 10.38 SaF classes of the *economicSAFSubcat*

SAFClass	Definition
businessAgricultureSAF	An economicSAFSubcat that is an agriculture organization
businessContractorSAF	An economicSAFSubcat that is a contractor organization
businessEnergySAF	An economicSAFSubcat that is an energy business organization
businessFinancialServicesSAF	An economicSAFSubcat that is a financial services organization
businessFishingSAF	An economicSAFSubcat that is a fishing business organization
businessManufacturingSAF	An economicSAFSubcat that is a manufacturing business organization
businessMediaSAF	An economicSAFSubcat that is a media business organization
businessMiningSAF	An economicSAFSubcat that is a mining business organization
businessServiceSAF	An economicSAFSubcat that is a service business organization
businessCulturalSAF	An economicSAFSubcat that is a cultural business organization
businessTimberSAF	An economicSAFSubcat that is a timber business organization
businessTourismSAF	An economicSAFSubcat that is a tourism business organization
businessTransportSAF	An economicSAFSubcat that is a transportation business organization
laborSAF	An economicSAFSubcat that is a labor organization
criminalSAF	An economicSAFSubcat that is a criminal business organization

Table 10.39 SaF classes of the *otherPeopleSAFSubcat*

SAFClass	Definition
NGOOrgSAF	An otherPeopleSAFSubcat that is an NGO organization
factionSocialSAF	An otherPeopleSAFSubcat that is a social faction organization
factionPoliticalSAF	An otherPeopleSAFSubcat that is a political faction organization
factionReligiousSAF	An otherPeopleSAFSubcat that is a religious faction organization

10.3 Concept Ontologies Relationships

The only relationships used in by the Semantic Concept Ontology are the *is-a* and *isSimilarTo* relationships, explained as follows:

is-a: Each instance of A is an instance of B and each property of B is a property of A. Its inverse relationship is *superClassOf*.
isSimilarTo: A is similar to B. Its inverse relationship is *isSimilarTo*.

Several relationships are used by the SaF Ontology. These are the formal relationships in the ontology that implement the SAF Roles, most of which were represented by unlabeled relationships in the diagrams. They are explained as follows:

is-a: Each instance of A is an instance of B and each property of B is a property of A. Its inverse relationship is *superClassOf*.
hasOrgPart: A has B as a part (composition) and B is an organization. Its inverse relationship is *isOrgPartOf*.
hasOrgCreator: A has B as a part (composition) and B is an Action that creates A. Its inverse relationship is *isOrgCreatorOf*.
hasPeopleTrainer: A has B as a part (composition) and B is an Action that trains A's people. Its inverse relationship is *isPeopleTrainerOf*.
hasPopTrainer: A has B as a part (composition) and B is an Action that trains A's population. Its inverse relationship is *isPopTrainerOf*.
hasOrgImpacter: A has B as a part (composition) and B is an Action that impacts its organization entity. Its inverse relationship is *isOrgImpacterOf*.
hasPopImpacter: A has B as a part (composition) and B is an Action that impacts its population. Its inverse relationship is *isPopImpacterOf*.
hasPopPart: A has B as a part (composition) and B is a population. Its inverse relationship is *isPopPartOf*.
hasRelatedPop: A has B as a part (composition) and B is a related population. Its inverse relationship is *isRelatedPopOf*.
hasKeyPerson: A has B as a part (composition) and B is a key person. Its inverse relationship is *isKeyPersonOf*.
hasRelatedActor: A has B as a part (composition) and B is a related Actor. Its inverse relationship is *isRelatedActorOf*.
hasItemPart: A has B as a part (composition) and B is an Object. Its inverse relationship is *isItemPartOf*.
hasOrgObject: A has B as a part (composition) and B is an Object related to its organization entity. Its inverse relationship is *isOrgObjectOf*.
hasPopObject: A has B as a part (composition) and B is an Object related to its population. Its inverse relationship is *isPopObjectOf*.
hasTypicalAction: A has B as a part (composition) and B is a typical action. Its inverse relationship is *isTypicalActionOf*.
hasOrgIncreaser: A has B as a part (composition) and B is an Action that increases its number of organizations. Its inverse relationship is *isOrgIncreaserOf*.

hasOrgDecreaser: A has B as a part (composition) and B is an Action that decreases its number of organizations. Its inverse relationship is *isOrgDecreaserOf*.

hasPeopleIncreaser: A has B as a part (composition) and B is an Action that increases its number of people. Its inverse relationship is *isPeopleIncreaserOf*.

hasPeopleDecreaser: A has B as a part (composition) and B is an Action that decreases its number of people. Its inverse relationship is *isPeopleDecreaserOf*.

hasNonhumanIncreaser: A has B as a part (composition) and B is an Action that increases its number of nonhuman Actors. Its inverse relationship is *isNonhumanIncreaserOf*.

hasNonhumanDecreaser: A has B as a part (composition) and B is an Action that decreases its number of nonhuman Actors. Its inverse relationship is *isNonhumanDecreaserOf*.

hasItemIncreaser: A has B as a part (composition) and B is an Action that increases its number of items. Its inverse relationship is *isItemIncreaserOf*.

hasItemDecreaser: A has B as a part (composition) and B is an Action that decreases its number of items. Its inverse relationship is *isItemDecreaserOf*.

hasPopIncreaser: A has B as a part (composition) and B is an Action that increases its population. Its inverse relationship is *isPopIncreaserOf*.

hasPopDecreaser: A has B as a part (composition) and B is an Action that decreases its population. Its inverse relationship is *isPopDecreaserOf*.

10.4 Recapitulation of the Concept Ontologies

The Actor, Action, and Object classes are related within their separate ontologies in their hierarchies of parent-child relationships. However, there are also relationships that cross the hierarchies in each ontology and cross among the ontologies. Similarity of meaning is one such relationship and it defines the Semantic Concept Ontology. Commonality of use is another such relationship and it defines the Stocks-and-Flows Ontology.

The Semantic Concept Ontology provides a means for computer connections of things that are related semantically by the *isSimilarTo* relationship. These associations of the Semantic Concept Ontology collectively include most, but not all, of the Actors, Actions, and Object classes, defined by similarities of meaning.

The SaF Ontology provides connections among things that are used together. It provides a set of connections among the elements that is both computer-readable, supporting virtual connections between elements, and human-readable, reminding humans about the elements that are connected in the face of large numbers of possible connections. These associations of the Stocks-and-Flows Ontology collectively include most, but not all, of the Actors, Actions, and Object classes, defined by the roles that the elements may play in interacting with each other. The associations are divided into organization-oriented, population-oriented, and environment-oriented SaF classes.

Chapter 11
MCO Theories Ontology

The transitions from one state to another are "simple" when recording what has happened in a real conflict – the transitions simply exist. However, when building a model of conflict, the transitions must occur, and there must be a reason for choosing a particular change rather than other possible changes. We will call these reasons "theories," even though some are not theories in a classical sense.

In one sense, theories are independent of the Modern Conflict Ontology (MCO). They are not needed in a purely descriptive use but only in building a computer model of conflict. Theories are included as part of the ontology structure because they are needed in "explaining" and understanding modern conflict and because a useful ontology of theories is not available elsewhere.

Figure 3.10 showed how the Theories Ontology fits into the overall MCO structure (with Fig. 11.1 illustrating its icon).

Fig. 11.1 Theories icon

D. S. Hartley III, *An Ontology of Modern Conflict*, Understanding Complex Systems, https://doi.org/10.1007/978-3-030-53214-7_11

11.1 Overview of Theories

There are five discipline categories of theories: social science theories, hard science theories, applied science theories, formal theories, and other theories. Figure 11.2 shows the top level of the Theories Ontology. The cartoon of the evolution of man in the figure represents all of the possible theories.

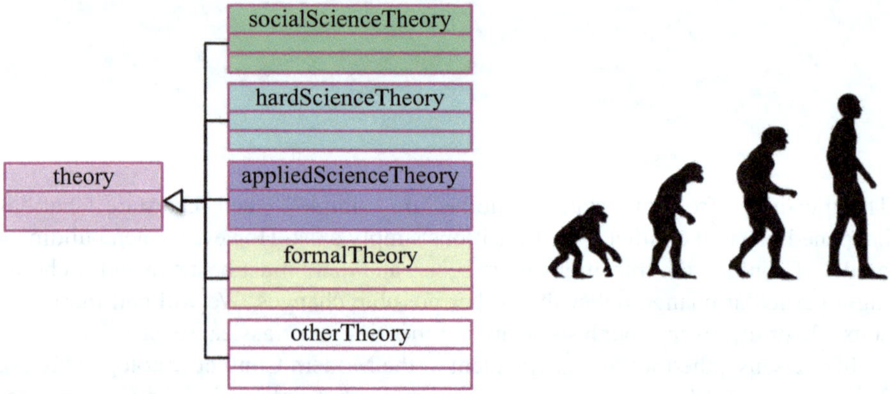

Fig. 11.2 Theories ontology overview

Changes to Theories from the UCO

In Chap. 3, we discussed the general changes from the Unconventional Conflict Ontology (UCO) to the Modern Conflict Ontology (MCO). Most of these changes are formal, e.g., changing spelling conventions and definition standards. The tables of the theories contained in the book about the UCO describe the content adequately (Hartley 2018). Therefore, its structure will be described here, but the lists of theories will be omitted. However, the treatment of validity has been modified in the MCO and will be discussed.

An Insight into Theories

This ontology is not a taxonomy, because many of the theory classes are subclasses of more than one subcategory. Currently there are 392 theories; however, there are 487 entries when the ontology is listed as a flat file. For example, *game-Theory* is found in *sociology*, *operationsResearch*, *microEconomics*, and *mathematics* and is thus a "subclass" of each of those subcategories. (It is technically not a subclass because the relationship is not *is-a*, but *hasPart*.) Its description says it covers "various types of games, with various rules." Its citations are abbreviated and truncated due to variable length restrictions in the database:

> Bennett, Peter G. 1995. "Modeling decisions in international relations: Game theory and beyond." Mershon International Studies Review 39:19-52; Von Neumann, John and Oskar Morgenstern. 1944. Theory of Games and Economic Behavior. Princeton, NJ: Princeton University Press; Parsons, Simon and Michael Wooldridge. 2002. "Game agents." in Game Theoretic and Decision Theoretic Agents: AAAI Spring Symposium, edited by S. Parsons and P. Gmytrasiewicz. Menlo Park, CA: American Association for Artificial Intelligence; Nash …

In most contexts, game theory is rated as having validity at the peerReview+ level (defined below).

The theories are preliminary. They need to be addressed by experts in the various fields as to which theories should be included, how the theories should be broken up or combined, and how the theories should be stated. Further, the Theories Ontology is built as a pointer to theories, not as a complete description of them.

11.2 Social Science Theories

A *socialScienceTheory* is a Theory from the social sciences. This discipline category of Theory is comprised of 12 disciplines or subcategories: political science, sociology, psychology, anthropology, history, human geography, law, linguistics, education, religion, demography, and criminology. Figure 11.3 shows the social science

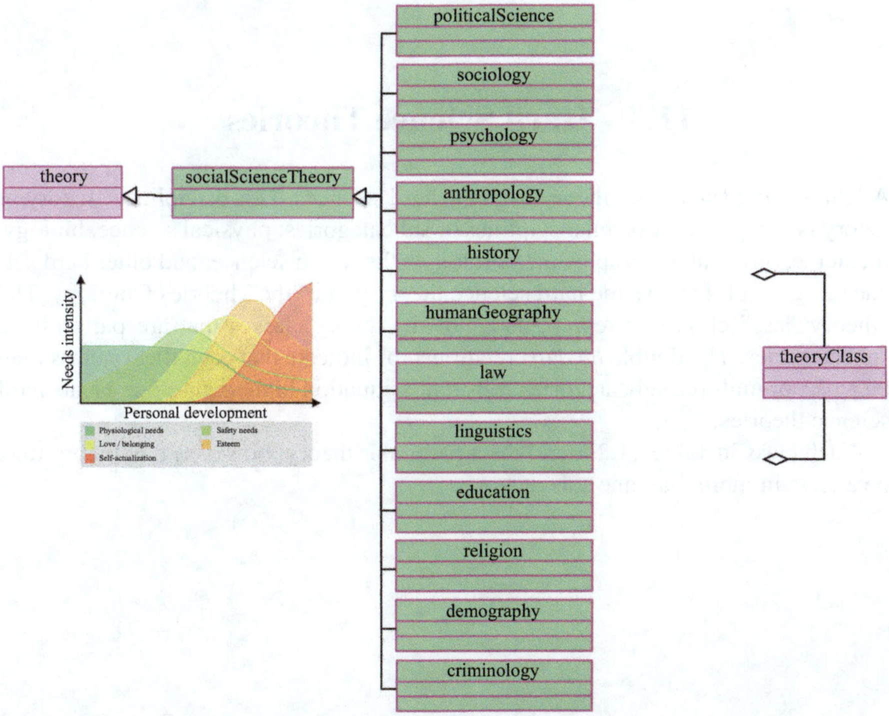

Fig. 11.3 Social science theories ontology

Theory part of the Theories Ontology. The "theoryClass" class box represents all of the Theory classes that are parts of the subcategories. The double *hasPart* relationships indicate that each Theory class may be a part of multiple subcategories. The graph in the figure represents Maslow's Theory of a hierarchy of needs.

Each class in Table 11.1 identifies a particular theory; however, that theory may be a class in more than one subcategory.

Table 11.1 Social science subcategories and classes

Subcategory	Discipline	# of classes
politicalScience	Political science	40
sociology	Sociology	74
psychology	Psychology	105
anthropology	Anthropology	25
history	History	3
humanGeography	Human geography	3
law	Law	6
linguistics	Linguistics	3
education	Education	10
religion	Religion	2
demography	Demography	1
criminology	Criminology	1

11.3 Hard Science Theories

A *hardScienceTheory* is a theory from the hard sciences. This discipline category of theory is comprised of seven disciplines or subcategories: physical science, biology, medicine, physical geography, meteorology, other earth science, and other hard science. Figure 11.4 shows the hard science theory part of the Theories Ontology. The "theoryClass" class box represents all of the theory classes that are parts of the subcategories. The double *hasPart* relationships indicate that each theory class may be a part of multiple subcategories. Einstein's equation in the figure represents hard science theories.

Each class in Table 11.2 identifies a particular theory; however, that theory may be a class in more than one subcategory.

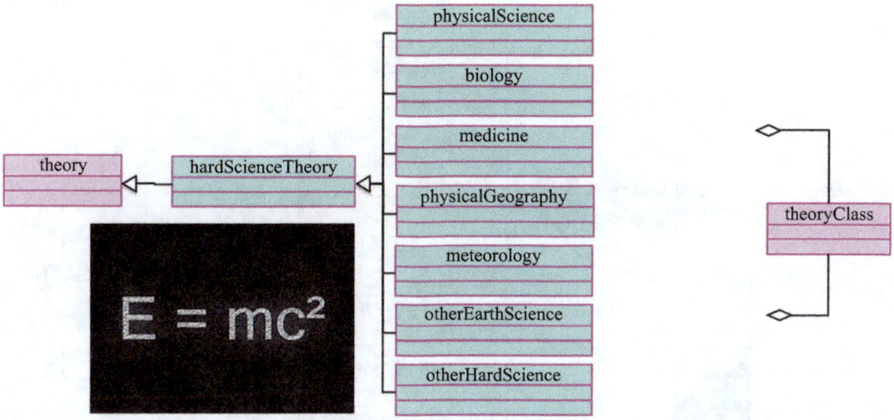

Fig. 11.4 Hard science theories ontology

Table 11.2 Hard science subcategories and classes

Subcategory	Discipline	# of classes
physicalScience	Physical science	6
biology	Biology	2
medicine	Medicine	5
physicalGeography	Meteorology	2
meteorology	History	3
otherEarthScience	Other earth sciences	2
otherHardScience	Other hard sciences	1

11.4 Applied Science Theories

An *appliedScienceTheory* is a Theory from the applied sciences. This discipline category of theory is comprised of 10 disciplines or subcategories: operations research, military science, business management, macroeconomics, microeconomics, civil engineering, other engineering, computer science, data issues, and other communications. Figure 11.5 shows the applied science theory part of the Theories Ontology. The "theoryClass" class box represents all of the theory classes that are parts of the subcategories. The double *hasPart* relationships indicate that each theory class may be a part of multiple subcategories. The graph in the figure illustrates linear programming, a theory in operations research.

Each class in Table 11.3 identifies a particular theory; however, that theory may be a class in more than one subcategory.

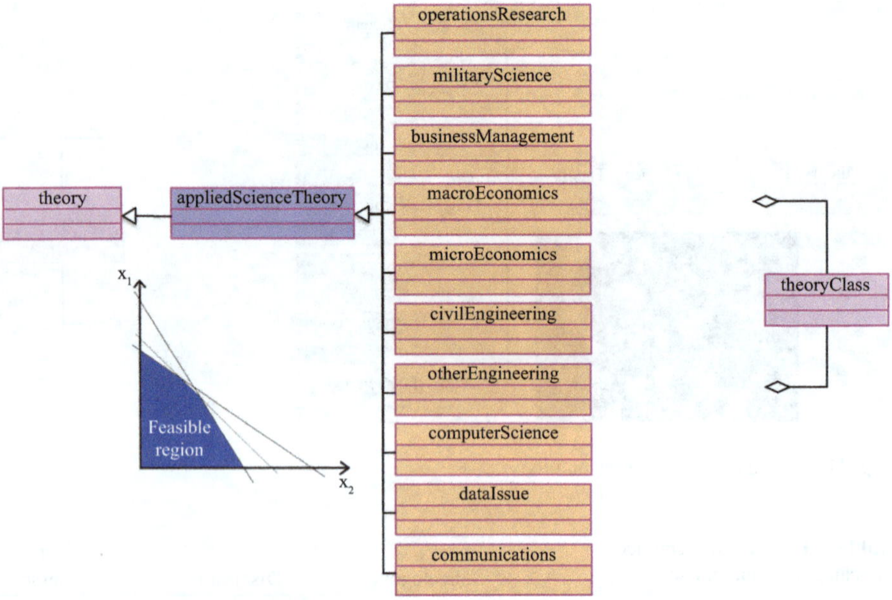

Fig. 11.5 Applied science theories ontology

Table 11.3 Applied science subcategories and classes

Subcategory	Discipline	# of classes
operationsResearch	Operations research	32
militaryScience	Military science	30
businessManagement	Business management	10
macroEconomics	Macroeconomics	31
microEconomics	Microeconomics	21
civilEngineering	Civil engineering	17
otherEngineering	Other engineering	2
computerScience	Computer science	15
dataIssue	Data sciences	2
communications	Communications	6

11.5 Formal Theories

A *formalTheory* is a theory from the formal disciplines. This discipline category of theory is comprised of three disciplines or subcategories: logic, mathematics, and statistics. Figure 11.6 shows the formal Theory part of the Theories Ontology. The "theoryClass" class box represents all of the theory classes that are parts of the subcategories. The double *hasPart* relationships indicate that each theory class may be a part of multiple subcategories. The three squares and the triangle in the figure illustrate the Pythagorean Theorem of mathematics.

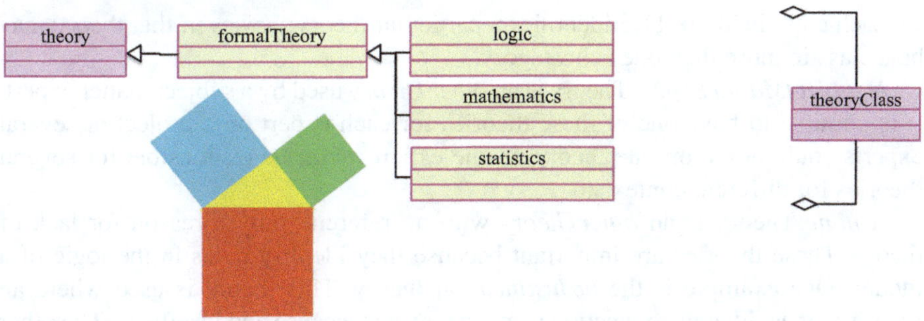

Fig. 11.6 Formal Theories Ontology

Table 11.4 Formal theories subcategories and classes

Subcategory	Discipline	# of classes
logic	Logic	1
mathematics	Mathematics	8
statistics	Statistics	2

Each class in Table 11.4 identifies a particular theory; however, that theory may be a class in more than one subcategory.

11.6 Other Theories

An *otherTheory* is a theory from other areas. This discipline category of theory is comprised of two disciplines or subcategories: subject matter expert and "none." Figure 11.7 shows the formal theory part of the Theories Ontology. The "theoryClass" class box represents all of the theory classes that are parts of the subcategories. The double *hasPart* relationships indicate that each theory class may be a part of multiple subcategories. The photograph of Commodore Grace Hopper (who created the term "debugging" a computer) in the figure represents a subject matter expert.

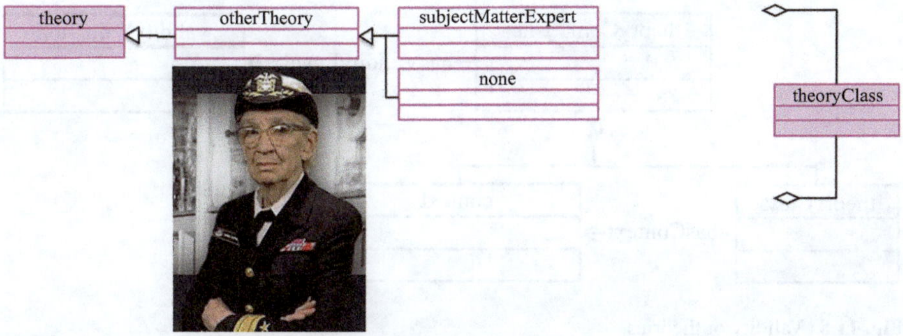

Fig. 11.7 Other Theories Ontology

Each class in Table 11.5 identifies a particular theory; however, that theory may be a class in more than one subcategory.

A *subjectMatterExpert* Theory is an *otherTheory* used by a subject matter expert. It is possible to have one of these theories for each expert on a project or several experts could agree on one theory or one expert might be responsible for several theories for different contexts.

A *none* Theory is an *otherTheory* with no referent, only a reason for lack of theory. These theories are important because they identify holes in the logic of a model. One example is the *noJustification* theory. This theory is used where an arbitrary or accidental connection is made between a cause and its effects. The other example is the *not-Modeled* theory that is used to identify a cause and effect relationship that is not modeled. This is used in verification and validation of a model.

Table 11.5 Other theories subcategories and classes

Subcategory	Discipline	# of classes
subjectMatterExpert	Subject matter expert	6
none	No referent	2

11.7 Validity of Theories

The word "Theory" has a number of connotations. In some sciences, a (proven within some context) statement of the results of some set of conditions is labeled a "law." A similar statement with greater generality will be accorded the label of "theory." In other domains, the word "theory" will be interchangeable with "hypothesis," indicating that it remains to be proved or that it has not been finally proven. We use the word "theory" in a loose way and add an estimate of validity, based on context, to differentiate among the various usages.

Figure 11.8 shows the relationship among theories, contexts, and validity. Each *theoryClass* has one or more *contexts*. Each pair forms a *theoryComposite*, which has a *validityValuation*.

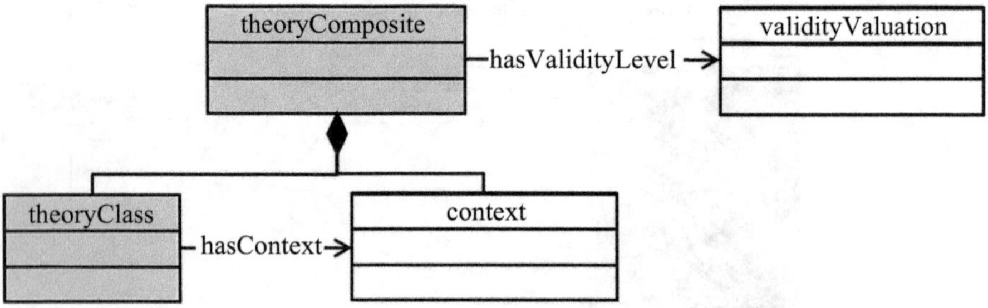

Fig. 11.8 Validity of theories

Contexts

A theory may be valid within one context and invalid within another. For example, Newton's laws are valid within everyday contexts, but not for systems approaching the speed of light. An economic theory might be valid within a capitalist economy, but not in a socialist economy. Another economic theory might be valid within both types of economies.

Validity Valuation Levels

The validity of the theories in the ontology should be treated as placeholders. In each case a general context was assumed, but not stated. The experts will need to decide what contexts are germane to each theory and how the theory should be assessed within each context. The validity assessment codes are shown in Table 11.6. They represent text anchors for a Likert-type assessment of the validity of the theories.

Note that "disproved" theories are accorded some value, as sometimes "disproved" theories are only out of favor and could later be accorded higher values. (Bayesian probability is an example.) Also, a "scientifically proved theory" may later undergo revisions; however, for these purposes we will regard both Newton's and Einstein's theories of gravitation as examples of Proved theories within appropriate domains of applicability (contexts).

Table 11.6 Theory validity codes

Type	Description	Value
Nil	Uncodified	0.0
Nil+	Codified but amorphous	0.5
WAG	"Wild Assed Guess" (WAG) or "disproved" theory	1.0
WAG+	WAG plus some science	1.5
SWAG	"Scientific Wild Assed Guess" (SWAG)	2.0
SWAG+	SWAG with some review or peer-reviewed with some negatives	2.5
PeerRvw	Peer-reviewed theory or well-reviewed with some negatives	3.0
PeerRvw+	Well-reviewed theory	3.5
Accepted	Generally accepted theory	4.0
Proved-	Close to proven theory	4.5
Proved	Scientifically proved theory	5.0

11.8 Theory Uses

There are three points of connection between the Theories Ontology and the Modern Conflict Ontology, all dealing with the Metrics.

Direct Connection to Metrics

As stated earlier, theories are called for to explain changes in Metric values resulting from Actions (or to justify creating those changes in a computer model). This requirement for theory is especially evident where changes to opinions are involved. The connections between Actions and results will be discussed in the Actor-Action-Result (AAR) section of Chap. 12.

Implicit Metric Models

Chapter 12 will also introduce the concept of an Actor's agenda. This agenda is logical, but not necessarily correct. It is based in part on a mental model of cause and effect and of the connections among state variables (Metrics). Each agenda is based on an implicit metric model. The instantiation of an agenda represents an instantiation of the Actor's beliefs concerning how Actions relate to Metrics and how the Metrics relate to the Actor's goals. These connections represent the Actor's implicit theories. By defining the Actor as a subject matter expert, these implicit theories could be codified as theories falling into the Other Theories category.

Figure 11.9 illustrates the structure of an Actor's Agenda. The Actor decomposes the Agenda into a set of Task-Goal pairs and decomposes each of these into Subtask-Subgoal pairs. Each Subtask is decomposed into Actions, and each Subgoal is decomposed into Metrics.

The Actor believes that accomplishing each Task will result in achieving its Goal and that accomplishing all of the Tasks will achieve the overall Intent. Similarly, the Actor believes that accomplishing each Subtask will result in achieving its Subgoal, and accomplishing all of the Subtasks of a Task will accomplish the Task and achieve that Task's Goal. The Actor believes that accomplishing all of the Actions of a Subtask will accomplish that Subtask. The Actor believes that the Metrics of a Subgoal will measure the achievement of the Subgoal.

Collectively, these beliefs can be stated by saying the Actor believes that a particular strategy will achieve the Actor's Intent and that this achievement can be followed through the values of the Metrics in his implicit metric model. Figure 11.10 shows a piece of such a metric model. The numbers in the boxes refer to the ID numbers of Metrics. Here the top box contains the Metric directly associated with a Goal. The gray boxes contain the Metrics directly associated with the Subgoals (divided by numbered dotted line boxes). The green boxes represent some of the other Metrics of the Subgoals (shown as metric boxes in Fig. 11.9). Thus the dia-

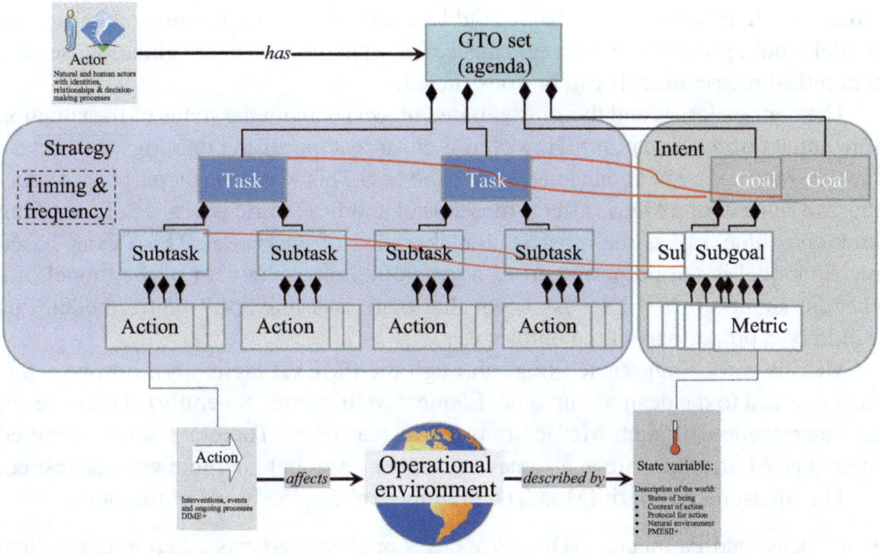

Fig. 11.9 Actor's Agenda

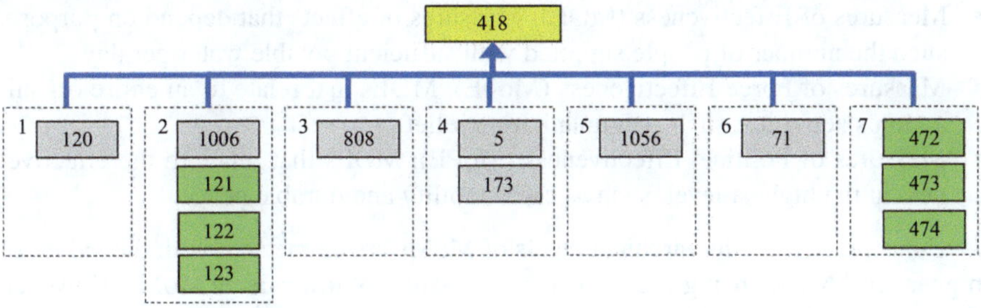

Fig. 11.10 Example of a partial implicit metric model

gram states that Metrics 1006 can be inferred from the values of Metrics 121, 122, and 123 and that Metric 418 can be inferred from the values of Metrics 120, 1006, 808, (5 and 173), 1056, 71, and (472, 473, and 474). In developing his Agenda, the Actor has picked these Metrics and associated them with his Goals and Subgoals, creating an implicit metric model.

Theoretical Metric Models

A theoretical metric model is similar to an Actor's metric model in structure; however, rather than describing the beliefs of one of the parties to the conflict, it uses the Metrics in the ontology and some explicit theoretical basis to describe the inferential connections among the Metrics for a generic modern conflict (or for some specific situation). If this theoretical basis could be validated, then the theoretical metric model would provide the "correct" answer. Absent any absolute validation, several theoretical metric models can be constructed.

The metric classes and their instantiated objects contain the status of the situation (or outputs of a simulation). However, a close reading of the ontology will reveal that the Metrics are not connected to one another! This was done on purpose. Saying that the value of one Metric affects the value of another Metric is a model. Describing how one value affects the other is an elaboration of the model. This theory-based metric model is not properly part of a general ontology but part of the model of a specific situation. If it later is shown that there is a universal metric model, this would be a proper part of the ontology.

Metrics can be aggregated (e.g., through the PMESII hierarchy); however, this may not lead to the desired outcome. Element Metrics are (generally) at a low level, and aggregation of such Metrics remains at that level. There are some specified aggregate Metrics that are at a somewhat higher level, but not quite what is desired.

The Measures of Merit (MoMs) hierarchy explains the levels of metrics:

- Dimensional Parameters (DP): Properties or characteristics inherent in physical entities, e.g., flow rate of a water purification unit under ideal conditions.
- Measures of Performance (MoP): Measures of direct results, e.g., flow rate of water purification unit in the field.
- Measures of Effectiveness (MOE): Measures of effects that depend on purpose, such the number of people supplied with sufficient potable water per day.
- Measures of Force Effectiveness (MoFE): MOEs that relate to an entire organization's activities, such as humanitarian relief.
- Measures of Political Effectiveness (MoPE): MOEs that relate to the effectiveness at the highest level, such as civil stability and durable peace.

Conceptually, the hierarchical levels of MoMs are connected, with the values of higher-level MoMs being determined by the values of lower-level MoMs. However, there are practical problems:

- The values of MoPs may (sometimes) be either deterministically or probabilistically calculable from a set of related DPs. For example, there may be a theory that connects the flow rate of a water purification unit in the field that is based on the flow rate under ideal conditions and the different conditions that obtain in the field. Alternatively, there may be a set of charts that give ranges of values that are based on testing. Probabilistic predictions can be derived from these charts.
- Inferring the values of MOEs from MoPs becomes more problematic, partly because the identification of the proper MoPs that should be used is often difficult and partly because the theoretical bases for making the connections are more difficult to determine and apply.
- Inferring MoFE values from MOEs and MoPE values from MoFE values is even more difficult. MoFEs and MoPEs depend on human factors to a very large extent, contributing to the difficulty of inferring values.

These examples show the need for theory-based inferences. However, some of the "theories" may not be codified in textbooks. This is the reason that the subject matter expert subcategory of theories exists. Some inferences will be based on what an expert says is the case, not on some peer-reviewed theory.

The theoretical metric model differs from the implicit metric models of the agendas. Those metric models represent the beliefs of the Actors, not necessarily the true inferential relations among the Metrics. The theoretical metric model is meant to represent the true relations, based on theories and, ultimately, on reality.

Figure 11.11 shows a partial theoretical metric model, asserted in the Interim Semi-static Stability Model (ISSM) (Hartley 2006b):

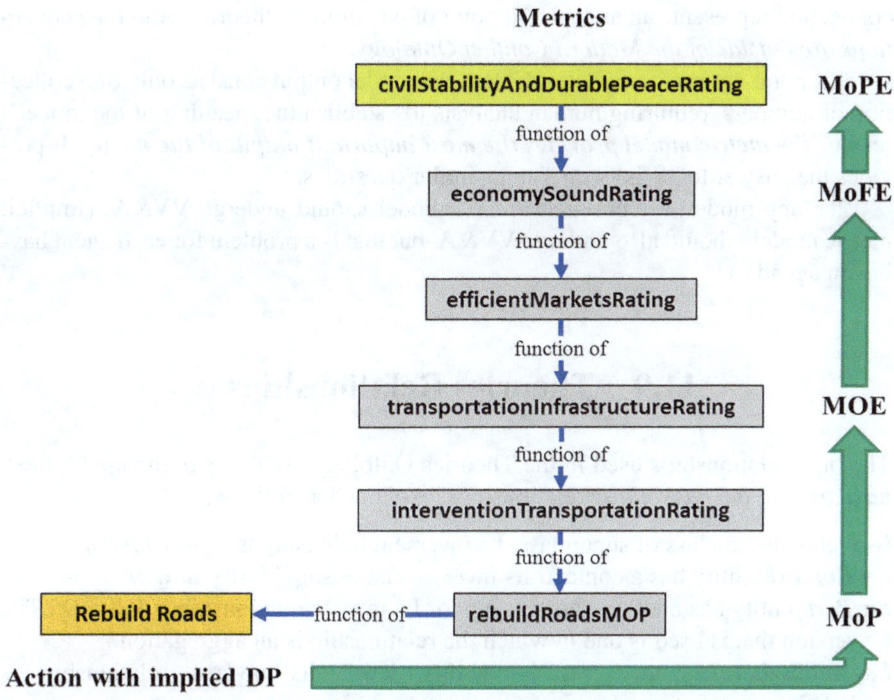

Fig. 11.11 Example of a partial theoretical metric model

- The Metric (an MoPE), which measures the success of the model (from the US point of view), *CivilStabilityAndDurablePeaceRating*, is a function of
- The MoFE *EconomySoundRating*, which is a function of
- The MOE *EfficientMarketsRating*, which is a function of
- The MOE *InterventionTransportationRating*, which is a function of
- The MOE *InterventionTransportationRating*, which is a function of
- The MOP *RebuildRoadsMOP*, which is a function of
- The DP of the Action, "Rebuild Roads"

The import of this model is that you can infer the value of the highest-level metrics, measures of political effectiveness (MoPEs), from the values of more accessible metrics in a real situation. And you can do the same thing in a computer model where the calculated values of the "accessible" metrics can be based on theories with relatively high validity. The MoPEs are the Metrics that represent answers to the questions of those responsible for dealing with a situation (or that the model is designed to answer).

Creating a theoretical metric model requires discipline and subject matter experts in several fields. It is not technically difficult because the ontology provides all the elements that need to be connected. However, it is very important because this is where the assertion is made that a certain state of the situation is better or worse than another state. Each connection requires a call to a specific theory or set of theories as justification. ***This is a key part of model design***. That means that each metric connection represents an attachment point of one or more theories. *And the connections are outside of the Modern Conflict Ontology.*

This point deserves emphasis. Unless the model output consists only of a collection of numbers, requiring human analysis to establish the meaning of the model's results, ***the metric model provides the most important output of the model***. It provides the answer to "Who won?" and similar questions.

Like any model, a theoretical metric model should undergo VV&A. (Implicit metric models should also undergo VV&A, but that is a problem for each agent having an agenda.)

11.9 Theories Relationships

The only relationships used in the Theories Ontology are the *is-a*, the *hasContext*, *hasPart*, and the *hasValidity* relationships, explained as follows:

is-a: class is subclass of superclass. Its inverse relationship is *superClassOf.*
hasContext: entity has a context. Its inverse relationship is *isContextOf.*
hasPart: entity has another entity as a part. Its inverse relationship is *isPartOf.* The version that is used is one in which the relationship is an aggregation.
hasValidityLevel: entity has a validity level. Its inverse relationship is *isValidityLevelOf.*

11.10 Theories Recapitulation

In this chapter we described the structure of the ontology, its contents and their validity, and its uses with respect to model validity. The Theories Ontology is a work in progress. It contains a structure and a set of classes. However, the contents require review, emendation, and additions by experts in the various disciplines before it can be regarded as nearly complete.

Chapter 12
MCO Scenario Ontologies

In the Modern Conflict Ontology (MCO), there are parts that don't depend on the scenario or situation. These situation-independent parts apply regardless of the situation. For example, the *genericAnimal* class is available for use in every scenario, even if it is never instantiated. This chapter deals with the parts of the MCO that are situation-dependent, those that depend on and describe the scenario.

The elements of the MCO, the Actor, Object, and Action classes, are its infrastructure; however, the heart of the description of conflict lies in the Goal-Task-Owner (GTO) sets of the ontology, described in this chapter. Conflicts arise between two or more Actors – but there are reasons for the conflicts. Perhaps the agendas of the parties have areas that are mutually incompatible, or perhaps the Actors don't like each other and develop agendas to do something about the mutual dislike. The GTO sets represent these agendas.

12.1 Overview of Scenario Ontologies

In the MCO there are five Scenario Ontologies under the *scenarioStructure* class: scenario identification, GTO sets, scenario rules, Actor-Action-Result (AAR) sets, and scenario relations, as illustrated in Fig. 12.1.

Three concepts are necessary to understanding the Scenario Ontologies: scenarios, owners, and entity-entity relationships.

D. S. Hartley III, *An Ontology of Modern Conflict*, Understanding Complex Systems, https://doi.org/10.1007/978-3-030-53214-7_12

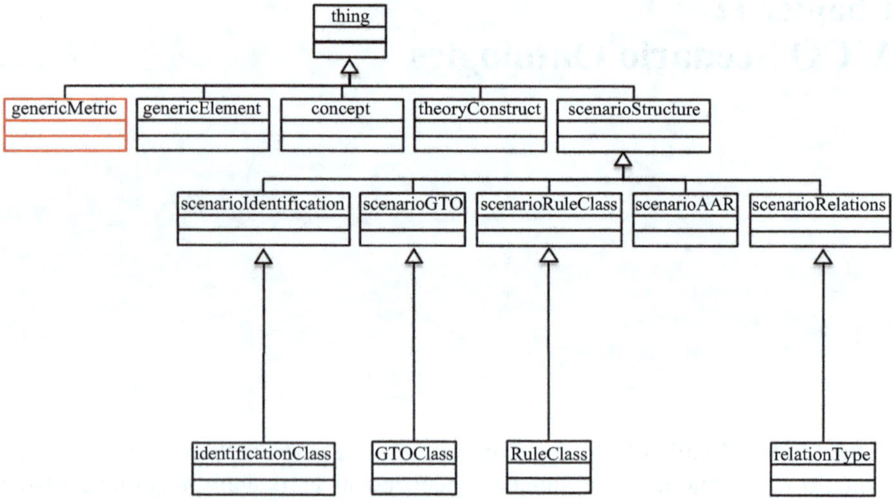

Fig. 12.1 The Scenario Ontologies

Scenarios

In this chapter, we will use the words "scenario" and "model" interchangeably. A prospective description of a conflict situation is more comfortably labeled a "scenario," while a description of an ongoing conflict is a "model," which emphasizes the imperfection of the description.

Owners

An Owner is an Actor with so much influence or power in the scenario that his desires need to be accounted for. An Owner is generally an organization, such as Al Qaeda, the Afghan National Police (or a subset thereof), the US State Department, or the Host Nation Military; however, the Owner could be a particular person. In the MCO, the identity is provided by text labels that are tied to a particular Actor class of the appropriate type, making it an instance of the Actor class. The Owner also has an *Intent* – another text label. This *Intent* is defined in the ontology by an instance of the GTO set structure. This all means that a particular owner can appear in multiple scenarios. Alternatively, a particular organization (or person) can be represented by several different owners in different scenarios. This latter case allows for representing different *Intents*.

Entity-Entity Relationships

Between the instances of two entities, there can be a relationship (see Chap. 13); however, there is more than just the bare relationship. The instances require a context to be meaningful. Entity-entity relationships provide the context and name the relationship.

Changes to Scenario Ontologies from the UCO

In the Unconventional Conflict Ontology (UCO), the Scenario Ontologies were referred to as situation-dependent ontologies. The concept is the same, only the name is different. In the MCO, the (very brief) Scenario Identification Ontology has been added to provide an externally defined scenario that can be used in the other Scenario Ontologies so that each can refer to the same scenario.

In the MCO, many of the diagrams have been changed from those in the UCO to provide a better representation of the concepts involved and, in some cases, to represent the structural changes to the ontologies. Further, sample instantiations have been added to support a better understanding of the ontologies.

An Insight into Scenario Ontologies

We will consider three scenarios to illustrate the concepts in this chapter. The first is a simple war, which introduces some of the concepts without too much detail. The second is a complex irregular war, which illustrates just how complex such conflicts can be. The third is a simple disaster relief scenario, which introduces nature as a party to a conflict.

12.2 Scenario Identification Ontology

The first of the five scenario ontologies is the Scenario Identification Ontology. Figure 12.1 showed how the Scenario Identification Ontology fits into the overall MCO structure (with Fig. 12.2 illustrating its icon as used in the context diagram, Fig. 3.5).

Fig. 12.2 Scenario identification icon

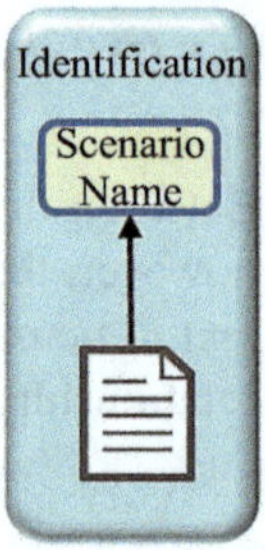

The Scenario Identification Ontology is a very simple ontology. Its purpose is to provide the structure and referent to instantiate scenarios.

Each model or scenario requires identifying information – at a minimum a unique identifying number, called the *ModelID*. We provide other information that is useful to humans, not just computers. This information includes a name, such as "*notionalModelSimpleWar*"; a sequence number to ensure that the different scenarios appear in a desired order; the scenario authors; a (possibly) more descriptive title, such as "Hartley Notional Simple War"; and a date. This information is part of the *scenarioID* class shown in Fig. 12.3. This class has a relationship (*hasCitation*) with the other *scenarioIdentification* class, *citationMetadata*. The citation contains information about the person responsible for the scenario, a (possibly) longer title for the scenario, a date for the citation, and other information.

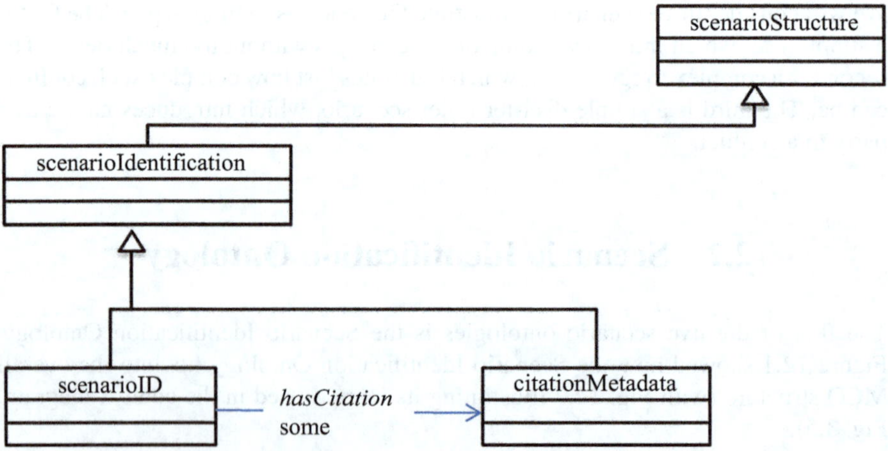

Fig. 12.3 Scenario Identification Ontology

Table 12.1 shows a set of scenario instances, with part of the information contained about each. The cite column contains the link to the citation data. The last three instances are the scenarios referred to in the "insight" subsection, above.

Table 12.1 ScenarioID ontology instances

Model ID	ScenarioName	ScenarioDate	Seq	Cite
1	*IWOntologyTRAC2011Workshop2*	4/27/2011	1	*OntWS2*
2	*notionalModelHartley001*	5/31/2011	2	*Hartley001*
3	*IWOntologyTRAC2011Workshop3*	7/15/2011	3	*OntWS3*
11	*notionalModelHartley002*	11/23/2011	4	*Hartley002*
12	*IWOntologyTRAC2012*	1/1/2013	5	*TRACOnt2*
13	*notionalModelSimpleConventionalWar*	9/10/2018	6	*Hartley003*
14	*notionalModelComplexIrregularWar*	9/10/2018	7	*Hartley004*
15	*notionalModelSimpleDisasterRelief*	9/10/2018	8	*Hartley005*

12.3 GTO Sets Ontology

The second of the five scenario ontologies is the GTO Sets Ontology. Figure 12.1 shows how the GTO Sets Ontology fits into the overall MCO structure (with Fig. 12.4 illustrating its icon).

GTO sets define the agendas of the various parties to a conflict. The "O" stands for the owner of the agenda (generally an organization). The "G" and "T" stand for the goals and tasks the owner believes will accomplish the goals that together comprise the owner's agenda. GTO sets are a generalization of the Line of Effort (LOE) concept described in Joint Pub 5-0 (Chairman of the Joint Chiefs of Staff 2017). LOEs are defined for US joint operations, whereas GTO sets are defined for any party to a conflict.

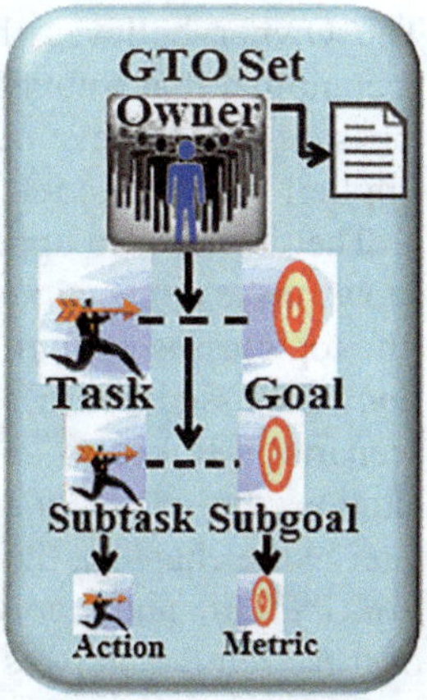

Fig. 12.4 GTO sets icon

Figure 12.5 illustrates the structure of an Owner's Agenda. The Owner decomposes the Agenda into a set of Task-Goal pairs and decomposes each of these into Subtask-Subgoal pairs. Each Subtask is decomposed into Actions and each Subgoal is decomposed into Metrics. The Task decomposition, along with timing and frequency of implementation, constitutes the Owner's **Strategy**.

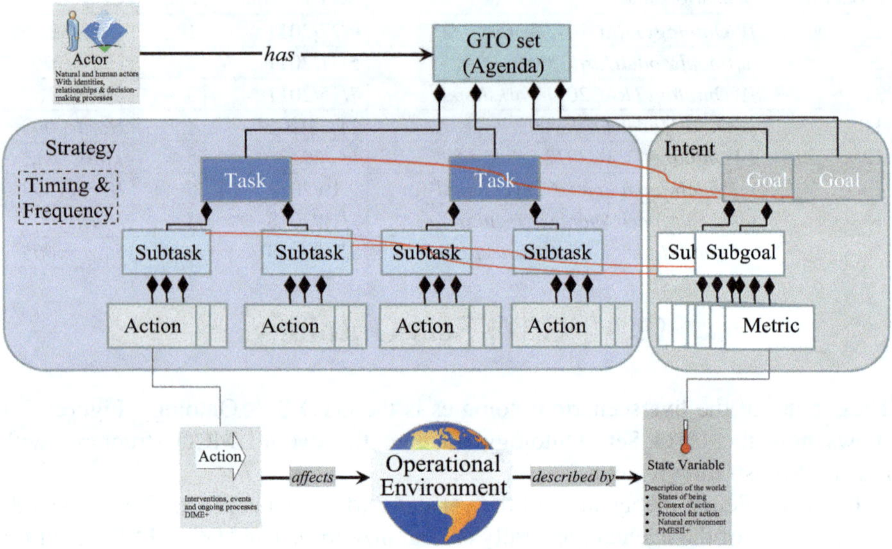

Fig. 12.5 Owner's agenda

The Owner believes that each accomplishing each Task will result in achieving the corresponding Goal and that accomplishing all of the Tasks will achieve the overall **Intent**. Similarly, the Owner believes that accomplishing each Subtask will result in achieving the corresponding Subgoal and accomplishing all of the Subtasks of a Task will accomplish the Task and achieve that Task's Goal. The Owner believes that accomplishing all of the Actions of a Subtask will accomplish that Subtask. The Owner believes that the Metrics of a Subgoal will measure the achievement of the Subgoal. Together, these beliefs constitute an Implicit Metric Model, described in Chap. 11.

The Owner also appears in other parts of the Scenario Ontologies. Figure 12.6 shows these connections. In this figure, four of the subclasses of *scenarioStructure* (the top classes of four of the scenario ontologies) are shown: *scenarioIdentification* (discussed above), *scenarioRuleClass*, *scenarioGTO*, and *scenarioAAR*. The scenario is linked to an owner by a relationship. The owner is also linked to the citation, the *ownerRuleSet* class, a subclass of the *scenarioRuleClass* (discussed later in the chapter), the *scenarioAAR* class (discussed later in the chapter), and the *taskGoalPair* and *actorReference* classes (discussed in this section). The meaning of these relationships is that a particular owner depends on the particular scenario and the owner has rule sets, GTO sets, and *scenarioAARs*.

Agendas are not programming prescriptions for a computer model. They do (at the finest level) describe actions that can be implemented in computer code; how-

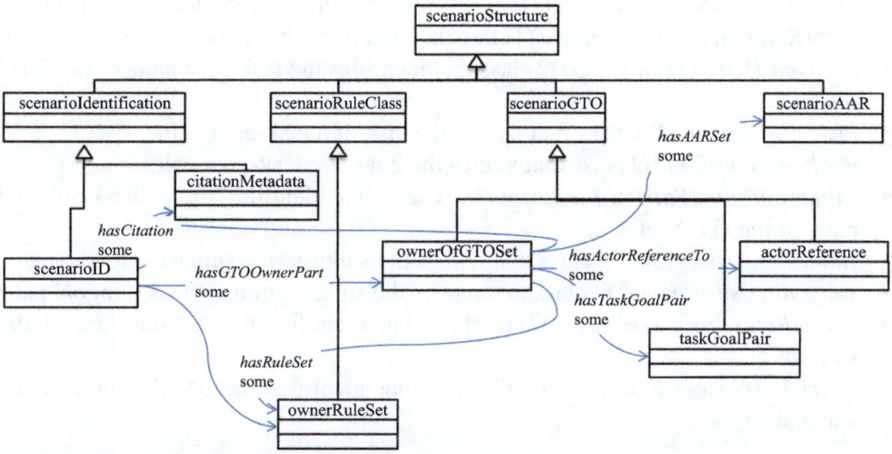

Fig. 12.6 Scenario GTO sets, owner-centered

ever, they are not meant to describe a set of simultaneous or sequential actions that the owner will take. Rather, they are a menu of possible actions, some of which will never take place because the unfolding situation will not require them. Some may be undertaken multiple times in multiple locations, mixed in with other actions, as determined by the evolving situation. Further, the owner may revise his agenda, partially or completely, at any time.

Figure 12.7 shows the details of the GTO Sets Ontology. The ten subclasses of the *scenarioGTO* class are the following:

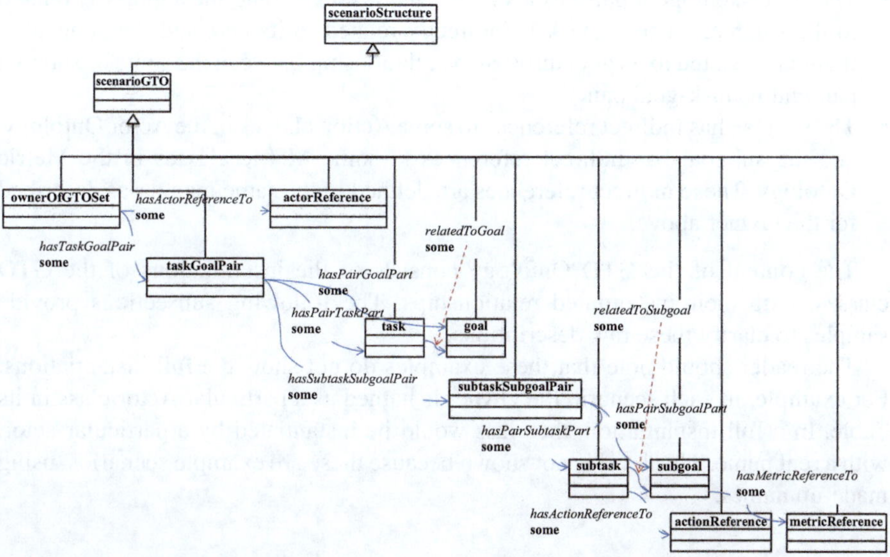

Fig. 12.7 GTO Ontology

- *ownerOfGTOSet*: a *scenarioGTO* class that identifies the owner of the GTO set.
- *actorReference*: a *scenarioGTO* class that identifies the Actor class of the owner.
- *taskGoalPair*: a *scenarioGTO* class that identifies the task-goal pairs of the GTO set.
- *task*: a *scenarioGTO* class that names the task of a task-goal pair.
- *goal*: a *scenarioGTO* class that names the goal of a task-goal pair.
- *subtaskSubgoalPair*: a *scenarioGTO* class that identifies the subtask-subgoal pairs of the GTO set.
- *subtask*: a *scenarioGTO* class that names the subtask of a subtask-subgoal pair.
- *subgoal*: a *scenarioGTO* class that names the subgoal of a subtask-subgoal pair.
- *actionReference*: a *scenarioGTO* class that identifies the Action class of the subtask.
- *metricReference*: a *scenarioGTO* class that identifies the Metric class of the subgoal.

The *is-a* relationships connect the *scenarioGTO* subclasses to the *scenarioGTO* class and are the primary structural relationships. The other relationships shown in the figure complete the definition of the GTO Ontology at the class level.

- The Owner has an indirect reference to an Actor class in the Actor Ontology. This is created by a *hasActorReferenctTo* relationship with the *actorReference* class. Upon instantiation, the *actorReference* instance points to an instance of an Actor class. This indirection is necessary because the particular Actor class that will be specified is only known for a given instance and thus cannot be specified in a class diagram, such as Fig. 12.7. The Owner also has some task-goal pairs.
- The task-goal pair has a task and a goal and the task is related to the goal. The task-goal pair also has some subtask-subgoal pairs.
- The subtask-subgoal pair has a subtask and a subgoal and the subtask is related to the subgoal. (The subtask is indirectly related to its task and the subgoal is indirectly related to its goal through the relationship between the subtask-subgoal pair and its task-goal pair.)
- The subtask has indirect references to some Action classes in the Actor Ontology, and the subgoal has indirect references to some Metric classes in the Metric Ontology. These indirect references are defined in the same manner as described for the Owner above.

The content of the GTO Ontology consists of the instantiations of the GTO classes, with their instantiated relationships. The following subsections provide samples to clarify these raw descriptions.

The reader should note that these examples do not show the full instantiations. For example, in each scenario the Owner is named as a particular Actor class in its Table. In a full instantiation, the class would be instantiated by a particular actor, with a real name. This level is not shown because these are example scenarios, using made-up names.

The Simple War Scenario

The simple war scenario has only two parties to the conflict. The owner of the first party is labeled the "*InterventionForce*." Because this is a notional model, the *ownerName* is simply "Intervention Force"; however, this could be something describing a particular coalition of nations or a particular nation's military. In this scenario, the Actor class that is cited is the simple *significantGroupActor*, → *armedForce*, →*interventionForceOrganization*, where "→" represents the subclass relationship. The other scenarios will refer to more complicated *hierarchicalActor* classes. The owner of the second party to the conflict is labeled the "*HNMilitary*," with ownerName "Host Nation Military," and refers to the simple *govtMilitaryForceOrganization* class (with the same superclasses).

As shown in Table 12.2, the intent of the first party is to invade the second party's country, defeat its military, and pacify the country. The intent of the second party is to resist and defeat the first party. This scenario models the initial, combat phase of the second Iraq War.

Table 12.2 Simple war scenario instantiation

Scenario	Owner	Intent
Conventional War Scenario (sketch)	*InterventionForce*	Defeat HN and establish security
	HNMilitary	Defeat military adversary

The InterventionForce

The *InterventionForce* has divided its agenda into two Goal-Task pairs:

1. **Goal:** Intervention defeats the opponent, the Host Nation (HN).
   Task: Intervention conducts general combat operations (against HN). Description: Intervention conducts offensive and defensive combat operations of modern conflict; opponent is called HN.
2. **Goal:** Safe, secure, and sTable environment in HN is established under Intervention control.
   Task: Intervention establishes civil security in HN (secure area of operation (AO)). Description: Intervention conducts combat operations to secure the area of operations.

These tasks and goals are subdivided into subtasks and subgoals, which are subdivided into Actions and Metrics. The *InterventionForce* believes that accomplishing these tasks, subtasks, and Actions will result in achieving the stated goals.

The HNMilitary

The *HNMilitary* has also divided its agenda into two Goal-Task Pairs:

1. **Goal:** Safe and secure military environment advanced under HN control.
   Task: HN conducts military operations (unique independent of intervention subtasks). Description: HN conducts military and intelligence operations independent of any intervention forces.
2. **Goal:** Safe and secure environment advanced.
   Task: Conduct military operations (core subtasks). Description: Conduct military and intelligence operations irrespective of independence of HN.

These two tasks and their descriptions are worded somewhat strangely. The reason is that the MCO contains instances of Tasks (and Goals, Subtasks, and Subgoals) that are derived from the original inputs of the people of TRAC who helped develop the Irregular Warfare (IW) Ontology (IWO) (Hartley and Lacy, Irregular Warfare (IW) Metrics Ontology Final Report, TRAC-H-TR-13-020, 2011, IW Ontology Final Report, 2013b). The first instance definitions were for a scenario in which an Intervention Force is helping to develop and train the Host Nation military. Subsequently, the Task was divided into a Task with Subtasks which the HN performed together with the Intervention, a Task with Subtasks that each party performed (core Subtasks), and a Task with Subtasks which the HN performed independently. The Task wording in parentheses was added to distinguish the three parts. A similar division was performed with respect to law enforcement Tasks.

This will be seen in the complex irregular war scenario in which there is a Task for the Intervention to conduct operations, a Task that is a set of core Subtasks that both the Intervention and the HN military each have, and a Task for the HN military to conduct operations independent of the Intervention. These latter two Tasks consist of the Subtasks that are necessary in this scenario and are thus used here.

These Tasks and Goals are subdivided into Subtasks and Subgoals, which are subdivided into Actions and Metrics. The *HNMilitary* believes that accomplishing these tasks, subtasks, and Actions will result in achieving the stated goals.

The Simple Disaster Relief Scenario

The simple disaster relief scenario has four parties, each with its own intent, as shown in Table 12.3. Three of these parties are human organizations and one is a natural (disaster) force.

Table 12.3 Simple disaster relief scenario instantiation

Scenario	Owner	Intent
Disaster Scenario (sketch)	*Disaster*	Create disruption
	Intervention	Help HN respond to disaster
	HNLawEnforcement	Conduct operations in response to HN disaster
	InternationalNGO	Provide social services in response to HN disaster

The Disaster

The *Disaster* has a simple agenda: create destruction. It might be a purely natural disaster, such as a hurricane, or a disaster that has a human cause, but acts through natural forces, such as a wildfire or chemical spill. The owner is represented by a *sideInConflict* class for this general scenario. In a specific instantiation, it might be a *storm* class, a *fireOrWildfire* class, a *manmadeDisaster* class, etc.

The *Disaster* has one Goal-Task pair.

1. **Goal:** Create destruction and inconvenience.
   Task: Natural forces conduct events. Description: Natural forces conduct disruptive events.

The InterventionForce

The agenda of the *InterventionForce* is to restore order, restore essential services, and communicate with the public. Its owner is represented by an *interventionForce-Organization* class.

The *InterventionForce* has divided its agenda into three Goal-Task pairs:

1. **Goal:** Safe, secure, and sTable environment in HN is established under Intervention control.
   Task: Intervention establishes civil security in HN (secure AO). Description: Intervention conducts combat operations to secure the area of operations.
2. **Goal:** Essential services are restored.
   Task: Intervention restores essential services. Description: Intervention restores sewage, water, electrical power, academic institutions, trash removal, and medical services.
3. **Goal:** Increased support for HN government and Intervention is established.
   Task: Intervention gathers and disseminates information about HN. Description: Tell the Story to the US public; marginalize insurgent influence (if any); isolate insurgents; provide context for HN operations; reinforce HN legitimacy; and create division between insurgent leadership and armed insurgents.

The HNLawEnforcement

In this scenario, the *HNLawEnforcement*, which includes the Host Nation Law Enforcement and First Responders, has a policing agenda in concert with the Intervention Force. The owner is represented by an *organizationalUnit* class. This class is a *hierarchicalActor* that has a *keyLeader* class and a *significantGroup* class and may have *organizationalUnit* classes as components. That is, the entire hierarchy of the Host Nation Law Enforcement and First Responders can be represented to the desired level of granularity.

The *HNLawEnforcement* has divided its agenda into two Goal-Task pairs. These parallel the first one of the *InterventionForce*, aimed at establishing civil security.

1. **Goal:** Rule of Law advanced under joint HN and Intervention control.
   Task: HN conducts law enforcement and first responders operations (unique in concert with intervention force subtasks). Description: HN conducts law enforcement and first responders working with intervention forces.
2. **Goal:** Rule of Law advanced under HN control.
   Task: HN conducts effective and independent law enforcement and first responders operations (unique independent of intervention subtasks). Description: HN conducts law enforcement and first responders operations independent of any intervention forces.

The InternationalNGO

The *InternationalNGO* represents the many non-governmental organizations that arrive in conflicts with the agenda of helping the people of the Host Nation. The owner is represented by an *NGOOrganization* class.

The *InternationalNGO* has one Goal-Task pair.

1. **Goal:** Humanitarian emergency is ended.
   Task: Distribute food, shelter, and health services. Description: Provide short-term immediate needs in emergencies.

The Complex Irregular War Scenario

The complex irregular war scenario has 15 parties, each with its own agenda, as shown in Table 12.4. Each agenda seems logical in itself; however, remember that the agendas of some parties will run in concert (at times) and the agendas of some parties will be in conflict. Thus, it is virtually impossible for any agenda to be accomplished precisely as envisioned by its owner.

Table 12.4 Complex irregular war scenario instantiation

Scenario	Owner	Intent
Irregular War Scenario (detailed)	*InterventionForce*	Standard US intent for unsTable HN
	HNMilitary	Work with Intervention Force and Intervention Organization to provide HN military functions
	HNLawEnforcement	Work with Intervention Force and Intervention Organization to provide HN law enforcement and first responder functions
	ProvincialForce	Secure local autonomy
	InterventionOrganization	Coordinate intervention operations (diplomatic and military) to "make things better" and gain personal political credit
	HNGovernment	Ensure HN government exists and functions in any way necessary and obtain personal advantages
	HNPoliticalReformer	Get elected and reform HN government
	InternationalNGO	Provide social services and increase sponsor donations
	InternationalContractor	Make money by repairing infrastructure and providing other services
	InternationalCorporation	Vulture Capitalist makes money by taking advantage of the situation
	AdversaryNGO	Provide social services and promote ideology
	HNCriminalOrganization	Make as much money as possible, as fast as possible, through means fair and foul
	HNSeparatistOrganization	Split off part of HN and form own state
	HNRevolutionaryOrganization	Take over the HN government by any means, including revolutionary war
	PatronCountry	Turn HN into client country
	TerroristOrganization	Take over the HN government by subversion, terror, etc., with the possible future goal of combining with other subverted areas

The InterventionForce

In this scenario, the Intervention Force has a peace operation agenda, together with humanitarian assistance and nation-building elements. The owner is represented by an *armedForceUnit* class, which is a *hierarchicalActor* that has a *keyLeader* class, an *armedForce* class, and a *milSystemsGroup* class and may have *armedForceUnit* classes as components. That is, the entire hierarchy of the Intervention Force, together with military systems, can be represented to the desired level of granularity. Note that is party is different from the *InterventionOrganization*, which controls it and which is described later.

The *InterventionForce* has divided its agenda into 11 Goal-Task pairs. The first goal is to ensure an environment that supports the rest of the goals.

1. **Goal:** Safe, secure, and sTable environment in HN is established under Intervention control.
   Task: Intervention establishes civil security in HN (secure AO). Description: Intervention conducts combat operations to secure the area of operations.

The next two goals concern establishing a useful HN military and intelligence service, with an intervention-unique task and a joint task with the HN military.

2. **Goal:** HN military and intelligence services functioning and compliant with both domestic laws and international human rights norms.
   Task: Intervention supports development of HN military and intelligence services (intervention unique subtasks). Description: Intervention supports the HN in developing HN military and intelligence services.
3. **Goal:** HN military and intelligences services are created and maintained.
   Task: Create and maintain HN military and intelligence services (core subtasks). Description: Create or reform and maintain the HN military and intelligence services.

The next goal is analogous to the first goal, but pertains to law enforcement.

4. **Goal:** Rule of Law established in HN under Intervention control.
   Task: Intervention establishes civil control in HN (external police-type operations). Description: Intervention conducts police-type operations.

The next two goals are analogous to goals 2 and 3, but pertain to HN law enforcement and first responders.

5. **Goal:** HN law enforcement functioning and compliant with both domestic laws and international human rights norms.
   Task: Intervention supports development of HN law enforcement and first responders (intervention unique subtasks). Description: Intervention supports the HN in developing HN law enforcement and first responders.
6. **Goal:** HN law enforcement and first responders are created and maintained.
   Task: Create and maintain HN law enforcement and first responders (core subtasks). Description: Create or reform and maintain the HN law enforcement and first responders.

Goals 7 and 8 cover serving the immediate needs of the populace.

7. **Goal:** Humanitarian emergency is ended.
   Task: Distribute food, shelter, and health services. Description: Provide short-term immediate needs in emergencies.
8. **Goal:** Essential services are restored.
   Task: Intervention restores essential services. Description: Intervention restores sewage, water, electrical power, academic institutions, trash removal, and medical services.

Goals 9 and 10 cover the intervention part in supporting HN economic and social development.

9. **Goal:** Economic foundation with sufficient infrastructure established.
   Task: Intervention supports economic, social, and infrastructure development (intervention unique subtasks). Description: Intervention provides public works, commerce, civilian supply, civilian healthcare, and agriculture support and coordinates civic assistance programs.
10. **Goal:** HN economy is functioning well.
   Task: Promote economic, social, and infrastructure development (core subtasks). Description: Assure that the economy is strong and that social development is possible.

The final goal covers the need for popular support for the intervention efforts.

11. **Goal:** Enjoy a positive perception of the force by the populace.
   Task: Develop/keep trust of population. Description: Develop and keep the trust of the population in the organization.

These tasks and goals are subdivided into subtasks and subgoals, which are subdivided into Actions and Metrics. The *InterventionForce* believes that accomplishing these tasks, subtasks, and Actions will result in achieving the stated goals.

The HNMilitary

In this scenario, the *HNMilitary*, which includes the Host Nation Military and Intelligence Services, has a military control agenda in concert with the Intervention Force. It also has a private sub-agenda of self-aggrandizement and preparing to lose. The owner is represented by an *armedForceUnit* class, just as the Intervention Force is.

The *HNMilitary* has divided its agenda into nine Goal-Task pairs. The first two goals parallel goals 2 and 3 of the *InterventionForce*, aimed at creating/reforming the military and intelligence services and ensuring civilian control. (Note that Goal 2 is the same as the *InterventionForce* Goal 3, the difference being who is acting.)

1. **Goal:** HN military and intelligence services are professional, effective, under civilian control, and able to conduct full spectrum of military operations without significant outside support.
   Task: HN controls effective, independent, and professional military and intelligence services (HN unique subtasks). Description: HN controls armed forces and intelligence services that are effective, independent, and professional that can support the country's needs.
2. **Goal:** HN military and intelligences services are created and maintained.
   Task: Create and maintain HN military and intelligence services (core subtasks). Description: Create or reform and maintain the HN military and intelligence services.

The next three goals concern military operations. Task 3 contains subtasks that are performed in concert with the intervention. Task 4 contains subtasks that are performed whether or not they are in concert with the intervention. And Task 5 contains subtasks that are performed by the HN independently. The concept is that at first Task 3 and Task 4 subtasks are performed. Once the *HNMilitary* is competent, this shifts to subtasks from Task 4 and Task 5. The corresponding *InterventionForce* subtasks will be largely from its Task 1.

3. **Goal:** Safe and secure military environment advanced under joint HN and Intervention control.
   Task: HN conducts military operations (unique in concert with intervention force subtasks). Description: HN conducts military and intelligence operations working with intervention forces.
4. **Goal:** Safe and secure environment advanced.
   Task: Conduct military operations (core subtasks). Description: Conduct military and intelligence operations irrespective of independence of HN.
5. **Goal:** Safe and secure military environment advanced under HN control.
   Task: HN conducts military operations (unique independent of intervention subtasks). Description: HN conducts military and intelligence operations independent of any intervention forces.

The next two *HNMilitary* goals concern obtaining and keeping popular and tribal support.

6. **Goal:** Enjoy a positive perception of the organization by the populace.
   Task: Develop/keep trust of population. Description: Develop and keep the trust of the population in the organization.
7. **Goal:** No effective tribal threat to entity organization government.
   Task: Support friendly tribal elements. Description: Enable entity-friendly tribal elements to maintain power and influence.

The final two goals reflect the reality of venal personnel and organizations and the fear of losing to the adversary.

8. **Goal:** Ensure resources to accommodate accepTable level of patronage.
   Task: Promote patronage network. Description: Accumulate resources to help patronage network.
9. **Goal:** Ensure permanent seat at the Table no matter who is in charge.
   Task: Prepare for possible adversary organization takeover. Description: Prepare for possible adversary organization takeover of HN.

The HNLawEnforcement

In this scenario, the *HNLawEnforcement*, which includes the Host Nation Law Enforcement and First Responders, has a policing agenda in concert with the Intervention Force. It also has a private sub-agenda of self-aggrandizement and preparing to lose. The owner is represented by an *organizationalUnit* class, which is similar to that of the two military forces above, but doesn't have a military systems component.

The *HNLawEnforcement* has divided its agenda into seven Goal-Task pairs. The first two goals parallel goals 5 and 6 of the *InterventionForce*, aimed at creating/reforming law enforcement and first responders and ensuring civilian control. (Note that Goal 2 is the same as the *InterventionForce* Goal 6, the difference being who is acting.)

1. **Goal:** HN law enforcement and first responders are effective, under civilian control.
   Task: HN controls effective law enforcement and first responders (HN unique subtasks). Description: HN controls law enforcement and first responders that are effective.
2. **Goal:** HN law enforcement and first responders are created and maintained.
   Task: Create and maintain HN law enforcement and first responders (core subtasks). Description: Create or reform and maintain HN law enforcement and first responders.

The next two goals concern operations of law enforcement and first responders. Task 3 includes subtasks done in concert with the Intervention, and Task 4 includes those done independently.

3. **Goal:** Rule of Law advanced under joint HN and Intervention control.
   Task: HN conducts law enforcement and first responders operations (unique in concert with intervention force subtasks). Description: HN conducts law enforcement and first responders working with intervention forces.
4. **Goal:** Rule of Law advanced under HN control.
   Task: HN conducts effective and independent law enforcement and first responders operations (unique independent of intervention subtasks). Description: HN conducts law enforcement and first responders operations independent of any intervention forces.

The fifth goal covers the need for popular support for the law enforcement and first responder efforts.

5. **Goal:** Enjoy a positive perception of the organization by the populace.
   Task: Develop/keep trust of population. Description: Develop and keep the trust of the population in the organization.

The final two goals reflect the reality of venal personnel and organizations and the fear of losing to the adversary. (These are the same as the ones for the *HNMilitary*. The only difference is who is being venal.)

6. **Goal:** Ensure resources to accommodate accepTable level of patronage.
   Task: Promote patronage network. Description: Accumulate resources to help patronage network.
7. **Goal:** Ensure permanent seat at the Table no matter who is in charge.
   Task: Prepare for possible adversary organization takeover. Description: Prepare for possible adversary organization takeover of HN.

The ProvincialForce

The *ProvincialForce*, or Host Nation Provincial Force, consists of a provincial governor or warlord who has decided he needs a private militia and private security force. He might later decide to become a revolutionary, taking over the government, or a separatist, creating his own state. However, neither option is included here. The owner is represented by an *organizationalUnit* class.

The *ProvincialForce* has only one Goal-Task pair.

1. **Goal:** Ensure Actor security.
   Task: Create and Maintain Organizational or Personal Security. Description: Hire or organize private security organization or militia.

The InterventionOrganization

In this scenario, the *InterventionOrganization*, which consists of the Intervention Diplomatic Organization, has the political agenda of creating/reforming the Host Nation Government, providing humanitarian and development assistance, and eliminating the narcotics industry. It also has a private sub-agenda of self-promotion. The owner is represented by an *organizationalUnit* class.

The *InterventionOrganization* has divided its agenda into nine Goal-Task pairs. The first goal is to ensure a Host Nation governmental environment that supports the rest of the goals.

1. **Goal:** HN government functioning as a sovereign, democratic country, responsive to population and respectful of international human rights law (IHRL) norms that does not require external support.
   Task: Intervention supports establishment and development of HN good governance, rule of law, and human rights. Description: Intervention supports the establishment and development of HN government practices, leading to good governance, rule of law, and human rights.

The second goal covers the need for popular support for the *InterventionOrganization* efforts.

2. **Goal:** Enjoy a positive perception of the organization by the populace.
   Task: Develop/keep trust of population. Description: Develop and keep the trust of the population in the organization.

The next four goals include the tasks needed for humanitarian assistance.

3. **Goal:** Healthcare is adequate.
   Task: Provide healthcare and reinforce medical reforms. Description: Provide various types of healthcare both in immediate crises and over the long term.
4. **Goal:** Education meets entity goals.
   Task: Provide education. Description: Provide education over the long term.

5. **Goal:** Move beyond humanitarian crisis stage; provide continuing humanitarian assistance.
   Task: Provide humanitarian assistance (long term). Description: Provide humanitarian assistance to population over the long term.
6. **Goal:** HN economy is functioning well.
   Task: Promote economic, social, and infrastructure development (core subtasks). Description: Assure that the economy is strong and that social development is possible.

The seventh goal is to eliminate the narcotics industry in the Host Nation.

7. **Goal:** Narcotics industry no longer fuels corruption or adversary organizations.
   Task: Eliminate narcotics industry. Description: Eliminate narcotics industry in all of its parts.

The last two goals involve information control and use.

8. **Goal:** Adversaries of Intervention become less effective.
   Task: Conduct information operations against adversaries of intervention. Description: Use information to degrade adversaries of Intervention.
9. **Goal:** Protect organizational reputation and its perquisites, become information power broker, and obtain promotion.
   Task: Control spin and promote self. Description: Recast (spin) problems into successes.

The HNGovernment

In this scenario, the *HNGovernment*, which consists of the Host Nation Whole of Government, has the political agenda of defending its government, attaining legitimacy, and providing humanitarian and development assistance. It also has a private sub-agenda of self-aggrandizement and preparing to lose. The owner is represented by a *wholeOfGovt* class, which is a *hierarchicalActor*, composed of a *demographicGroupActor* class, its *keySocialLeader*, and a possible *populationUnit* class.

The *HNGovernment* has divided its agenda into nine Goal-Task pairs. The first goal is to defend its governance.

1. **Goal:** HN achieves functional independence.
   Task: HN defends its constitutional governance, rule of law, and human rights. Description: HN assures that the machinery of government is in place and supports democratic institutions.

The Goals 2 and 3 cover the need for popular support and legitimacy of the *HNGovernment*.

2. **Goal:** Enjoy a positive perception of the organization by the populace.
   Task: Develop/keep trust of population. Description: Develop and keep the trust of the population in the organization.

3. **Goal:** HN achieves international and domestic legitimacy.
   Task: HN achieves domestic and international legitimacy. Description: HN assures that social and cultural policies and services are in place to ensure domestic and international legitimacy.

The next four goals include the tasks needed for humanitarian assistance.

4. **Goal:** Healthcare is adequate.
   Task: Provide healthcare and reinforce medical reforms. Description: Provide various types of healthcare both in immediate crisis and over the long term.
5. **Goal:** Education meets entity goals.
   Task: Provide education. Description: Provide education over the long term.
6. **Goal:** Move beyond humanitarian crisis stage; provide continuing humanitarian assistance.
   Task: Provide humanitarian assistance (long term). Description: Provide humanitarian assistance to population over the long term.
7. **Goal:** HN economy is functioning well.
   Task: Promote economic, social, and infrastructure development (core subtasks). Description: Assure that the economy is strong and that social development is possible.

The final two goals reflect the reality of venal personnel and organizations and the fear of losing to the adversary. (These are the same as the ones for the *HNMilitary* and the *HNLawEnforcement*. The only difference is who is being venal.)

8. **Goal:** Ensure resources to accommodate accepTable level of patronage.
   Task: Promote patronage network. Description: Accumulate resources to help patronage network.
9. **Goal:** Ensure permanent seat at the Table no matter who is in charge.
   Task: Prepare for possible adversary organization takeover. Description: Prepare for possible adversary organization takeover of HN.

The HNPoliticalReformer

The *HNPoliticalReformer*, or Host Nation Political Reform Organization, consists of a key leader who has decided the Host Nation needs to be reformed in some respect. He might later decide to become a revolutionary, taking over the government, or a separatist, creating his own state. However, neither option is included here. The owner is represented by an *organizationalUnit* class.

The *HNPoliticalReformer* has only one Goal-Task pair.

1. **Goal:** HN policies and personnel are changed.
   Task: Reform HN Government. Description: Change HN government to fit own vision.

The InternationalNGO

The *InternationalNGO* represents the many non-governmental organizations that arrive in conflicts with the agenda of helping the people of the Host Nation. (If there are multiple NGOs with similar characteristics, these may be represented by a single owner. However, multiple owners can be instantiated, each with separate goals and tasks, at the user's discretion.) The NGO represented here avoids or downplays any contact with the military to avoid being linked to it in people's minds. It also has a sub-agenda of self-development. The owner is represented by an *organizationalUnit* class.

The *InternationalNGO* has six Goal-Task pairs. The first three relate to its humanitarian efforts.

1. **Goal:** Humanitarian emergency is ended.
   Task: Distribute food, shelter, and health services. Description: Provide short-term immediate needs in emergencies.
2. **Goal:** Move beyond humanitarian crisis stage; provide continuing humanitarian assistance.
   Task: Provide humanitarian assistance (long term). Description: Provide humanitarian assistance to population over the long term.
3. **Goal:** Enable population to work cohesively to solve their own problems and complete sustainable development.
   Task: Promote and operate small-scale, self-reliant, grassroots economic and social development. Description: Provide development that is focused at the local level to support small-scale, self-reliant, grassroots development.

The last three relate to its operational preferences and need for continued existence.

4. **Goal:** Enforce mission division between NGO and military and perception of agreements.
   Task: NGO avoids military entanglements. Description: Avoid military contacts and public agreements with the military.
5. **Goal:** NGO has access to needy.
   Task: Ensure NGO access to needy. Description: Work with power structure to ensure that projects can be accomplished at the local level.
6. **Goal:** Establish/maintain niche of the NGO; enable follow-on work.
   Task: Promote NGO organization and engage with donors/constituents. Description: Advertise and promote the NGO brand. Engage with NGO home country, Intervention and HN donors, and constituents.

The InternationalContractor

The *InternationalContractor*, or International Contractor Organization, represents a contractor hired by the Intervention to construct or repair the Host Nation infrastructure, with that as its agenda. The owner is represented by an *organizationalUnit* class.

The *InternationalContractor* has four Goal-Task pairs. Each relates to one category of infrastructure.

1. **Goal:** Energy infrastructure is established.
   Task: Construct/repair energy infrastructure. Description: Execute construction/repair/maintenance work in the energy field.
2. **Goal:** Transportation infrastructure is established.
   Task: Construct/repair transportation infrastructure. Description: Execute construction/repair/maintenance work in the transportation field.
3. **Goal:** Water and sanitation infrastructure is established.
   Task: Construct/repair water and sanitation infrastructure. Description: Execute construction/repair/maintenance work in the water and sanitation field.
4. **Goal:** Buildings and other infrastructure is established.
   Task: Construct/repair buildings and other infrastructure. Description: Execute construction/repair/maintenance work in the other infrastructure areas.

The InternationalCorporation

The *InternationalCorporation* represents an external company that wants to take advantage of the unsTable situation in the Host Nation. The owner is represented by an *organizationalUnit* class.

The *InternationalContractor* has four Goal-Task pairs. Each relates to a part of the business process.

1. **Goal:** Resources, equipment, and companies are acquired.
   Task: Acquire natural resources and equipment and acquire/divest companies. Description: Acquire companies and restructure things needed for successful operations.
2. **Goal:** Business capacities are increased.
   Task: Increase business capacities. Description: Increase the capabilities needed for successful operations.
3. **Goal:** Operations produce gross profits.
   Task: Conduct business operations. Description: Conduct the production and trading operations.
4. **Goal:** Operations produce net profits.
   Task: Manage operations. Description: Manage the operations, workforce, and logistics.

The AdversaryNGO

The *AdversaryNGO* represents the non-governmental organizations that arrive in conflicts with the agenda of promoting their ideology by helping the people of the Host Nation. This NGO avoids any contact with the military to avoid being linked to it in people's minds. The owner is represented by an *organizationalUnit* class.

The *AdversaryNGO* has three Goal-Task pairs. The first relates to its propaganda and humanitarian efforts.

1. **Goal:** Social needs are met and political/religious purposes are advanced.
   Task: Provide social support. Description: Provide social support for both immediate and long-term needs in order to advance political/religious views.

   The last two relate to its operational preferences and need for continued existence.

2. **Goal:** Enforce mission division between NGO and military and perception of agreements.
   Task: NGO avoids military entanglements. Description: Avoid military contacts and public agreements with the military.
3. **Goal:** Establish/maintain niche of the NGO; enable follow-on work.
   Task: Promote NGO organization and engage with donors/constituents. Description: Advertise and promote the NGO brand; engage with NGO home country, Intervention and HN donors, and constituents.

The HNCriminalOrganization

The *HNCriminalOrganization* represents the criminal element in the Host Nation. Its agenda consists of making money through criminal activities, which necessitates obtaining a certain amount of political influence. The owner is represented by an *organizationalUnit* class.

The *ProvincialForce* has three Goal-Task pairs.

1. **Goal:** Uninterrupted growth, production, and export cycle of criminal operations.
   Task: Develop secure criminal environment and co-opt local actors. Description: Develop a secure environment that enables long-term operations of the criminal enterprise.
2. **Goal:** Consistent and abundant criminal income flow.
   Task: Conduct profitable criminal operations. Description: Conduct profiTable criminal enterprises.
3. **Goal:** Organization has political power.
   Task: Obtain political power for organization. Description: Translate organization success into political power.

The HNSeparatistOrganization

The *HNSeparatistOrganization*, or Host Nation Separatist Faction, has the agenda of separating a portion of the Host Nation from it and creating a new state. As part of its strategy, it may ally with other anti-government organizations, although its ultimate goal is to defeat or subsume them. The owner is represented by an *armedForceUnit* class.

The *HNSeparatistOrganization* has ten Goal-Task pairs. The first four are organizational.

1. **Goal:** Adversary organization is strong and effective.
   Task: Create and maintain the adversary organization. Description: The adversary organization is organized and working.
2. **Goal:** Organization has political power.
   Task: Obtain political power for organization. Description: Translate organization success into political power.
3. **Goal:** Adversary coalition is strong and effective.
   Task: Create and maintain the coalition of adversary organizations. Description: Create a coalition of adversary organizations with common interests; maintain its coherence; and dominate the coalition in the end-game.
4. **Goal:** No effective tribal threat to entity organization government.
   Task: Support friendly tribal elements. Description: Enable entity-friendly tribal elements to maintain power and influence.

The last six are operational in nature.

5. **Goal:** Control all aspects of HN society.
   Task: Expand shadow government at the expense of the HN government. Description: Expand reach of shadow government, decreasing reach of the HN government.
6. **Goal:** Cause covered by mass media for worldwide audience.
   Task: Conduct information operations for adversary. Description: Conduct information operations in support of cause.
7. **Goal:** Loss of local and international confidence in HN and Intervention efforts to provide security.
   Task: Undermine perception of security. Description: Undermine local and international perceptions of security.
8. **Goal:** HN legitimacy is reduced.
   Task: Reduce HN legitimacy. Description: Act to reduce legitimacy of the HN.
9. **Goal:** Entity is in control of sectors of HN.
   Task: Undermine and destabilize HN government. Description: Undermine the authority and effectiveness of HN government institutions.
10. **Goal:** Organization control is greater than HN government control of region.
    Task: Reduce the HN government control over territory. Description: Localized actions reduce HN government control over selected areas.

The HNRevolutionaryOrganization

The *HNRevolutionaryOrganization*, or Host Nation Revolutionary Organization, has the agenda of replacing the government of the Host Nation with itself. As part of its strategy, it may ally with other anti-government organizations, although its ultimate goal is to defeat or subsume them. The owner is represented by an *armedForceUnit* class.

The *HNSeparatistOrganization* has 12 Goal-Task pairs. The first four are organizational.

1. **Goal:** Adversary organization is strong and effective.
   Task: Create and maintain the adversary organization. Description: The adversary organization is organized and working.
2. **Goal:** Organization has political power.
   Task: Obtain political power for organization. Description: Translate organization success into political power.
3. **Goal:** Adversary coalition is strong and effective.
   Task: Create and maintain the coalition of adversary organizations. Description: Create a coalition of adversary organizations with common interests; maintain its coherence; and dominate the coalition in the end-game.
4. **Goal:** No effective tribal threat to entity organization government.
   Task: Support friendly tribal elements. Description: Enable entity-friendly tribal elements to maintain power and influence.

The last eight are operational in nature.

5. **Goal:** Cause covered by mass media for worldwide audience.
   Task: Conduct information operations for adversary. Description: Conduct information operations in support of cause.
6. **Goal:** Political/religious view accepted.
   Task: Promote political/religious view. Description: Use position as aid to the populace to support a particular political or religious view.
7. **Goal:** Loss of local and international confidence in HN and Intervention efforts to provide security.
   Task: Undermine perception of security. Description: Undermine local and international perceptions of security.
8. **Goal:** HN legitimacy is reduced.
   Task: Reduce HN legitimacy. Description: Act to reduce legitimacy of the HN.
9. **Goal:** Entity is in control of sectors of HN.
   Task: Undermine and destabilize HN government. Description: Undermine the authority and effectiveness of HN government institutions.
10. **Goal:** Organization control is greater than HN government control of region.
    Task: Reduce the HN government control over territory. Description: Localized actions reduce HN government control over selected areas.
11. **Goal:** No Intervention present.
    Task: Expel intervention forces. Description: Expel foreign forces (Intervention) from HN.
12. **Goal:** HN government is defeated militarily.
    Task: Adversary defeats HN government. Description: Adversary defeats HN government forces militarily.

The PatronCountry

The *PatronCountry* has the agenda of converting the Host Nation into a client state, joining it to an existing patronage network. For this scenario, we assume that the Intervention is an alliance of countries, one of which has this additional agenda.

The *PatronCountry* has five Goal-Task pairs.

1. **Goal:** Tie the HN to self economically and then politically, leading to military bases in the HN.
   Task: Promote economic, social, and infrastructure development that ties HN to self. Description: Promote and complete expensive developments with loans that the HN will not be able to repay.
2. **Goal:** Ensure sufficient resources to accommodate accepTable level of patronage.
   Task: Promote patronage network. Description: Accumulate resources to help patronage network.
3. **Goal:** Adversaries of Intervention become less effective.
   Task: Conduct information operations against adversaries of intervention. Description: Use information to degrade adversaries of the Intervention.
4. **Goal:** Protect national reputation and its perquisites, become information power broker, and obtain promotion.
   Task: Control spin and promote self. Description: Recast (spin) problems into successes.
5. **Goal:** Patron country has political power.
   Task: Obtain political power for country. Description: Translate national success into political power.

The TerroristOrganization

The *TerroristOrganization*, or International Terrorist Organization, has the agenda of creating terror and destruction. The reason is unspecified (compare Al Qaeda versus ISIS and its desired Caliphate). As part of its strategy, it may ally with other anti-government organizations, although its ultimate goal is to defeat or subsume them. The owner is represented by an *armedForceUnit* class.

The *HNSeparatistOrganization* has 12 Goal-Task pairs. The first two are organizational.

1. **Goal:** Adversary organization is strong and effective.
   Task: Create and maintain the adversary organization. Description: The adversary organization is organized and working.
2. **Goal:** Adversary coalition is strong and effective.
   Task: Create and maintain the coalition of adversary organizations. Description: Create a coalition of adversary organizations with common interests; maintain its coherence; and dominate the coalition in the end-game.

The last ten are operational in nature.

3. **Goal:** Control all aspects of HN society.
   Task: Expand shadow government at the expense of the HN government. Description: Expand reach of shadow government, decreasing reach of the HN government.
4. **Goal:** Cause covered by mass media for worldwide audience.
   Task: Conduct information operations for adversary. Description: Conduct information operations in support of cause.
5. **Goal:** Political/religious view accepted.
   Task: Promote political/religious view. Description: Use position as aid to the populace to support a particular political or religious view.
6. **Goal:** Loss of local and international confidence in HN and Intervention efforts to provide security.
   Task: Undermine perception of security. Description: Undermine local and international perceptions of security.
7. **Goal:** HN legitimacy is reduced.
   Task: Reduce HN legitimacy. Description: Act to reduce legitimacy of the HN.
8. **Goal:** Entity is in control of sectors of HN.
   Task: Undermine and destabilize HN government. Description: Undermine the authority and effectiveness of HN government institutions.
9. **Goal:** Area of operations is terrorized.
   Task: Conduct terror campaign. Description: Conduct acts to terrorize all elements.
10. **Goal:** Organization control is greater than HN government control of region.
    Task: Reduce the HN government control over territory. Description: Localized actions reduce HN government control over selected areas.
11. **Goal:** No Intervention present.
    Task: Expel intervention forces. Description: Expel foreign forces (Intervention) from HN.
12. **Goal:** HN government is defeated militarily.
    Task: Adversary defeats HN government. Description: Adversary defeats HN government forces militarily.

Interacting Agendas

The parties' agendas are not immaculate stand-alone expressions of intent: they interact with each other over time and space. For example, in the complex irregular war scenario, we see the following interactions:

- The *InterventionForce* Tasks 2 and 3 and the *HNMilitary* Tasks 1 and 2 interact explicitly in establishing the Host Nation military and intelligence services. The

InterventionForce Task 1 and the *HNMilitary* Tasks 3, 4, and 5 interact explicitly in conducting operations.

- The *InterventionForce* Tasks 5 and 6 and the *HNLawEnforcement* Tasks 1 and 2 interact explicitly in establishing the Host Nation law enforcement and first responders. The *InterventionForce* Task 4 and the *HNLawEnforcement* Tasks 3 and 4 interact explicitly in conducting operations.
- The *InterventionOrganization* Tasks 3, 4, 5, and 6 and the *HNGovernment* Tasks 4, 5, 6, and 7 interact explicitly in providing humanitarian assistance and development support.
- The *InterventionForce* and the *InterventionOrganization* interact implicitly, as the former is controlled by the latter.
- The *HNMilitary*, the *HNLawEnforcement*, and the *HNGovernment* interact implicitly, as the first two are nominally controlled by the *HNGovernment*.
- The *InternationalNGO*, the *InterventionForce*, the *InterventionOrganization*, and the *HNGovernment* interact explicitly in their various humanitarian assistance tasks (although some of the interactions may be tacit, rather than overt). The *InternationalNGO* Task 3 and the *InterventionOrganization* Task 6 may interact by conflicting and neither may align precisely with the *HNGovernment* Task 7, all of which concern economic and social development.
- The *AdversaryNGO* relates to the *InternationalNGO* implicitly by competing for "customers."
- The *InternationalContractor* interacts explicitly with the *InterventionForce* and the *InterventionOrganization*, whichever hired it, and interacts implicitly with the other and with the *HNGovernment*.
- The *InternationalCorporation* interacts with the Host Nation economic community (not represented as a party in this scenario) and implicitly with the *HNGovernment*.
- The *HNPoliticalReformer* interacts regularly with the *HNGovernment*.
- The *ProvincialForce* may interact explicitly with the *InterventionForce* Task 1 and with the *HNMilitary* Tasks 3, 4, and 5, depending on its position regarding the *HNGovernment*.
- The *HNCriminalOrganization* Tasks 1 and 2 interact explicitly with the *InterventionForce* Task 4 and with the *HNLawEnforcement* Tasks 3 and 4. The *HNCriminalOrganization* Task 3 interacts explicitly with the *HNGovernment*.
- The *HNSeparatistOrganization*, the *HNRevolutionaryOrganization*, and the *TerroristOrganization* interact with each other, either as part of a coalition or through competition. They share a number of tasks; however, that does not imply that they act together on these tasks, only that they perform similar actions. They each also interact explicitly with the *InterventionForce* and the *HNMilitary* as they are in mutual conflict and combat. One or more of them may be allied with the *AdversaryNGO* in tasks to promote political/religious views. They each may also interact implicitly with other parties by attacking them from time to time.

- The *PatronCountry* (not shown separately in the figure below) operates as part of the *InterventionOrganization*, interacting with the *HNGovernment*. The *PatronCountry* also interacts with the *InternationalCorporation* in furtherance of its separate agenda.

The actual unfolding of events, when tied back to the parties' Goal-Task pairs, can be viewed as braided intentions. This view is not an artifact of the ontology but a reality of modern conflict and is a key to understanding modern conflict. Generally, the population will provide the background for these braided intentions. However, there are situations (not included in this scenario) in which the general population becomes an active party. Portions of the general population, for example, the Shi'a and Sunni populations in Arab states, might each become an active party. Participation of major portions of these populations would probably be preceded by Shi'a and Sunni Factions becoming active parties, drawing in larger and larger portions of the respective populations.

Figure 12.8 illustrates the concept of the braided intentions. The sequence of interactions from top to bottom of the figure is illustrative, based here on the sequence of the descriptions above, not on an attempt to show when interactions might actually occur.

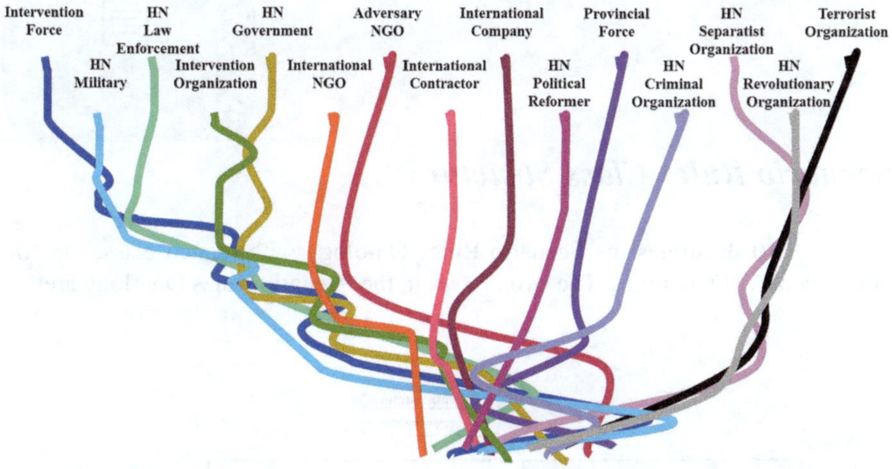

Fig. 12.8 Interacting agendas

GTO Set Instances

The MCO contains a large set of Goal-Task pairs, Subtask-Subgoal pairs, and referents to Actors and Metrics. The complex irregular war scenario exposes most, although not all, of these at the Goal-Task pair level. These are available for use or modification, as needed.

12.4 Scenario Rules Ontology

The third of the five scenario ontologies is the Scenario Rules Ontology. Figure 12.1 shows how the Scenario Rule Classes Ontology fits into the overall MCO structure (with Fig. 12.9 illustrating its icon).

Scenario rules implement the concept of formal "rules of engagement (ROE)" that are created for military forces. They also extend the concept to other Actors who may have much looser, informal rules of engagement; however, these rules can still be represented.

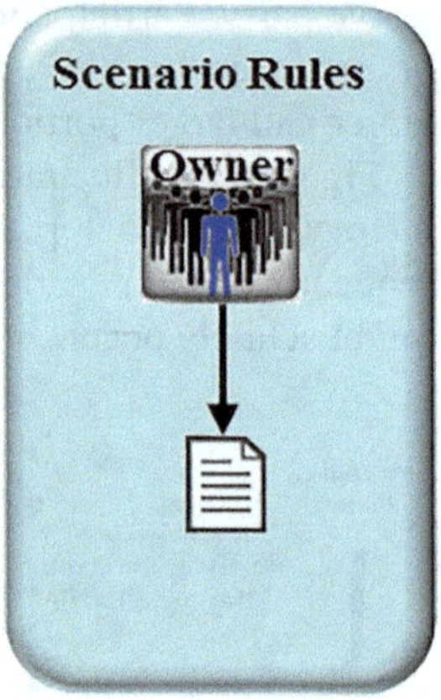

Fig. 12.9 Scenario rule classes icon

Scenario Rules Class Structure

Figure 12.10 illustrates the Scenario Rules Ontology with related classes in two other Scenario Ontologies. The two classes in the Scenario Rules Ontology are:

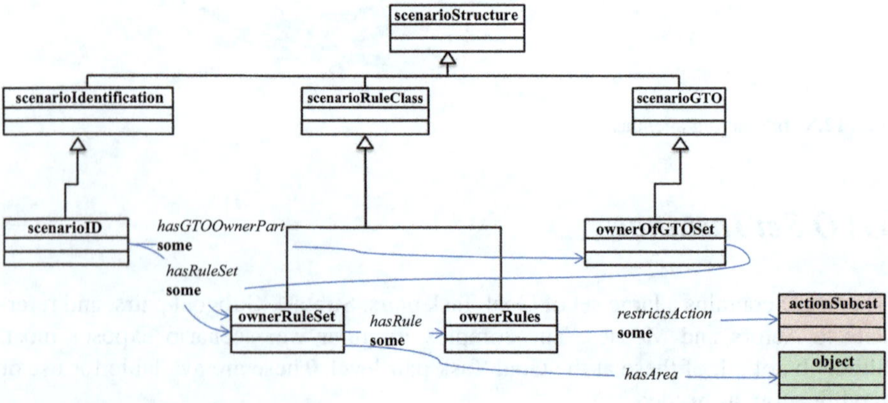

Fig. 12.10 Scenario Rules Ontology

- *ownerRuleSet*: a scenarioRuleClass that identifies the rule set by owner, date, and location
- *ownerRules*: a scenarioRuleClass that contains the rules used in *ownerRuleSets*

The *ownerRuleSet* is related to the *scenarioID* and the *ownerOfGTOSet* classes, identifying the scenario and the owner of the rule set. The *ownerRuleSet* is also related to *ownerRules*, which provides the actual rules in the rule set, and to the Object class, which provides the area in which the rule set pertains. The *ownerRules* class is also related to each of the Action subcategory classes, as these contain all of the Action classes that may be restricted by the rules.

Scenario Rules Instances

Table 12.5 illustrates two instances of *ownerRuleSets*. One is a rule set for restricted combat and the other is for unrestricted combat. They are each attached to a scenario and an owner.

Table 12.6 lists a number of rule instances. Naturally, these could be worded differently and other rules can be created; however, they provide examples that support

Table 12.5 *ownerRuleSet* instances

RuleSetName	StartDate	Relationship	entityName
ruleSet001restrictedCombat001	27-Apr-11	hasArea	ProvinceAGeographicalSubdivision
ruleSet002unrestrictedCombat001	18-Dec-17	hasArea	ProvinceAGeographicalSubdivision

Table 12.6 *ownerRules* instances

RuleName	Rule
noCollateralDamage	Avoid damage to civilian infrastructure, friendly force infrastructure, Host Nation infrastructure
unrestrictedDamage	Damage is only restricted by resources and capabilities
noCollateralCasualties	Avoid casualties to civilians, friendly forces, no chemical, biological, or radiological weapons
unrestrictedCasualties	Casualties are only restricted by resources and capabilities
noEnvironmentDamage	Do not damage the environment
noInjuries	Do not injure anyone
selfDefense	Use weapons only when attacked; avoid civilian casualties
unrestrictedConflict	Conduct conflict by any means available
noConflict	Avoid all conflict
noTerrorismOrPiracy	Do not participate in terrorism or piracy
limitedCriminalActivities	Engage in limited extrajudicial, criminal, and suppression activities
unlimitedCriminalActivities	Engage in unlimited extrajudicial, criminal, and suppression activities

understanding. Each rule set instance is composed of a subset of the rule instances, and each rule instance is attached to the appropriate Action subcategory class.

Table 12.7 shows the rules that are connected to each rule set by the *hasRule* relationship.

Table 12.7 Rule set to rule connections

RuleSetName	RuleName
ruleSet001restrictedCombat001	noCollateralDamage
ruleSet001restrictedCombat001	noCollateralCasualties
ruleSet001restrictedCombat001	selfDefense
ruleSet001restrictedCombat001	noTerrorismOrPiracy
ruleSet002unrestrictedCombat001	unrestrictedDamage
ruleSet002unrestrictedCombat001	unrestrictedCasualties
ruleSet002unrestrictedCombat001	unrestrictedConflict
ruleSet002unrestrictedCombat001	unlimitedCriminalActivities

Table 12.8 lists the Action subcategories connected to each of the rules that was used by the *restrictsAction* relationship.

Thus, an Owner in a particular scenario will have one or more rule sets, each of which has a start date and an effective area. The rule set will have a set of rules that restrict the Action classes contained in one or more Action subcategories.

Table 12.8 Rule to action subcategory connections

RuleName	ActionSubcategory
noCollateralDamage	*strikeTargetAction*
noCollateralDamage	*strikeMethodAction*
noCollateralDamage	*generalConflictAction*
unrestrictedDamage	*strikeTargetAction*
unrestrictedDamage	*strikeMethodAction*
unrestrictedDamage	*generalConflictAction*
noCollateralCasualties	*strikeTargetAction*
noCollateralCasualties	*strikeMethodAction*
noCollateralCasualties	*generalConflictAction*
noCollateralCasualties	*criminalAction*
unrestrictedCasualties	*strikeTargetAction*
unrestrictedCasualties	*strikeMethodAction*
unrestrictedCasualties	*generalConflictAction*
unrestrictedCasualties	*criminalAction*
selfDefense	*strikeTargetAction*
selfDefense	*strikeMethodAction*
selfDefense	*generalConflictAction*
unrestrictedConflict	*generalConflictAction*
limitedCriminalActivities	*criminalAction*

For a real-world scenario, the rules affect the actions of the participants more or less successfully. In a model, the restrictions will have to be coded.

12.5 Scenario AAR Ontology

The fourth of the five scenario ontologies is the Scenario AAR Ontology. Figure 12.1 shows how the Actor-Action-Result (AAR) Sets Ontology fits into the overall MCO structure (with Fig. 12.11 illustrating its icon) .

Fig. 12.11 AAR sets icon

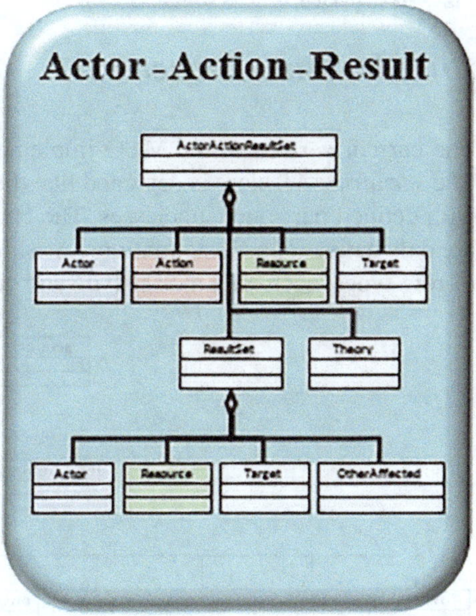

AAR sets implement the concept that actions have consequences. These consequences begin with a description of the environment of the action and include all of the actors and objects that are affected by the action.

AAR Class Structure

The scenario AAR Ontology has only one class, *scenarioAAR*, because all of the content consists of instantiations. As shown in Fig. 12.12, this class is related to the scenario and to the owner.

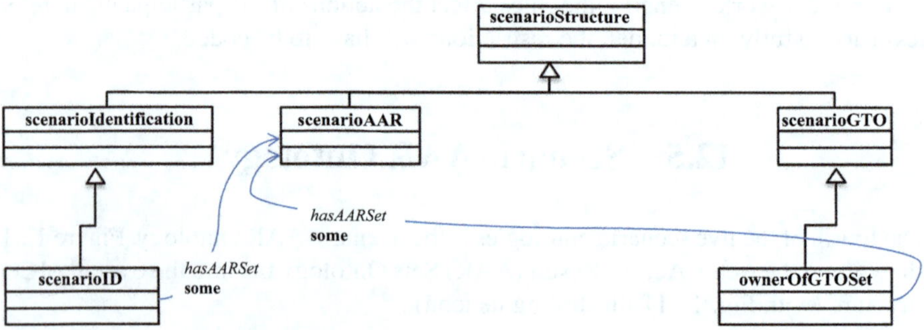

Fig. 12.12 Scenario AAR Ontology

AAR Instance Structure

The current version of the MCO implements this ontology through instantiations. The *scenarioAAR* class is designed like the composite classes described in Chap. 7, with defined parts, not subclasses. The class is internally composed of links, which are only active in the instantiations. As shown in Fig. 12.13, the links are comprised of role-types, each with a class type and cardinality restriction.

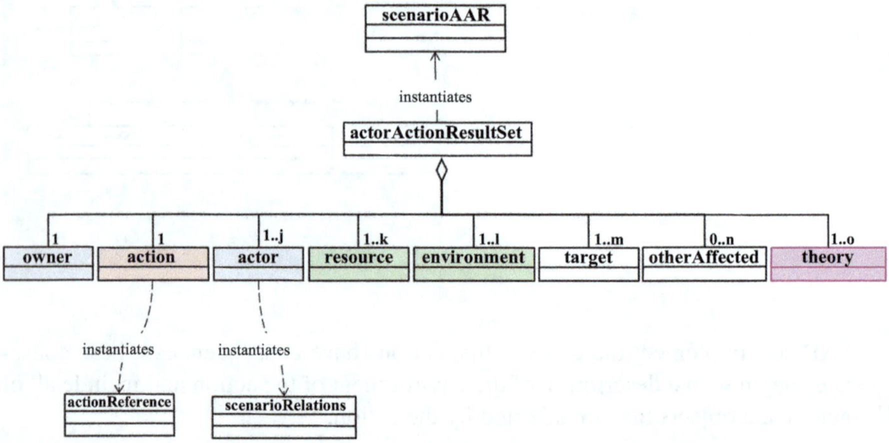

Fig. 12.13 AAR instantiations

The blue-filled boxes represent instantiations of Actor classes. The pink-orange-filled box represents the instantiation of an Action class. The green-filled boxes represent instantiations of Object classes. The white (or "unfilled") boxes may be instantiations of Actor classes, Object classes, or a mixture of the two. The magenta-filled box represents instantiations of Theory classes.

Figure 12.12 indicates that each *scenarioAAR* class is connected to a scenario and to an owner. The instantiation of this owner fills the first role (the ultimate instigator) in an instantiation of the *scenarioAAR* class.

The second role is the Action class instance that is at the heart of the Actor-**Action**-Result concept. As shown in Fig. 12.13, the action is an instantiation of one *actionReference* class in the owner's GTO set.

The third role is filled by a set of one to *j* Actor instantiations, also at the heart of the **Actor**-Action-Result concept. These are the ones who carry out the Action. This set could be an instantiation of one of the subclasses of the *scenarioRelations* class (described in the next section), perhaps the *groupRelations* class or the *networkRelations* class. It could also be the instantiation of an ad hoc set of Actor classes.

The fourth role is filled by a set of one to *k* resources. These are Object class instances that are needed to complete the action, such as weapons, explosives, vehicles, etc.

The fifth role is filled by a set of one to *l* (the letter) descriptions of the environment in which the action takes place.

The sixth role is filled by a set of one to *m* targets of the action. These might be Actor instances, Object instances, or both.

The seventh role is filled by a set of zero to *n* other-affected Actor or Object instances. The difference between a target and an other-affected instance is that the target is meant to be affected and the other-affected instance either is not meant to be affected or is irrelevant to the intention. An example would be civilians in the area who are affected, but not specifically targeted.

The eighth role is filled by a set of one to *o* (the letter) Theory instances. These theories justify the changes of state of any or all of the Actor and Object instances.

Figure 12.14 illustrates the Results part of the Actor-Action-Result concept. The left side of the figure illustrates the situation in the real world: an Action *hasInfluenceOn* an element (Actor or Object), with a concomitant change in its Metric. (A short-hand description might also be given that the Action *hasInfluenceOn* a Metric.) The right side of the figure expands this by adding the time, location, and environ-

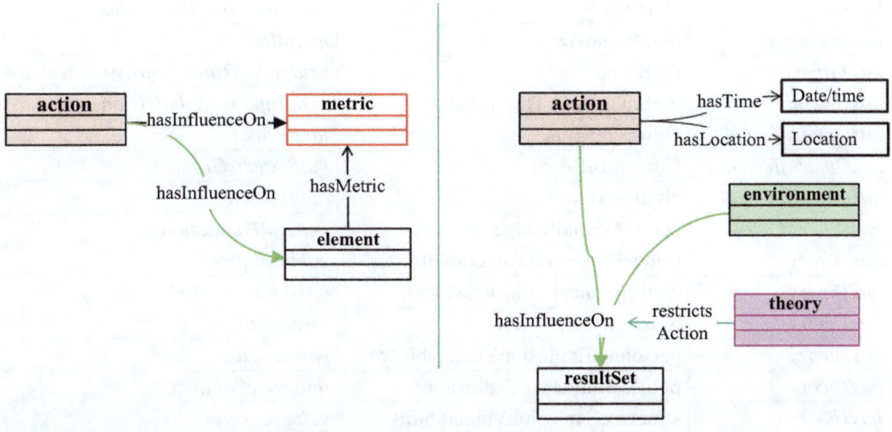

Fig. 12.14 AAR results justification

ment and by including all affected elements in a *resultSet*. It also adds the Theory that *restricts* the effect of the Action. In the real world, the Theory is added as an explanation of what happened, whereas, in a model, the Theory is used to create the effect.

The MCO contains one instantiation to provide an example. Table 12.9 displays the major contents of a pre-defined query showing the *actorActionResultSet* instance of the *scenarioAAR* class. This is an action to blow up a particular bridge by Afghan terrorists.

Table 12.10 shows the details of this action. The Action class is *strikeWithHighYieldExplosives*, and the name of the instance is blowUp8thAveBridge. Seven Actor instances are named, instantiated from two Actor classes. We will see these same instances in a network in the next section. We could have just named the network as the sole entry as an actor; however, this would have required another level

Table 12.9 *actorActionResultSet* instance

Data name	Data contents
Scenario	notionalModelHartley002
Owner	Afghan Terrorists
AAR Set Name	blowUpBridge

Table 12.10 AAR instance

Data name	Data contents	Class
AARDate	12/20/17	
hasAction	blowUp8thAveBridge	*strikeWithHighYieldExplosives*
hasActor	AbduhltheBad	*keyNonGovtArmedOfficial*
hasActor	PersonofInterest01	*keyNonGovtArmedOfficial*
hasActor	PersonofInterest02	*keyNonGovtArmedOfficial*
hasActor	PersonofInterest03	*keyNonGovtArmedOfficial*
hasActor	Contact01	*nonGovtArmedIndividual*
hasActor	Contact02	*nonGovtArmedIndividual*
hasActor	Contact03	*nonGovtArmedIndividual*
hasResource	highExplosive	*munition*
hasTarget	theBridge	*bridgeAndTunnelInfrastructure*
hasGeoSubdivision	theEnvironsOfTheBridge	*geographicalSubdivision*
hasOtherAffected	Number19Bus	*landVehicle*
hasOtherAffected	Convoy14Bde	*milSystemsGroup*
hasOtherAffected	civilianAuto01	*landVehicle*
hasOtherAffected	peopeAroundBridge	*generalPopulation*
hasTheory	bridgeExplosionVulnerability	*vulnerability*
hasTheory	civilViolenceExplosionModel	*civilViolenceModel*
hasTheory	insecurityExplosion	*insecurity*
hasTheory	personnelExplosionVulnerability	*vulnerability*
hasTheory	politicalInfluenceExplosion	*influencePolitical*
hasTheory	structureExplosionVulnerability	*vulnerability*
hasTheory	vehicleExplosionVulnerability	*vulnerability*

of connection to show who is involved, and it would not have shown the feature of including multiple actors. To keep things simple, only one resource and one target are represented. However, four other-affected instances are included to illustrate both Actor and Object class instances. Seven theory instances from three Theory classes are included to cover the vulnerability of the classes (including the bridge's class, the Owner's class, and the perpetrators' Actor classes) and to cover possible political influence, insecurity, and civil violence resulting from the action.

Using AAR Sets

Building AAR sets clearly involves a lot of work. However, a model would have to calculate the results from every action instance, and this information would be required for each one. Depending on the granularity of the model, some of the details could be assumed by creating categories of vignettes and assessing by category, perhaps with random variations between instances.

Possible Future Enhancements of AAR Sets

The current version of the AAR Ontology has minimal class structure, proceeding directly to the instance structure. A possible future version of the MCO might use an approach more like that of the GTO sets, providing a class structure instead of concealing the details in the instance structure.

Currently, the relationship between the action instance and the instance in a GTO set is implied rather than implemented, in the MCO. A future version might make this relationship explicit. Similarly, the relationship between the actor instances and the Scenario Relations groups is implied, not implemented, in the MCO. This relationship could also be made explicit in a future version.

12.6 Scenario Relations Ontology

The fifth of the five scenario ontologies is the Scenario Relations Ontology. Figure 12.1 shows how the Scenario Relations Ontology fits into the overall MCO structure (with Fig. 12.15 illustrating its icon).

Scenario relations implement the concept that things have diverse relationships, including different relationships between the same two things. There are five scenario relations classes:

- *actorActorRelations*: a scenarioRelation that relates an Actor to another Actor
- *actorObjectRelations*: a scenarioRelation that relates an Actor to an Object

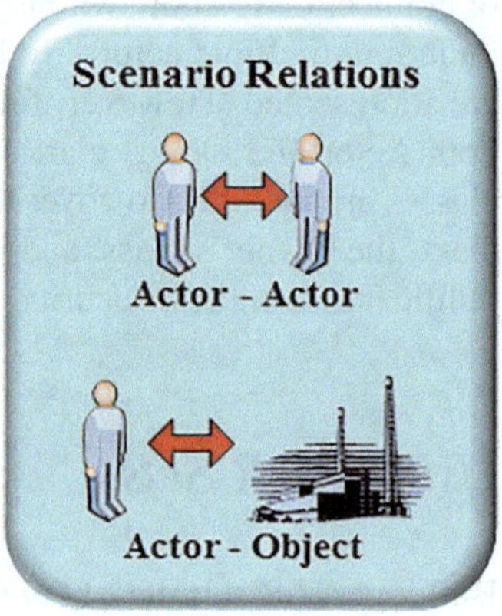

Fig. 12.15 Scenario relations icon

- *objectObjectRelations*: a scenarioRelation that relates an Object to an Object
- *groupRelations*: a scenarioRelation that collects relationship pairs into a group
- *networkRelations*: a scenarioRelation that collects relationship pairs into a network

Scenario Relations Class Structure

The first three define relationships between pairs of Actor classes, Actor and Object classes, and Object classes, respectively. The *groupRelations* class defines relationships in a set of pairs, each pair consisting of the group itself and an Actor or Object class. The *networkRelations* class defines relationships in a set of pairs of Actors, with the set of pairs defining a network. As shown in Fig. 12.16, each class is connected to the scenario.

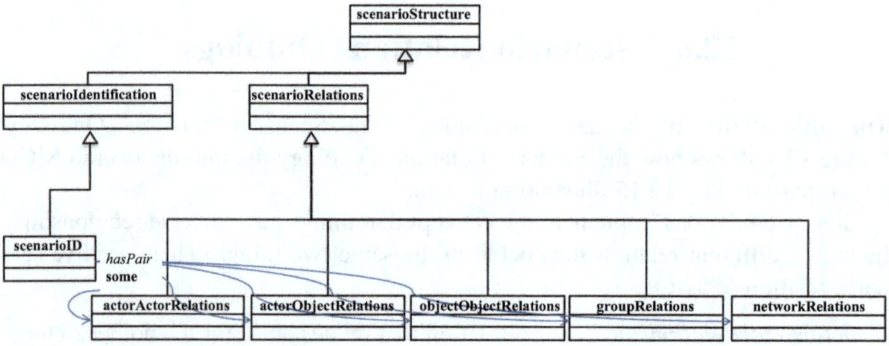

Fig. 12.16 Scenario Relations Ontology

Scenario Relations Instance Structure

Figure 12.17 illustrates the structure of the instances of the five scenario relations classes. In each case, the instance is named (e.g., *AAEntityEntityGroup*) and has relationships with one or more pairs (e.g., *<pair1stPartA, pair2ndPartA>*).

In the first three classes to the left of the figure (*actorActorRelations*, *actorObjectRelations*, and *objectObjectRelations*), exactly one pair is instantiated. Hence the relationships are *hasPair1stPart* and *hasPair2ndPart*, with singular "pair." (If the same pair of things have multiple relationships, multiple pairs can be created, each with the same things, but different relationships.) Note that the labels for the pairs-parts in the instances have the appropriate "A" or "O" suffix to indicate the type of element involved.

The last two classes on the right (*groupRelations* and *networkRelations*) have instantiations of multiple pairs. Hence the relationships are *hasPairs1stPart* and *hasPairs2ndPart*, with plural "pairs." The instance pair-parts of the *groupRelations* class have a "G" in the first part to indicate the group and an "E" in the second part to indicate either an Object or Actor instance. The instance pair-parts of the networkRelations class have "E" suffixes to allow for either Actor or Object entities, although probably only Actor entities will actually form a network.

In each case, there is a relationship between the first part of the pair and the second part of the pair, indicated by a generic "has relationship" label, as the actual relationship depends on what is being represented.

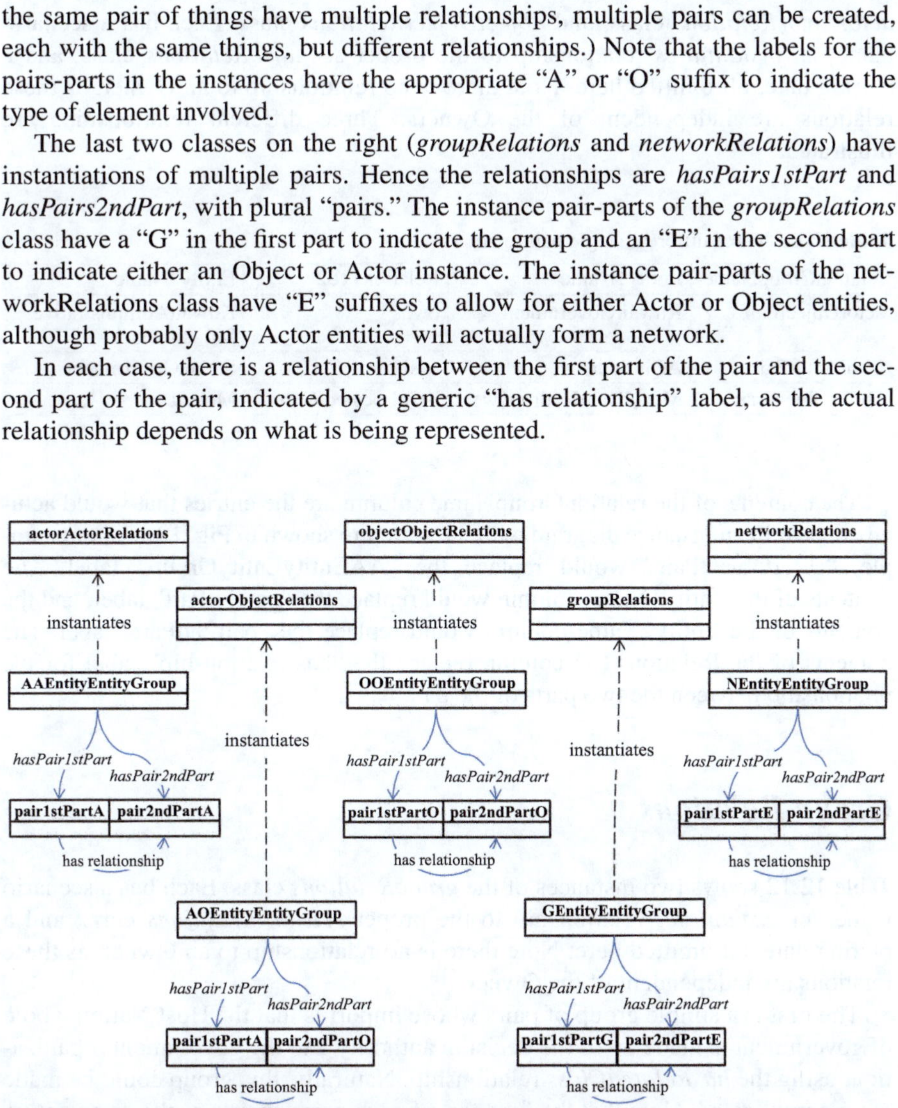

Fig. 12.17 Scenario relations instance structure

The examples below are intended to convert this abstract description into real examples in order to show how these relations can be used to describe this part of a scenario.

Single-Pair Relations

Table 12.11 shows three instances, one each for the *actorObjectRelations* class, *actorActorRelations* class, and *objectObjectRelations* class. Each has a scenario name, an *instantiates* relationship to the proper *scenarioRelations* class, and a pairing date, all omitted here. Note there is no relationship to an Owner, as these relations are independent of the Owners. Three different relationships are illustrated.

Table 12.11 Three different single pair relations

relationGroupName	Entity1Name	Relation1To2	Entity2Name
actorObjectPair	KuwaitGovernment	owns	KuwaitAdministrative Infrastructure
actorActorPair	AQinSudan	hasEnmityFor	SudanGovernment
objectObjectPair	MosulDamHydropower	isCoLocatedWith	MosulDam

The contents of the relationGroupName column are the entries that would actually appear in an instance diagram with the structure shown in Fig. 12.17. For example, "actorObjectPair" would replace the "AAEntityEntityGroup" label. The contents of the Entity1Name column would replace the "pair1stPart" label, and the contents of the Entity2Name column would replace the "pair2ndPart" label. The contents of the Relation1To2 column replace the "has relationship" label for the relationship between the two parts of the pair.

Group Relations

Table 12.12 shows two instances of the *groupRelations* class. Each has a scenario name, an *instantiates* relationship to the proper *scenarioRelations* class, and a pairing date, all omitted here. Note there is no relationship to an Owner, as these relations are independent of the Owners.

The first is a simple group of pairs whose import is that the Host Nation whole of government is made up of the decision authority and the government organization, using the *hasRelatedClass* relationship. Naturally, this group could be made more complicated. Note that the first part of each pair consists of the same entity. This entity's name is similar to the group name, but not identical, as they are two

Table 12.12 Two group relations

relationGroupName	Entity1Name	Relation1To2	Entity2Name
HNWholeGovt	theHNWholeOfGovt	hasRelatedClass	theHNDecisionAuthority
HNWholeGovt	theHNWholeOfGovt	hasRelatedClass	theHNGovtOrganization
HNGovtOrganization	theHNGovtOrganization	hasKeyPerson	Inam-ul-Haq
HNGovtOrganization	theHNGovtOrganization	hasKeyPerson	AbdManaf
HNGovtOrganization	theHNGovtOrganization	hasKeyPerson	BarkatAli
HNGovtOrganization	theHNGovtOrganization	hasKeyPerson	Fakhral-Din
HNGovtOrganization	theHNGovtOrganization	hasRelatedClass	theHNSupremeCourt
HNGovtOrganization	theHNGovtOrganization	hasRelatedClass	theDeptofState
HNGovtOrganization	theHNGovtOrganization	hasRelatedClass	theCabinet
HNGovtOrganization	theHNGovtOrganization	hasRelatedClass	theParliament
HNGovtOrganization	theHNGovtOrganization	hasSubCompositePart	aProvincialGovt

different entities. The group name refers to the artificial group, while the first part of each pair refers to the instantiation of an actual class.

The second instance is an expansion of the government organization; however, since it uses a different name from that of the first group, it represents a different entity. Here the government organization is defined to have four key persons, a Supreme Court, a Department of State, a Cabinet, and a Parliament, and one provincial government as a sub-composite part. Note that additional work using other groups would be required to identify which key person goes with which sub-organization. For example, a group for theHNSupremeCourt could identify Inam-ul-Haq as the Chief Justice and identify other individuals as Justices.

The representation of a group using the *groupRelations* class can be compared to the representation using composite classes that is discussed in Chap. 7. Each requires constructing a mental model of the group before creating the instance. The composite classes method uses pre-defined structures and is probably easier if the situation is a good fit with the pre-defined structure. The method of *groupRelations* classes is more flexible and may fit better with very simple groups or with very complicated groups that are not hierarchically structured.

Network Relations

Table 12.13 shows three instances of the *networkRelations* class. Each has a scenario name, an *instantiates* relationship to the proper *scenarioRelations* class, and a pairing date, all omitted here. Note there is no relationship to an Owner, as these relations are independent of the Owners.

The first instance, network001, shows the actors that were used in the AAR instance of Table 12.10. This is a relatively simple network. The second network,

Table 12.13 Three network relations

relationGroupName	Entity1Name	Relation1To2	Entity2Name
network001	AbduhltheBad	directsOrControls	Contact01
network001	Contact01	hasInfluenceOn	Contact02
network001	Contact01	isMemberOf	AQinSudan
network001	Contact02	isMemberOf	AQinSudan
network001	Contact03	isMemberOf	AQinSudan
network001	PersonofInterest01	isMemberOf	AQinSudan
network001	PersonofInterest02	directsOrControls	Contact02
network001	PersonofInterest03	directsOrControls	Contact03
aCrimeNetwork	Archibald	talksWith	Barry
aCrimeNetwork	Archibald	isMemberOf	DumbGang
aCrimeNetwork	Barry	talksWith	Celeste
aCrimeNetwork	Barry	isMemberOf	DumbGang
aCrimeNetwork	Celeste	talksWith	Dominic
aCrimeNetwork	Celeste	isMemberOf	DumbGang
aCrimeNetwork	Dominic	talksWith	Evelyn
aCrimeNetwork	Dominic	isMemberOf	DumbGang
aCrimeNetwork	Evelyn	talksWith	Archibald
aCrimeNetwork	Evelyn	isMemberOf	DumbGang
aCrimeNetwork	Evelyn	isMemberOf	CityCops
aCrimeNetwork	MasterMind	isMemberOf	DumbGang
aCrimeNetwork	MasterMind	sendsMessagesTo	Dominic
aCrimeNetwork	MasterMind	sendsMessagesTo	Archibald
aCrimeNetwork	MasterMind	sendsMessagesTo	Evelyn
aCrimeNetwork	MasterMind	sendsMessagesTo	Celeste
aCrimeNetwork	MasterMind	sendsMessagesTo	Barry
DominicFamily	Dominic	isParentOf	Joseph
DominicFamily	Dominic	isParentOf	Harvey
DominicFamily	Dominic	isParentOf	Irmentrude
DominicFamily	Dominic	isSpouseOf	Gertrude
DominicFamily	Dominic	isExSpouseOf	Francine
DominicFamily	Francine	isParentOf	Harvey
DominicFamily	Gertrude	isParentOf	Joseph
DominicFamily	Gertrude	isParentOf	Irmentrude
DominicFamily	Gertrude	isStepParentOf	Harvey
DominicFamily	Harvey	isSiblingOf	Joseph
DominicFamily	Harvey	isSiblingOf	Irmentrude
DominicFamily	Irmentrude	isSiblingOf	Joseph

aCrimeNetwork, is more complicated. Each entity has the relationship *isMemberOf* with DumbGang and each *talksWith* or *sendsMessagesTo* at least one other entity. Evelyn is a police plant and has the relationship *isMemberOf* with CityCops.

The third network, DominicFamily, is built to explore family relationships. Dominic has three children, one with his ex-wife, Francine, and two with his current wife, Gertrude. The children are siblings.

As it turns out, Dominic is also a member of the crime network, so that the two networks are linked through him. One advantage that a Web Ontology Language (OWL) implementation of the MCO using Protégé has over the Access database implementation is the facility to display diagrams. Figure 12.18 displays the combination of these two networks. The Dominic family network appears in the upper left portion of the figure, and the crime network appears in the lower right portion. Other than the links to the scenario through the network names (upper right box), the only connection between the two networks is Dominic (lower center of the figure). Note that the CityCops (center right) is only linked to Evelyn and the crime network name. The various colors of the links distinguish among the relationship types.

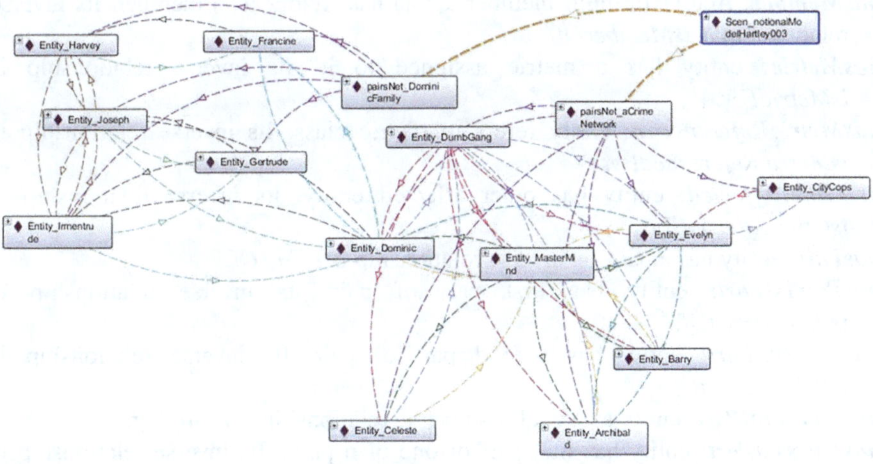

Fig. 12.18 Crime network and Dominic family

12.7 Scenario Ontologies Relationships

The relationships used in by the Scenario Ontologies are named and explained as follows:

directsOrControls: Actor1 directs or controls the actions of Actor2. Its inverse relationship is *isDirectedOrControlledBy*.
hasAARSet: entity has AAR Set. Its inverse relationship is *isAARSetOf*.
hasAction: entity commits action. Its inverse relationship is *isActionOf*.
hasActionReferenceTo: entity refers to action class. Its inverse relationship is *isActionReferencedBy*.

hasActor: entity has an actor. Its inverse relationship is *isActorOf*.
hasActorReferenceTo: entity refers to actor class. Its inverse relationship is *isActorReferencedBy*.
hasArea: entity affects action in the area. Its inverse relationship is *isAreaOf*.
hasCitation: entity has a citation (a set of data with authorship and dates). Its inverse relationship is *isCitationOf*.
hasEnmityFor: Actor1 feels hatred, intends injury to, or opposes the interests of Actor2. Its inverse relationship is *hasEnemy*.
hasGeoSubdivision: entity has *geographicSubdivision* with location. Its inverse relationship is *isGeoSubdivisionOf*.
hasGTOOwnerPart: scenario has another entity as a GTO owner. Its inverse relationship is *isGTOOwnerPartOf*.
hasInfluenceOn: entity has influence on another entity. Its inverse relationship is *isInfluencedBy*.
hasKeyPerson: entity has a key person. Its inverse relationship is *isKeyPersonOf*.
hasLocation: implied relationship, created by having a Metric.
hasMember: Actor1 (a multi-member actor) has Actor2 as a member. Its inverse relationship is *isMemberOf*.
hasMetric: entity has a metric assigned to it. Its inverse relationship is *isMetricFor*.
hasMetricReferenceTo: entity refers to metric class. Its inverse relationship is *isMetricReferencedBy*.
hasOtherAffected: entity has other affected entity. Its inverse relationship is *isOtherAffectedBy*.
hasPair: entity has a pair. Its inverse relationship is *isPairOf*.
hasPair1stPart: entity has first part of pair. Its inverse relationship is *isPair1stPartOf*.
hasPair2ndPart: entity has second part of pair. Its inverse relationship is *isPair2ndPartOf*.
hasPairGoalPart: entity as a goal. Its inverse relationship is *isGoalOf*.
hasPairs1stPart: entity has first part of one of n pairs. Its inverse relationship is *isPairs1stPartOf*.
hasPairs2ndPart: entity has second part of one of n pairs. Its inverse relationship is *isPairs2ndPartOf*.
hasPairSubgoalPart: entity has a subgoal. Its inverse relationship is *isSubgoalOf*.
hasPairSubtaskPart: entity has a subtask. Its inverse relationship is *isSubtaskOf*.
hasPairTaskPart: entity has a task. Its inverse relationship is *isTaskOf*.
hasRelatedClass: entity1 has entity2 as a related class – government for government composite, armed force for armed force composite, etc. Its inverse relationship is *isRelatedClassOf*.
hasResource: entity has a resource. Its inverse relationship is *isResourceOf*.
hasRule: entity has rule. Its inverse relationship is *isRuleOf*.
hasRuleSet: entity has rule set. Its inverse relationship is *isRuleSetOf*.
hasSubCompositePart: entity1 has as a part a composite entity, entity2, of the same type as entity1 as a part. Its inverse relationship is *isSubCompositePartOf*.

hasSubtaskSubgoalPair: entity has a subtask-subgoal pair. Its inverse relationship is *isSubtaskSubgoalOf*.

hasTarget: entity has target. Its inverse relationship is *isTargetOf*.

hasTaskGoalPair: entity owns a task-goal pair. Its inverse relationship is *isTaskGoalOf*.

hasTheory: entity has theory for computing results. Its inverse relationship is *isTheoryOf*.

hasTime: implied relationship, created by having a Metric.

is-a: class is subclass of superclass. Its inverse relationship is *superClassOf*.

isCoLocatedWith: entity1 and entity2 are collocated.

isExSpouseOf: Actor1 is ex-spouse of Actor2.

isInstanceOf (instantiates): entity is instance of a class. Its inverse relationship is *hasInstance*.

isParentOf: Actor1 has given birth to or nurtured and raised Actor2. Its inverse relationship is *isChildOf*.

isSiblingOf: Actor2 has one or both parents in common with Actor1.

isSpouseOf: Actor2 is married to Actor1.

isStepParentOf: Actor1 is stepparent of Actor2. Its inverse relationship is *isStepChildOf*.

owns: Actor owns or has some property rights to the thing. Its inverse relationship is *isOwnedBy*.

relatedToGoal: task is related to goal. Its inverse relationship is *relatedToTask*.

relatedToSubgoal: subtask is related to subgoal. Its inverse relationship is *relatedToSubtask*.

restrictsAction: entity restricts the action. Its inverse relationship is *isRestrictedBy*.

sendsMessagesTo: Actor1 sends messages to Actor2. Its inverse relationship is *receivesMessagesFrom*.

talksWith: Actor1 and Actor2 talk together.

12.8 Recapitulation of the Scenario Ontologies

In this chapter we described the structure of the Scenario Ontologies and their contents. An ontology of conflict is incomplete if it does not address scenarios. We have said that the point of the ontology is to describe what we can know about conflicts in general and to support recording what we know about a particular conflict. A scenario is the context that reifies the **concept** of conflict and defines the **particular** conflict.

The scenario defines which parts of the ontology are to be expressed (instantiated) in a particular situation and which parts will be suppressed as unneeded. Unsurprisingly, the Scenario Ontologies require instantiations for full understanding. That is, the structures are defined in the MCO; however, their value lies in describing a particular scenario and examples are required (and provided) to see how they work.

The Scenario Identification Ontology identifies a scenario. The GTO Sets Ontology defines the important parties and their agendas in a scenario. The Scenario Rules Ontology defines the restrictions the parties apply to their own actions in a scenario. The AAR Sets Ontology defines the connections of the parties' actions to results in a scenario. And the Scenario Relations Ontology defines the relationships between Actors [such as kinship], between Actors and Objects [such as ownership], between Objects [such as co-location], among entities in a group, and among entities in a network.

In a future version of the MCO, several changes might be made. The current version of the AAR Ontology has minimal class structure, proceeding directly to the instance structure. A possible future version of the MCO might use an approach more like that of the GTO sets, providing a class structure instead of concealing the details in the instance structure. Currently, the relationship between the action instance in an AAR Set and the action instance in a GTO set is implied, not implemented. A future version might make this relationship explicit. Similarly, the relationship between the actor instances in an AAR Set and the Scenario Relations groups is implied, not implemented. A future version might make this relationship explicit.

The next chapter defines and categorizes the relationships of the MCO.

Chapter 13
MCO Relationships

This chapter provides a review of the Modern Conflict Ontology (MCO) relationships.

13.1 Overview of Relationships

Previous chapters have concentrated on the classes of the ontology, with excursions into some instantiations to provide examples of the employment of the ontology. The goal has been to show the breadth and complexity of modern conflict and the ontology necessary to capture its content.

Changes to Relationships from the UCO

Although the UCO certainly had definite relationships between ontology elements and structures, they were not explicitly described. The MCO explicitly names and defines its relationships and describes a structure for them.

Insights into Relationships

Throughout these chapters, various relationships among the classes and instances have been asserted; however, they have not been the focus of the discussions. This chapter is devoted to displaying the rich diversity and scope of the MCO relationships.

D. S. Hartley III, *An Ontology of Modern Conflict*, Understanding Complex Systems, https://doi.org/10.1007/978-3-030-53214-7_13

There are three types of relationships: annotation, data, and object, as shown in Fig. 13.1. Within these, there are 41 annotation relationships, 2 data relationships, and 131 object relationships. The picture in the figure illustrates a family relationship, one of the object relationships.

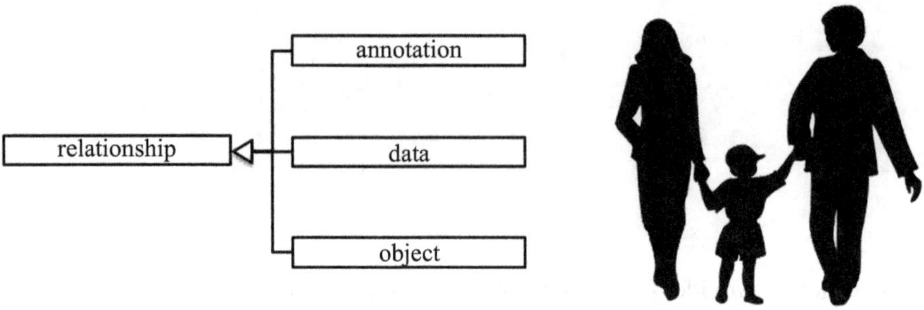

Fig. 13.1 Top-level relationships

13.2 Annotation Relationships

Annotations consist of text data about classes that are attached to the data by relationships that define the type of data. Thus, annotations are relationships between either a class or an instance and text data.

Annotation Relationship Categories

Figure 13.2 extends Fig. 13.1 by adding the 41 sub-relationships of the *annotationRelationship*. Each of the sub-relationships is defined in the next section. The picture of the calendar in the figure represents a date relationship.

Annotation Relationship Definitions

All annotation relationships are sub-relationships of *annotationRelationship* and have no inverse relationships.

AARDate: entity has actor-action-result date.
actionCatID: unique ID for action category.
actionSubCatID: unique ID for action subcategory.
actorCatID: unique ID for actor category.
actorSubCatID: unique ID for actor subcategory.

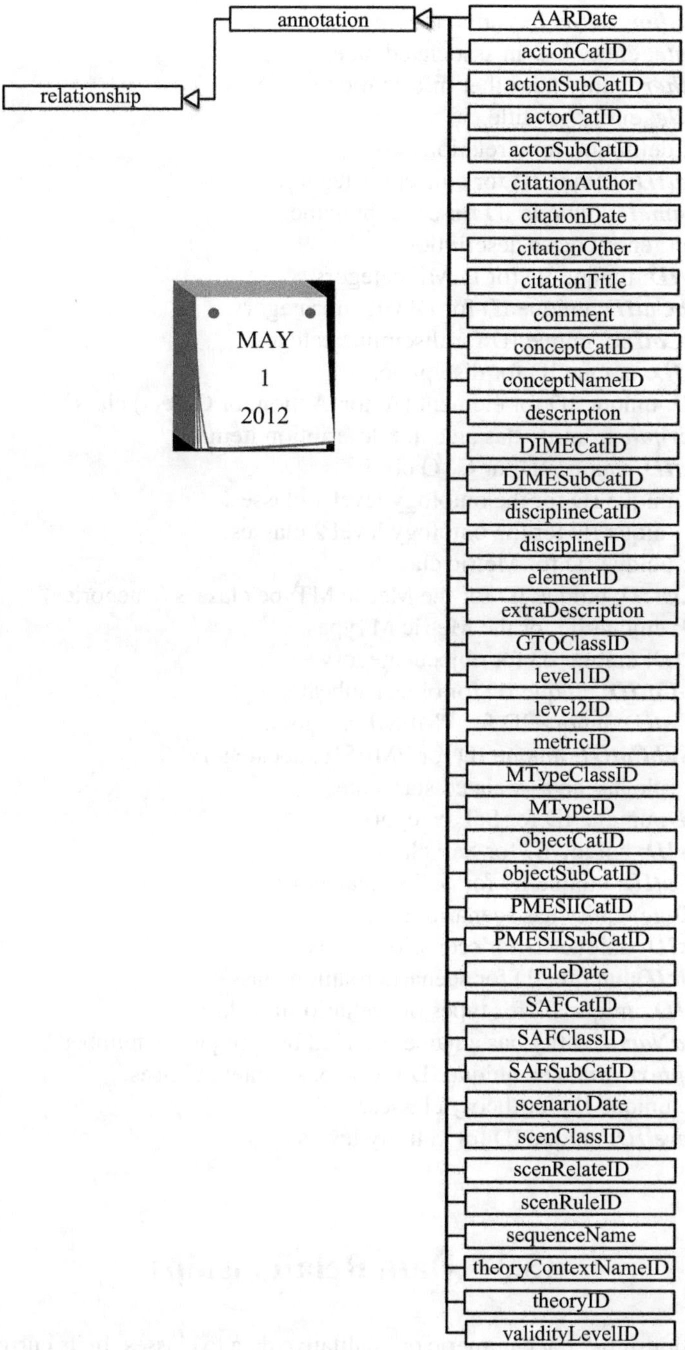

Fig. 13.2 *annotationRelationships*

citationAuthor: entity has authors.
citationDate: entity has an associated date.
citationOther: entity has other information.
citationTitle: entity has title.
comment: definition for a relationship.
conceptCatID: unique ID for concept category.
conceptNameID: unique ID for concept name.
description: entity has a description.
DIMECatID: unique ID for DIME category.
DIMESubCatID: unique ID for DIME subcategory.
disciplineCatID: unique ID for discipline category.
disciplineID: unique ID for discipline.
elementID: unique ID for element (Actor, Action, or Object) class.
extraDescription: entity has an extra description item.
GTOClassID: unique ID for GTO class.
level1ID: unique ID for the ontology level 1 classes.
level2ID: unique ID for the ontology level 2 classes.
metricID: unique ID for Metric class.
MTypeClassID: unique ID for the Metric MType classes (categories).
MTypeID: unique ID for the Metric MTypes.
objectCatID: unique ID for object category.
objectSubCatID: unique ID for object subcategory.
PMESIICatID: unique ID for PMESII category.
PMESIISubCatID: unique ID for PMESII subcategory.
ruleDate: rule has an associated start date.
SAFCatID: unique ID for SaF category.
SAFClassID: unique ID for SaF class.
SAFSubCatID: unique ID for SaF subcategory.
scenarioDate: entity has scenario date.
scenClassID: unique ID for scenario classes.
scenRelateID: unique ID for scenario relation classes.
scenRuleID: unique ID for types of scenario rule class.
sequencedName: entity has a name prefixed by a sequence number.
theoryContextNameID: unique ID for theory context classes.
theoryID: unique ID for theory classes.
validityLevelID: unique ID for validity levels.

13.3 Data Relationships

Data relationships attach numeric or qualitative data to classes. In its current design, the MCO places most of the numeric and qualitative data into the Metrics, leaving only these two relationships between classes or instances and data values.

Data Relationship Categories

Figure 13.3 extends Fig. 13.1 by adding the two sub-relationships of the *dataRelationship*. Each of the sub-relationships is defined in the next section. The words in the green box in the figure illustrate a data relationship of the *hasValidityValue* subcategory.

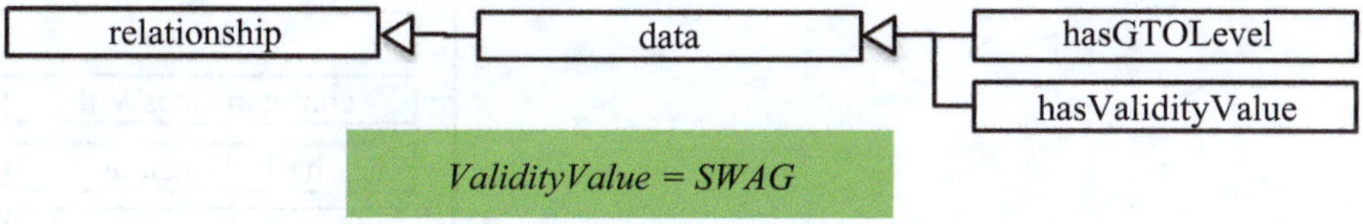

Fig. 13.3 *dataRelationships*

Data Relationship Definitions

All data relationships are sub-relationships of *dataRelationship*, and none has an inverse relationship.

hasGTOLevel: entity has importance level (this refers to the importance of the Action to a subtask or a Metric to a subgoal).
hasValidityValue: entity has a validity value.

Possible Future Enhancements to Data Relationships

It is likely that an effort to fully populate the MCO for a particular conflict might motivate the creation of relationships between the Metrics and various data types, requiring new data relationships.

13.4 Object Relationships

Object relationships connect classes and instances. (The word "object" in this usage has no connection to the Object classes of the MCO. This is the term used in Protégé for this type of relationship.) The MCO contains numerous types of relationships, with some being sub-relationships of other relationships. Object relationships are the most visible of the relationships and permit the representation of difficult features of the modern conflict domain. Most of these relationships are created specifically for the MCO because real-world relationships call for them. Their names, therefore, have been invented. As a result, it is possible that some names could be refined and some of the definitions could be clarified. However, they are presented here as they currently exist.

Object Relationship Categories

Figure 13.4 extends Fig. 13.1 by adding the 20 top-level sub-relationships of *objectRelationship*. The New York Yankee's baseball cap in the figure illustrates the *isAffliatedWith* subcategory.

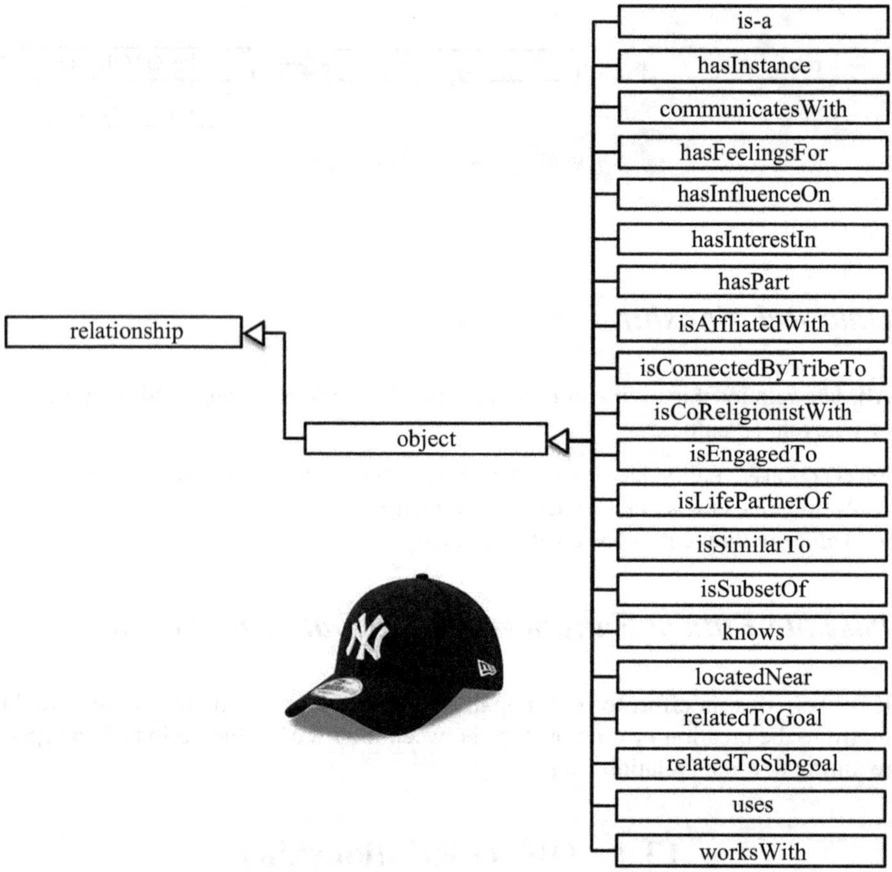

Fig. 13.4 Top-level *objectRelationships*

Some of these sub-relationships have their own sub-relationships. The following figures illustrate those with lower levels, described as families in the captions. The sub-relationships and their sub-sub-relationships are defined in a later section.

Figure 13.5 extends Fig. 13.4 by adding the two sub-relationships of the *communicatesWith* relationship. Figure 13.6 extends Fig. 13.4 by adding the seven sub-relationships of the *hasFeelingsFor* relationship. Figure 13.7 extends Fig. 13.4 by adding the one sub-relationship of the *hasInfluenceOn* relationship. Figure 13.8

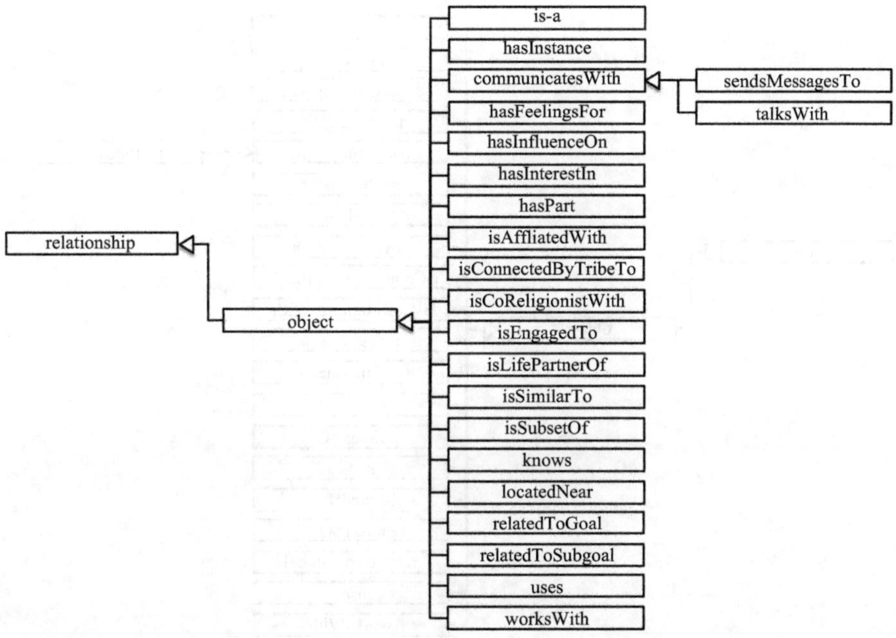

Fig. 13.5 *communicatesWith* relationship family

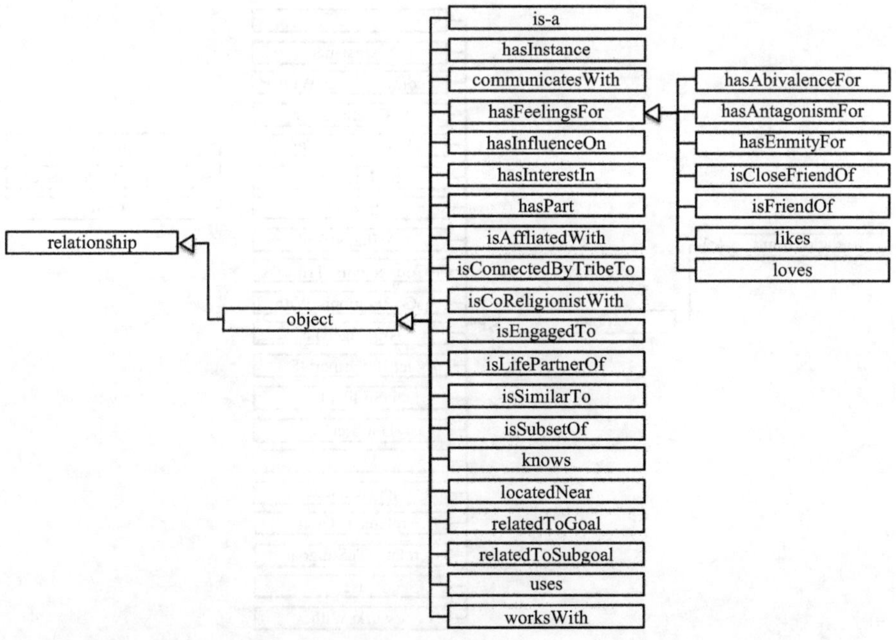

Fig. 13.6 *hasFeelingsFor* relationship family

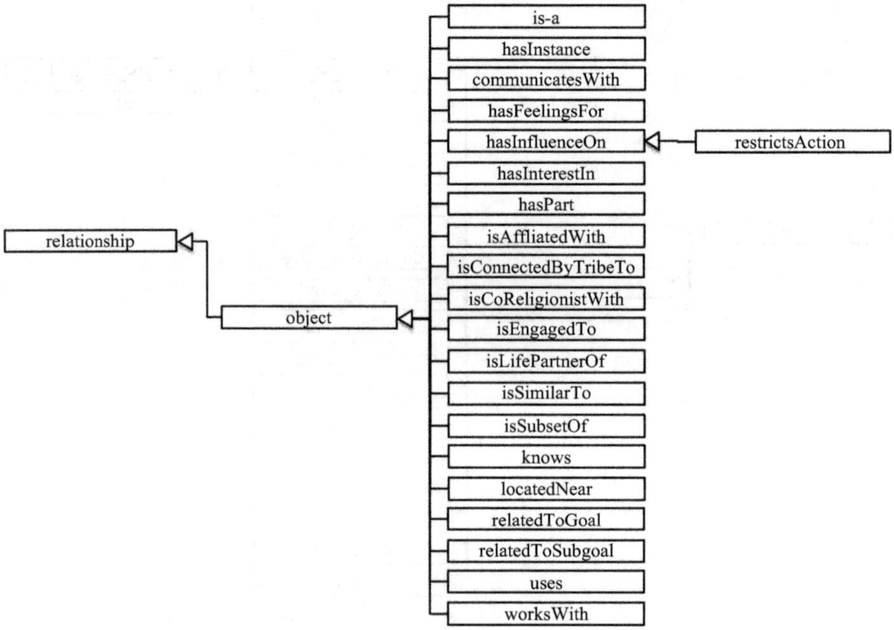

Fig. 13.7 *hasInfluenceOn* relationship family

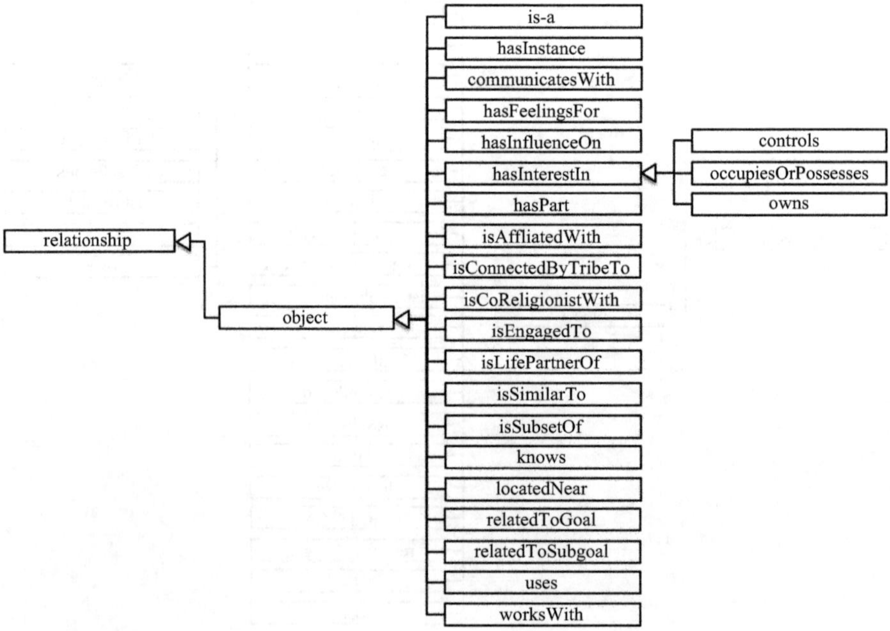

Fig. 13.8 *hasInterestIn* relationship family

extends Fig. 13.4 by adding the three sub-relationships of the *hasInterestIn* relationship.

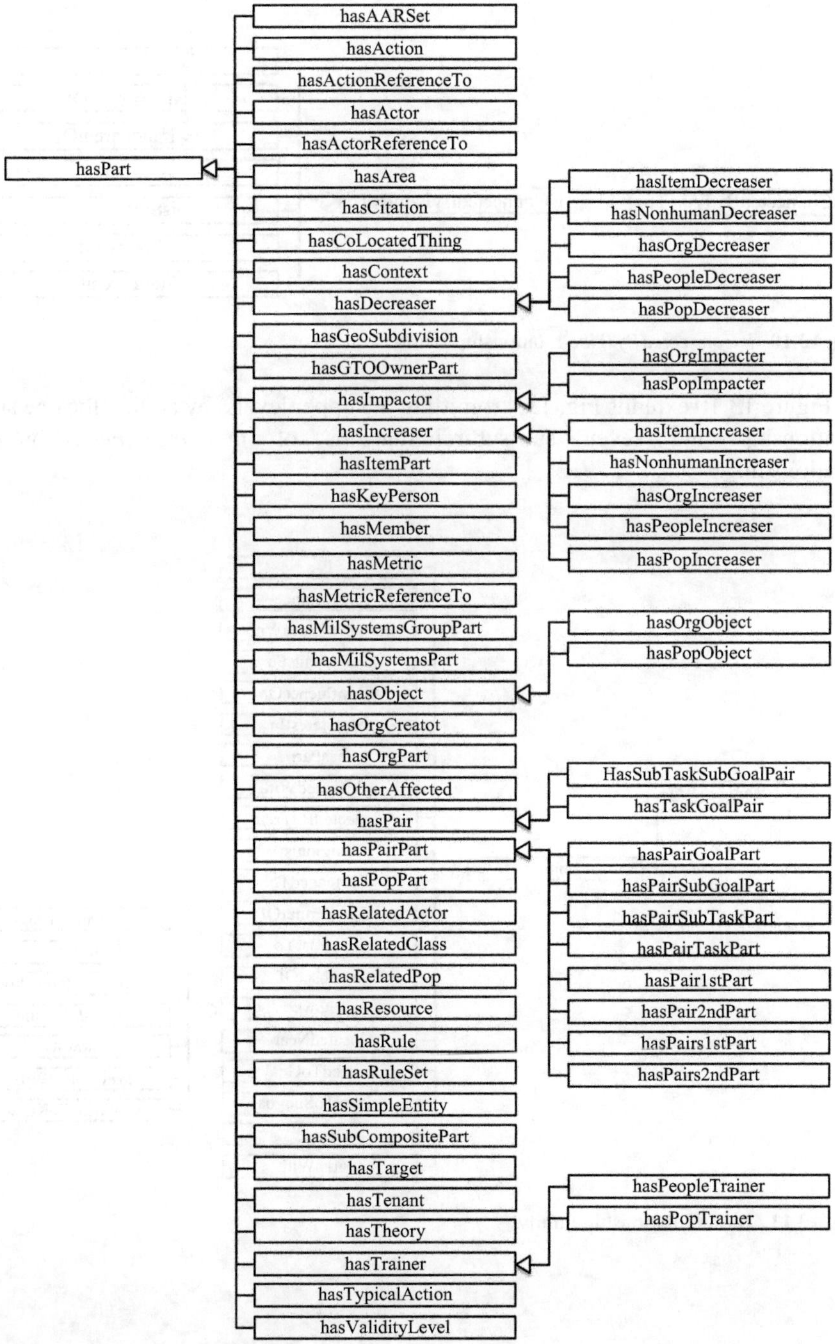

Fig. 13.9 *hasPart* relationship family

Figure 13.9 extends Fig. 13.4 (omitting the upper levels) by adding the 42 sub-relationships and 26 sub-sub-relationships of the *hasPart* relationship.

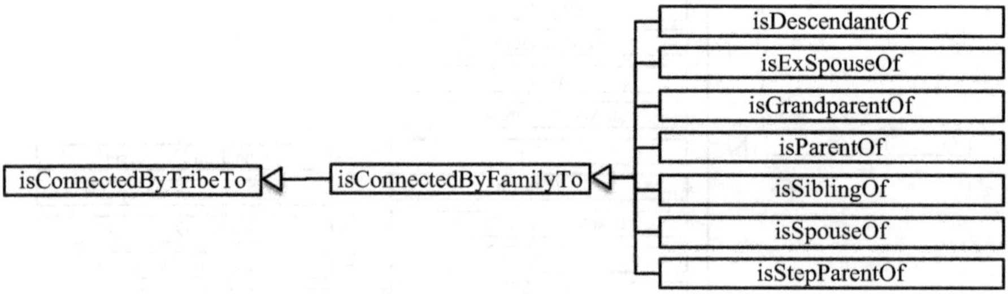

Fig. 13.10 *isConnectedByTribe* relationship family

Figure 13.10 extends Fig. 13.4 (omitting the upper levels) by adding the one sub-relationship and seven sub-sub-relationships of the *isConnectedByTribe* relationship.

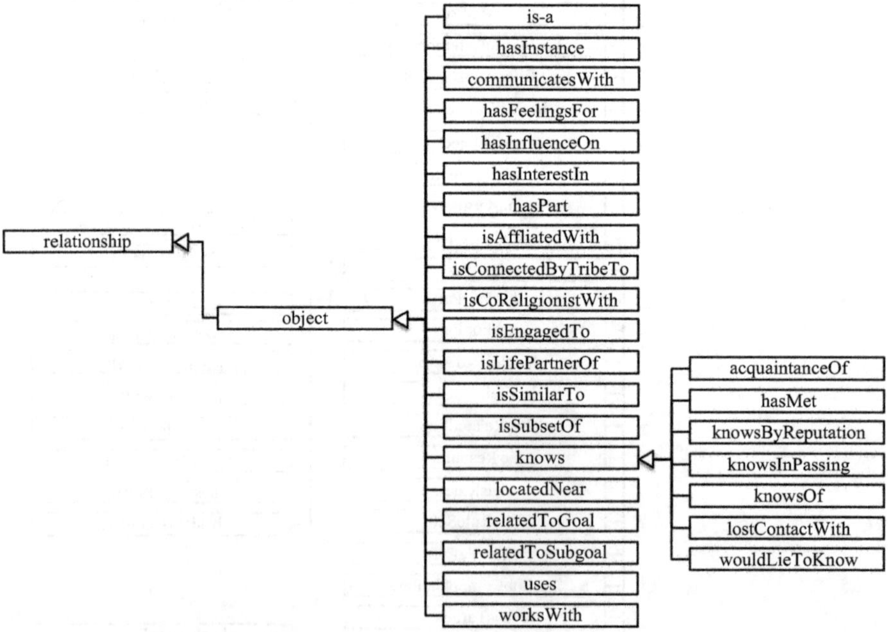

Fig. 13.11 *knows* relationship family

Figure 13.11 extends Fig. 13.4 by adding the seven sub-relationships of the *knows* relationship.

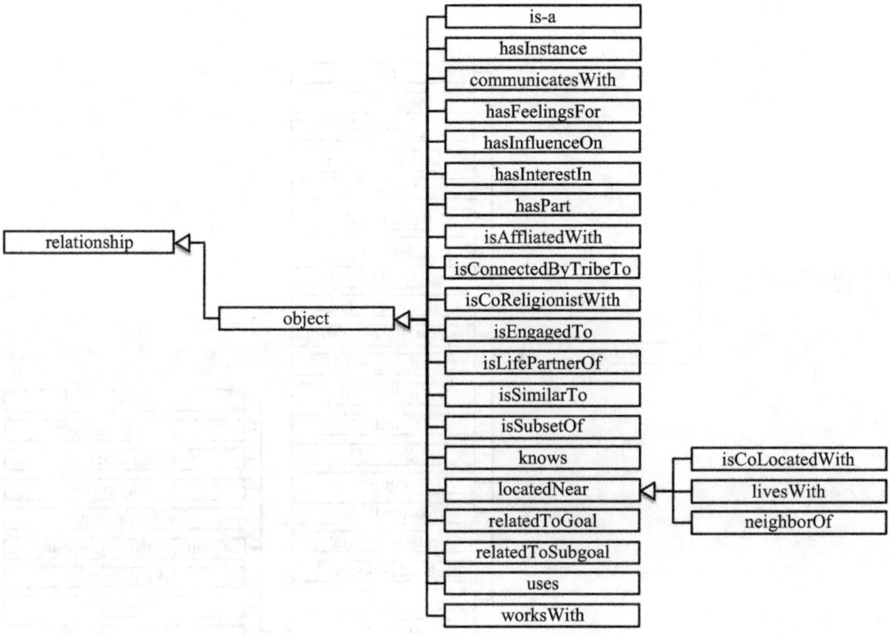

Fig. 13.12 *locatedNear* relationship family

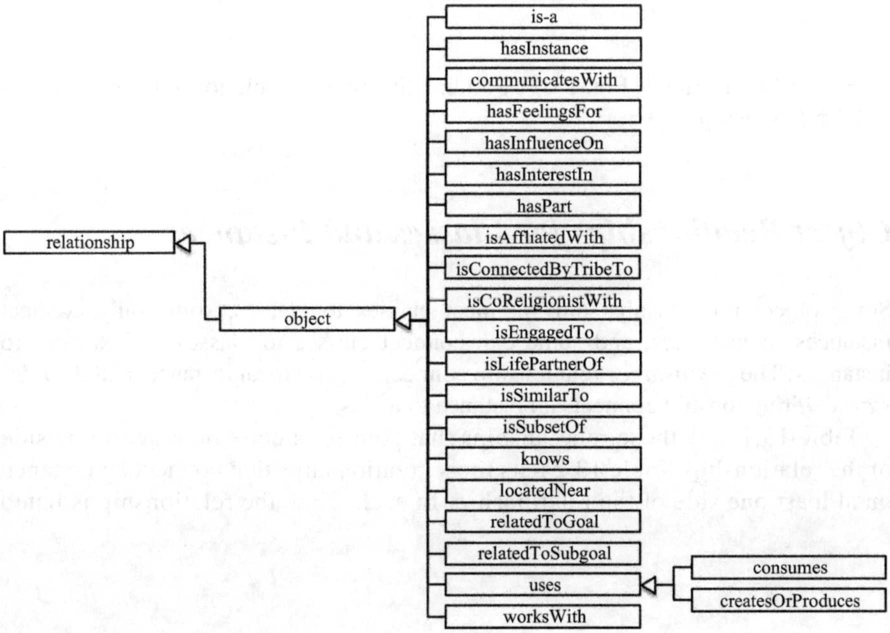

Fig. 13.13 *uses* relationship family

Figure 13.12 extends Fig. 13.4 by adding the three sub-relationships of the *locatedNear* relationship. Figure 13.13 extends Fig. 13.4 by adding the two sub-relationships of the *uses* relationship.

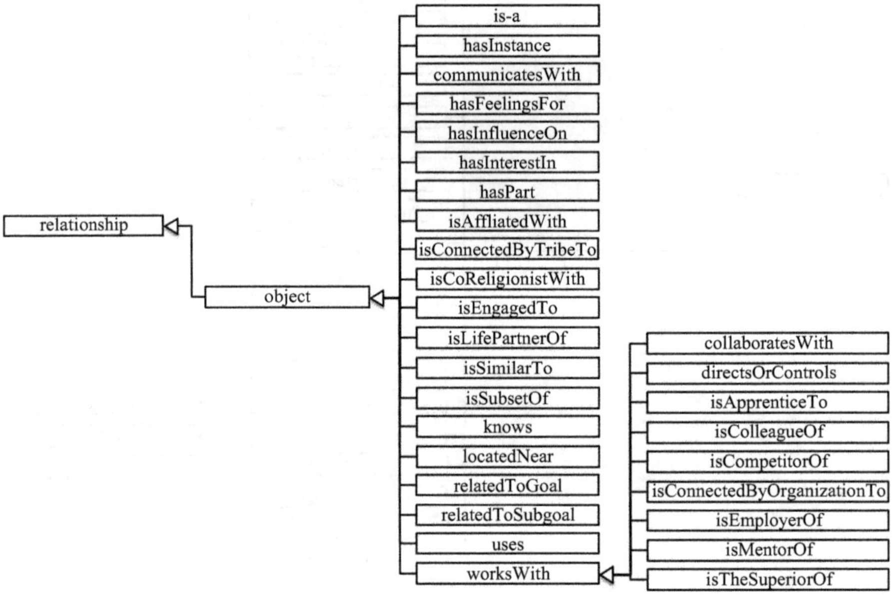

Fig. 13.14 *worksWith* relationship family

Figure 13.14 extends Fig. 13.4 by adding the nine sub-relationships of the *worksWith* relationship.

Object Relationships for Classes and Instances

Some object relationships only connect classes to classes; some only connect instances to instances; and some can connect classes to classes or instances to instances. The *hasInstance* relationship connects a class to an instance, and the *isInstanceOf* relationship connects an instance to a class.

Table 13.1 lists those relationships that connect a class on at least one side of the relationship. Table 13.2 lists those relationships that connect an instance on at least one side of the relationship. In each table, the relationship is listed

Table 13.1 Object relationships used with classes

Super-relationship	Relationship	Inverse
	is-a	*superClassOf*
	hasInstance	*isInstanceOf*
	hasInfluenceOn	*isInfluencedBy*
hasInfluenceOn	*restrictsAction*	*isRestrictedBy*
	hasPart	*isPartOf*
hasPart	*hasAARSet*	*isAARSetOf*
hasPart	*hasActionReferenceTo*	*isActionReferencedBy*
hasPart	*hasActorReferenceTo*	*isActorReferencedBy*
hasPart	*hasCitation*	*isCitationOf*
hasPart	*hasContext*	*isContextOf*
hasPart	*hasDecreaser*	*isDecreaserOf*
hasDecreaser	*hasItemDecreaser*	*isItemDecreaserOf*
hasDecreaser	*hasNonhumanDecreaser*	*isNonhumanDecreaserOf*
hasDecreaser	*hasOrgDecreaser*	*isOrgDecreaserOf*
hasDecreaser	*hasPeopleDecreaser*	*isPeopleDecreaserOf*
hasDecreaser	*hasPopDecreaser*	*isPopDecreaserOf*
hasPart	*hasGTOOwnerPart*	*isGTOOwnerPartOf*
hasPart	*hasImpacter*	*isImpacterOf*
hasImpacter	*hasOrgImpacter*	*isOrgImpacterOf*
hasImpacter	*hasPopImpacter*	*isPopImpacterOf*
hasPart	*hasIncreaser*	*isIncreaserOf*
hasIncreaser	*hasItemIncreaser*	*isItemIncreaserOf*
hasIncreaser	*hasNonhumanIncreaser*	*isNonhumanIncreaserOf*
hasIncreaser	*hasOrgIncreaser*	*isOrgIncreaserOf*
hasIncreaser	*hasPeopleIncreaser*	*isPeopleIncreaserOf*
hasIncreaser	*hasPopIncreaser*	*isPopIncreaserOf*
hasPart	*hasItemPart*	*isItemPartOf*
hasPart	*hasKeyPerson*	*isKeyPersonOf*
hasPart	*hasMetric*	*isMetricFor*
hasPart	*hasMetricReferenceTo*	*isMetricReferencedBy*
hasPart	*hasObject*	*isObjectOf*
hasObject	*hasOrgObject*	*isOrgObjectOf*
hasObject	*hasPopObject*	*isPopObjectOf*
hasPart	*hasOrgCreator*	*isOrgCreatorOf*
hasPart	*hasOrgPart*	*isOrgPartOf*
hasPart	*hasPair*	*isPairOf*
hasPair	*hasSubtaskSubgoalPair*	*isSubtaskSubgoalOf*
hasPair	*hasTaskGoalPair*	*isTaskGoalOf*
hasPart	*hasPairPart*	*isPartOfPairOf*
hasPairPart	*hasPairGoalPart*	*isGoalOf*
hasPairPart	*hasPairSubgoalPart*	*isSubgoalOf*
hasPairPart	*hasPairSubtaskPart*	*isSubtaskOf*

(continued)

Table 13.1 (continued)

Super-relationship	Relationship	Inverse
hasPairPart	*hasPairTaskPart*	*isTaskOf*
hasPart	*hasPopPart*	*isPopPartOf*
hasPart	*hasRelatedActor*	*isRelatedActorOf*
hasPart	*hasRelatedPop*	*isRelatedPopOf*
hasPart	*hasRule*	*isRuleOf*
hasPart	*hasRuleSet*	*isRuleSetOf*
hasPart	*hasTrainer*	*isTrainerOf*
hasTrainer	*hasPeopleTrainer*	*isPeopleTrainerOf*
hasTrainer	*hasPopTrainer*	*isPopTrainerOf*
hasPart	*hasTypicalAction*	*isTypicalActionOf*
hasPart	*hasValidityLevel*	*isValidityLevelOf*
	isSimilarTo	*isSimilarTo*
	relatedToGoal	*relatedToTask*
	relatedToSubgoal	*relatedToSubtask*

Table 13.2 Object properties used with instances

Super-relationship	Relationship	Inverse
	is-a	*superClassOf*
	isInstanceOf	*hasInstance*
	relatedToGoal	*relatedToTask*
	relatedToSubgoal	*relatedToSubtask*
	hasInfluenceOn	*isInfluencedBy*
	communicatesWith	*communicatesWith*
communicatesWith	*sendsMessagesTo*	*receivesMessagesFrom*
communicatesWith	*talksWith*	*talksWith*
	hasFeelingsFor	*isSubjectOfFeelingsBy*
hasFeelingsFor	*hasAmbivalenceFor*	*isViewedWithAmbivalenceBy*
hasFeelingsFor	*hasAntagonismFor*	*hasAntagonist*
hasFeelingsFor	*hasEnmityFor*	*hasEnemy*
hasFeelingsFor	*isCloseFriendOf*	*isCloseFriendOf*
hasFeelingsFor	*isFriendOf*	*isFriendOf*
hasFeelingsFor	*likes*	*isLikedBy*
hasFeelingsFor	*loves*	*isLovedBy*
	hasInterestIn	*engagesInterestOf*
hasInterestIn	*controls*	*isControlledBy*
hasInterestIn	*occupiesOrPossesses*	*isOccupiedOrPossessedBy*
hasInterestIn	*owns*	*isOwnedBy*
	hasPart	*isPartOf*
hasPart	*hasAARSet*	*isAARSetOf*
hasPart	*hasAction*	*isActionOf*
hasPart	*hasActionReferenceTo*	*isActionReferencedBy*

(continued)

Table 13.2 (continued)

Super-relationship	Relationship	Inverse
hasPart	*hasActor*	*isActorOf*
hasPart	*hasActorReferenceTo*	*isActorReferencedBy*
hasPart	*hasArea*	*isAreaOf*
hasPart	*hasCitation*	*isCitationOf*
hasPart	*hasCoLocatedThing*	*isCoLocatedWith*
hasPart	*hasGeoSubdivision*	*isGeoSubdivisionOf*
hasPart	*hasGTOOwnerPart*	*isGTOOwnerPartOf*
hasPart	*hasKeyPerson*	*isKeyPersonOf*
hasPart	*hasMember*	*isMemberOf*
hasPart	*hasMetricReferenceTo*	*isMetricReferencedBy*
hasPart	*hasMilSystemsGroupPart*	*isMilSystemsGroupOf*
hasPart	*hasMilSystemsPart*	*isMilSystemsPartOf*
hasPart	*hasOtherAffected*	*isOtherAffectedBy*
hasPart	*hasPair*	*isPairOf*
hasPair	*hasSubtaskSubgoalPair*	*isSubtaskSubgoalOf*
hasPair	*hasTaskGoalPair*	*isTaskGoalOf*
hasPart	*hasPairPart*	*isPartOfPairOf*
hasPairPart	*hasPair1stPart*	*isPair1stPartOf*
hasPairPart	*hasPair2ndPart*	*isPair2ndPartOf*
hasPairPart	*hasPairGoalPart*	*isGoalOf*
hasPairPart	*hasPairs1stPart*	*isPairs1stPartOf*
hasPairPart	*hasPairs2ndPart*	*isPairs2ndPartOf*
hasPairPart	*hasPairSubgoalPart*	*isSubgoalOf*
hasPairPart	*hasPairSubtaskPart*	*isSubtaskOf*
hasPairPart	*hasPairTaskPart*	*isTaskOf*
hasPart	*hasRelatedClass*	*isRelatedClassOf*
hasPart	*hasResource*	*isResourceOf*
hasPart	*hasRule*	*isRuleOf*
hasPart	*hasRuleSet*	*isRuleSetOf*
hasPart	*hasSimpleEntity*	*isSimpleEntityOf*
hasPart	*hasSubCompositePart*	*isSubCompositePartOf*
hasPart	*hasTarget*	*isTargetOf*
hasPart	*hasTenant*	*isTenantOf*
hasPart	*hasTheory*	*isTheoryOf*
	isAffiliatedWith	*hasAffiliate*
	isConnectedByTribeTo	*isConnectedByTribeTo*
isConnectedByTribeTo	*isConnectedByFamilyTo*	*isConnectedByFamilyTo*
isConnectedByFamilyTo	*isDescendantOf*	*isAncestorOf*
isConnectedByFamilyTo	*isExSpouseOf*	*isExSpouseOf*
isConnectedByFamilyTo	*isGrandparentOf*	*isGrandchildOf*
isConnectedByFamilyTo	*isParentOf*	*isChildOf*
isConnectedByFamilyTo	*isSiblingOf*	*isSiblingOf*

(continued)

Table 13.2 (continued)

Super-relationship	Relationship	Inverse
isConnectedByFamilyTo	*isSpouseOf*	*isSpouseOf*
isConnectedByFamilyTo	*isStepParentOf*	*isStepChildOf*
	isCoReligionistWith	*isCoReligionistWith*
	isEngagedTo	*isEngagedTo*
	isLifePartnerOf	*isLifePartnerOf*
	isSubsetOf	*isSupersetOf*
	knows	*isKnownBy*
knows	*acquaintanceOf*	*hasBeenMadeAcquaintanceBy*
knows	*hasMet*	*hasBeenMetBy*
knows	*knowsByReputation*	*isKnownByReputationBy*
knows	*knowsInPassing*	*isKnownInPassingBy*
knows	*knowsOf*	*isKnownBy*
knows	*lostContactWith*	*isNotInContactWith*
knows	*wouldLikeToKnow*	*isDesiredToBeKnownBy*
	locatedNear	*locatedNear*
locatedNear	*isCoLocatedWith*	*isCoLocatedWith*
locatedNear	*livesWith*	*livesWith*
locatedNear	*neighborOf*	*neighborOf*
	uses	*isUsedBy*
uses	*consumes*	*isConsumedBy*
uses	*createsOrProduces*	*isCreatedOrProducedBy*
	worksWith	*worksWith*
worksWith	*collaboratesWith*	*collaboratesWith*
worksWith	*directsOrControls*	*isDirectedOrControlledBy*
worksWith	*isApprenticeTo*	*hasApprentice*
worksWith	*isColleagueOf*	*isColleagueOf*
worksWith	*isCompetitorOf*	*hasCompetitor*
worksWith	*isConnectedByOrganizationTo*	*isConnectedByOrganizationTo*
worksWith	*isEmployerOf*	*isEmployedBy*
worksWith	*isMentorOf*	*isMenteeOf*
worksWith	*isTheSuperiorOf*	*isTheSubordinateOf*

in the center column (the "sorted" column), with its super-relationship (the relationship of which it is a sub-relationship) in the left column, and its inverse relationship in the right column. The center and right columns have the names indented to show sub-relationship associations. Where a super-relationship entry is blank, *objectRelationship* can be assumed to be the direct super-relationship.

All of the relationships are collected and ordered alphabetically in the next section. In that section, each relationship is named and defined. The inverse relationship and the super-relationships are also named.

Object Relationship Definitions

All object relationships are sub-relationships of *objectRelationship*, either directly or indirectly. All object relationships have inverse relationships, which must be explicitly defined in the ontology. In general, the inverse relationships are not included as separate entries in the definitions below but are named as the inverse in the definition of the dominant relationship.

Some relationships are symmetric: the inverse relationship is the same as the relationship. An example is *collaboratesWith*. Thus, if Actor2 *collaboratesWith* Actor1, then Actor1 *collaboratesWith* Actor2. Relationships that are optionally symmetric are specifically defined as non-symmetric. For example, if Actor1 *hasAmbivalenceFor* Actor2, then Actor2 *hasAmbivalenceFor* Actor1 might be true or false. Thus, the inverse is defined as *isViewedWithAmbivalenceBy*, which correctly expresses the relationship in the other direction: Actor2 *isViewedWithAmbivalenceBy* Actor1. The actual feelings of Actor2 toward Actor1 must be stated separately and explicitly.

acquaintanceOf: Actor2 has more than slight or superficial knowledge of Actor1 but short of friendship. Its inverse relationship is *hasBeenMadeAcquaintanceBy*. It is a sub-relationship of *knows*.

collaboratesWith: Actor2 works toward a common goal with Actor1. Its inverse relationship is *collaboratesWith*. It is a sub-relationship of *worksWith*.

communicatesWith: Actor1 communicates with Actor2. Its inverse relationship is *communicatesWith*. It is a sub-relationship of *objectRelationship*.

consumes: Actor consumes the thing or portions of the thing. Its inverse relationship is *isConsumedBy*. It is a sub-relationship of *uses*.

controls: Actor controls access or use of the thing. Its inverse relationship is *isControlledBy*. It is a sub-relationship of *hasInterestIn*.

createsOrProduces: Actor creates or produces the thing. Its inverse relationship is *isCreatedOrProducedBy*. It is a *sub-relationship* of *uses*.

directsOrControls: Actor1 directs or controls the actions of Actor2. Its inverse relationship is *isDirectedOrControlledBy*. It is a sub-relationship of *worksWith*.

hasAARSet: entity has AAR Set. Its inverse relationship is *isAARSetOf*. It is a sub-relationship of *hasPart*.

hasAction: entity commits action. Its inverse relationship is *isActionOf*. It is a sub-relationship of *hasPart*.

hasActionReferenceTo: entity refers to action class. Its inverse relationship is *isActionReferencedBy*. It is a sub-relationship of *hasPart*.

hasActor: entity has an actor. Its inverse relationship is *isActorOf*. It is a sub-relationship of *hasPart*.

hasActorReferenceTo: entity refers to actor class. Its inverse relationship is *isActorReferencedBy*. It is a sub-relationship of *hasPart*.

hasAmbivalenceFor: Actor1 has mixed feelings or emotions toward Actor2. Its inverse relationship is *isViewedWithAmbivalenceBy*. It is a sub-relationship of *hasFeelingsFor*.

hasAntagonismFor: Actor1 opposes and contends against Actor2. Its inverse relationship is *hasAntagonist*. It is a sub-relationship of *hasFeelingsFor*.

hasArea: entity affects action in the area. Its inverse relationship is *isAreaOf*. It is a sub-relationship of *hasPart*.

hasCitation: entity has a citation. Its inverse relationship is *isCitationOf*. It is a sub-relationship of *hasPart*.

hasCoLocatedThing: entity1 has a thing, entity2, co-located with it. Its inverse relationship is *isCoLocatedWith*. It is a sub-relationship of *hasPart*.

hasContext: entity has a context. Its inverse relationship is *isContextOf*. It is a sub-relationship of *hasPart*.

hasDecreaser: entity has an action that decreases it in number. Its inverse relationship is *isDecreaserOf*. It is a sub-relationship of *hasPart*.

hasEnmityFor: Actor1 feels hatred, intends injury to, or opposes the interests of Actor2. Its inverse relationship is *hasEnemy*. It is a sub-relationship of *hasFeelingsFor*.

hasFeelingsFor: Actor1 has feelings for Actor2. Its inverse relationship is *isSubjectOfFeelingsBy*. It is a sub-relationship of *objectRelationship*.

hasGeoSubdivision: entity has *geographicSubdivision* with location. Its inverse relationship is *isGeoSubdivisionOf*. It is a sub-relationship of *hasPart*.

hasGTOOwnerPart: scenario has another entity as a GTO owner. Its inverse relationship is *isGTOOwnerPartOf*. It is a sub-relationship of *hasPart*.

hasImpacter: entity has an action that impacts it. Its inverse relationship is *isImpacterOf*. It is a sub-relationship of *hasPart*.

hasIncreaser: entity has an action that increases it in number. Its inverse relationship is *isIncreaserOf*. It is a sub-relationship of *hasPart*.

hasInfluenceOn: entity has influence on another entity. Its inverse relationship is *isInfluencedBy*. It is a sub-relationship of *objectRelationship*.

hasInstance: class has an instance. Its inverse relationship is *isInstanceOf*. It is a sub-relationship of *objectRelationship*.

hasInterestIn: Actor is concerned about the thing. Its inverse relationship is *engagesInterestOf*. It is a sub-relationship of *objectRelationship*.

hasItemDecreaser: entity has an action that decreases its number of items. Its inverse relationship is *isItemDecreaserOf*. It is a sub-relationship of *hasDecreaser*.

hasItemIncreaser: entity has an action that increases its number of items. Its inverse relationship is *isItemIncreaserOf*. It is a sub-relationship of *hasIncreaser*.

hasItemPart: entity has an object as a part. Its inverse relationship is *isItemPartOf*. It is a sub-relationship of *hasPart*.

hasKeyPerson: entity has a key person. Its inverse relationship is *isKeyPersonOf*. It is a sub-relationship of *hasPart*.

hasMember: Actor1 (a multi-member actor) has Actor2 as a member. Its inverse relationship is *isMemberOf*. It is a sub-relationship of *hasPart*.

hasMet: Actor2 has met Actor1 whether in passing or longer. Its inverse relationship is *hasBeenMetBy*. It is a sub-relationship of *knows*.

hasMetric: entity has a metric assigned to it. Its inverse relationship is *isMetricFor*. It is a sub-relationship of *hasPart*.

hasMetricReferenceTo: entity refers to metric class. Its inverse relationship is *isMetricReferencedBy*. It is a sub-relationship of *hasPart*.

hasMilSystemsGroupPart: entity1 has entity2, which is a *milSystemsGroup*, as a part. Its inverse relationship is *isMilSystemsGroupOf*. It is a sub-relationship of *hasPart*.

hasMilSystemsPart: entity1 has a *milSystem* entity, entity2, as a part. Its inverse relationship is *isMilSystemsPartOf*. It is a sub-relationship of *hasPart*.

hasNonhumanDecreaser: entity has an action that decreases its number of *nonhumanActors*. Its inverse relationship is *isNonhumanDecreaserOf*. It is a sub-relationship of *hasDecreaser*.

hasNonhumanIncreaser: entity has an action that increases its number of *nonhumanActors*. Its inverse relationship is *isNonhumanIncreaserOf*. It is a sub-relationship of *hasIncreaser*.

hasObject: entity has a related object. Its inverse relationship is *isObjectOf*. It is a sub-relationship of *hasPart*.

hasOrgCreator: entity has an action that creates it. Its inverse relationship is *isOrgCreatorOf*. It is a sub-relationship of *hasPart*.

hasOrgDecreaser: entity has an action that decreases its number of organizations. Its inverse relationship is *isOrgDecreaserOf*. It is a sub-relationship of *hasDecreaser*.

hasOrgImpacter: entity has an action that impacts its organization entity. Its inverse relationship is *isOrgImpacterOf*. It is a sub-relationship of *hasImpacter*.

hasOrgIncreaser: entity has an action that increases its number of organizations. Its inverse relationship is *isOrgIncreaserOf*. It is a sub-relationship of *hasIncreaser*.

hasOrgObject: entity has an object related to its organization entity. Its inverse relationship is *isOrgObjectOf*. It is a sub-relationship of *hasObject*.

hasOrgPart: entity has an organization entity as a part. Its inverse relationship is *isOrgPartOf*. It is a sub-relationship of *hasPart*.

hasOtherAffected: entity has other affected entity. Its inverse relationship is *isOtherAffectedBy*. It is a sub-relationship of *hasPart*.

hasPair: entity has a pair. Its inverse relationship is *isPairOf*. It is a sub-relationship of *hasPart*.

hasPair1stPart: entity has first part of pair. Its inverse relationship is *isPair1stPartOf*. It is a sub-relationship of *hasPairPart*.

hasPair2ndPart: entity has second part of pair. Its inverse relationship is *isPair2ndPartOf*. It is a sub-relationship of *hasPairPart*.

hasPairGoalPart: entity as a goal. Its inverse relationship is *isGoalOf*. It is a sub-relationship of *hasPairPart*.

hasPairPart: entity has a part of a pair. Its inverse relationship is *isPartOfPairOf*. It is a sub-relationship of *hasPart*.

hasPairs1stPart: entity has first part of one of n pairs. Its inverse relationship is *isPairs1stPartOf*. It is a sub-relationship of *hasPairPart*.

hasPairs2ndPart: entity has second part of one of n pairs. Its inverse relationship is *isPairs2ndPartOf*. It is a sub-relationship of *hasPairPart*.

hasPairSubgoalPart: entity has a subgoal. Its inverse relationship is *isSubgoalOf*. It is a sub-relationship of *hasPairPart*.

hasPairSubtaskPart: entity has a subtask. Its inverse relationship is *isSubtaskOf*. It is a sub-relationship of *hasPairPart*.

hasPairTaskPart: entity has a task. Its inverse relationship is *isTaskOf*. It is a sub-relationship of *hasPairPart*.

hasPart: entity has another entity as a part. Its inverse relationship is *isPartOf*. It is a sub-relationship of *objectRelationship*.

hasPeopleDecreaser: entity has an action that decreases its number of people. Its inverse relationship is *isPeopleDecreaserOf*. It is a sub-relationship of *hasDecreaser*.

hasPeopleIncreaser: entity has an action that increases its number of people. Its inverse relationship is *isPeopleIncreaserOf*. It is a sub-relationship of *hasIncreaser*.

hasPeopleTrainer: entity has an action that trains its people. Its inverse relationship is *isPeopleTrainerOf*. It is a sub-relationship of *hasTrainer*.

hasPopDecreaser: entity has an action that decreases its population. Its inverse relationship is *isPopDecreaserOf*. It is a sub-relationship of *hasDecreaser*.

hasPopImpacter: entity has an action that impacts its population. Its inverse relationship is *isPopImpacterOf*. It is a sub-relationship of *hasImpacter*.

hasPopIncreaser: entity has an action that increases its population. Its inverse relationship is *isPopIncreaserOf*. It is a sub-relationship of *hasIncreaser*.

hasPopObject: entity has an object related to its population. Its inverse relationship is *isPopObjectOf*. It is a sub-relationship of *hasObject*.

hasPopPart: entity has a population entity as a part. Its inverse relationship is *isPopPartOf*. It is a sub-relationship of *hasPart*.

hasPopTrainer: entity has an action that trains its population. Its inverse relationship is *isPopTrainerOf*. It is a sub-relationship of *hasTrainer*.

hasRelatedActor: entity has a related Actor. Its inverse relationship is *isRelatedActorOf*. It is a sub-relationship of *hasPart*.

hasRelatedClass: entity1 has entity2 as a related class – government for government composite, armed force for armed force composite, etc. Its inverse relationship is *isRelatedClassOf*. It is a sub-relationship of *hasPart*.

hasRelatedPop: entity has a related population. Its inverse relationship is *isRelatedPopOf*. It is a sub-relationship of *hasPart*.

hasResource: entity has a resource. Its inverse relationship is *isResourceOf*. It is a sub-relationship of *hasPart*.

hasRule: entity has rule. Its inverse relationship is *isRuleOf*. It is a sub-relationship of *hasPart*.

hasRuleSet: entity has rule set. Its inverse relationship is *isRuleSetOf*. It is a sub-relationship of *hasPart*.

hasSimpleEntity: entity2 is the simple entity base of the composite entity1. Its inverse relationship is *isSimpleEntityOf*. It is a sub-relationship of *hasPart*.

hasSubCompositePart: entity1 has a composite entity, entity2, of the same type as entity1 as a part. Its inverse relationship is *isSubCompositePartOf*. It is a sub-relationship of *hasPart*.

hasSubtaskSubgoalPair: entity has a subtask-subgoal pair. Its inverse relationship is *isSubtaskSubgoalOf*. It is a sub-relationship of *hasPair*.

hasTarget: entity has target. Its inverse relationship is *isTargetOf*. It is a sub-relationship of *hasPart*.

hasTaskGoalPair: entity owns a task-goal pair. Its inverse relationship is *isTaskGoalOf*. It is a sub-relationship of *hasPair*.

hasTenant: entity1 has an organization, entity2, that is a tenant. Its inverse relationship is *isTenantOf*. It is a sub-relationship of *hasPart*.

hasTheory: entity has theory for computing results. Its inverse relationship is *isTheoryOf*. It is a sub-relationship of *hasPart*.

hasTrainer: entity has an action that trains it. Its inverse relationship is *isTrainerOf*. It is a sub-relationship of *hasPart*.

hasTypicalAction: entity has a typical action. Its inverse relationship is *isTypicalActionOf*. It is a sub-relationship of *hasPart*.

hasValidityLevel: entity has a validity level. Its inverse relationship is *isValidityLevelOf*. It is a sub-relationship of *hasPart*.

is-a: class is subclass of superclass. Its inverse relationship is *superClassOf*. It is a sub-relationship of *objectRelationship*.

isAffiliatedWith: Actor1 is affiliated with Actor2. Its inverse relationship is *hasAffiliate*. It is a sub-relationship of *objectRelationship*.

isApprenticeTo: Actor2 serves as a trusted counselor or teacher to Actor1. Its inverse relationship is *hasApprentice*. It is a sub-relationship of *worksWith*.

isCloseFriendOf: Actor2 shares a close mutual friendship with Actor1. Its inverse relationship is *isCloseFriendOf*. It is a sub-relationship of *hasFeelingsFor*.

isColleagueOf: Actor2 is a member of the same profession as Actor1. Its inverse relationship is *isColleagueOf*. It is a sub-relationship of *worksWith*.

isCoLocatedWith: entity1 is collocated with entity2. Its inverse relationship is *isCoLocatedWith*. It is a sub-relationship of *locatedNear*.

isCompetitorOf: Actor1 is a competitor of Actor2. Its inverse relationship is *hasCompetitor*. It is a sub-relationship of *worksWith*.

isConnectedByFamilyTo: Actor1 is related to Actor2 by familial relationships. Its inverse relationship is *isConnectedByFamilyTo*. It is a sub-relationship of *isConnectedByTribeTo*.

isConnectedByOrganizationTo: Actor1 is connected to Actor2 by a mutual organization or group. Its inverse relationship is *isConnectedByOrganizationTo*. It is a sub-relationship of *worksWith*.

isConnectedByTribeTo: Actor1 is related to Actor2 by tribal relationship. Its inverse relationship is *isConnectedByTribeTo*. It is a sub-relationship of *object Relationship*.

isCoReligionistWith: Actor1 and Actor2 belong to the same religion. Its inverse relationship is *isCoReligionistWith.* It is a sub-relationship of *object Relationship.*

isDescendantOf: Actor1 is descended of Actor2. Its inverse relationship is *isAncestorOf.* It is a sub-relationship of *isConnectedByFamilyTo*.

isEmployerOf: Actor1 has engaged the services of Actor2. Its inverse relationship is *isEmployedBy*. It is a sub-relationship of *worksWith*.

isEngagedTo: Actor1 is betrothed to Actor2. Its inverse relationship is *isEngagedTo.* It is a sub-relationship of *objectRelationship*.

isExSpouseOf: Actor1 is ex-spouse of Actor2. Its inverse relationship is *isExSpouseOf.* It is a sub-relationship of *isConnectedByFamilyTo*.

isFriendOf: Actor2 shares mutual friendship with Actor1. Its inverse relationship is *isFriendOf.* It is a sub-relationship of *hasFeelingsFor*.

isGrandparentOf: Actor2 is the parent of any of Actor1's parents. Its inverse relationship is *isGrandchildOf.* It is a sub-relationship of *isConnectedByFamilyTo*.

isInstanceOf: entity is instance of a class. Its inverse relationship is *hasInstance*. It is a sub-relationship of *objectRelationship*.

isLifePartnerOf: Actor2 has made a long-term commitment to Actor1. Its inverse relationship is *isLifePartnerOf.* It is a sub-relationship of *objectRelationship*.

isMentorOf: Actor1 serves as a trusted counselor or teacher to Actor2. Its inverse relationship is *isMenteeOf.* It is a sub-relationship of *worksWith*.

isParentOf: Actor1 has given birth to or nurtured and raised Actor2. Its inverse relationship is *isChildOf.* It is a sub-relationship of *isConnectedByFamilyTo*.

isSiblingOf: Actor2 has one or both parents in common with Actor1. Its inverse relationship is *isSiblingOf.* It is a sub-relationship of *isConnectedByFamilyTo*.

isSimilarTo: entity is similar to an entity. Its inverse relationship is *isSimilarTo*. It is a sub-relationship of *objectRelationship*.

isSpouseOf: Actor2 is married to Actor1. Its inverse relationship is *isSpouseOf.* It is a sub-relationship of *isConnectedByFamilyTo*.

isStepParentOf: Actor1 is stepparent of Actor2. Its inverse relationship is *isStepChildOf.* It is a sub-relationship of *isConnectedByFamilyTo*.

isSubsetOf: Actor1 is a subset (sub-group, sub-organization, etc.) of Actor2. Its inverse relationship is *isSupersetOf.* It is a sub-relationship of *objectRelationship*.

isTheSuperiorOf: Actor1 is the superior of Actor2 in some organization. Its inverse relationship is *isTheSubordinateOf.* It is a sub-relationship of *worksWith*.

knows: Actor1 knows Actor2. Its inverse relationship is *isKnownBy*. It is a sub-relationship of *objectRelationship*.

knowsByReputation: Actor2 is known by Actor1 primarily for a particular action, position, or field of endeavor. Its inverse relationship is *isKnownByReputationBy*. It is a sub-relationship of *knows*.

knowsInPassing: Actor1 has slight or superficial knowledge of Actor2. Its inverse relationship is *isKnownInPassingBy*. It is a sub-relationship of *knows*.

knowsOf: Actor2 has come to be known to Actor1 through his, her, or its actions or position. Its inverse relationship is *isKnownBy*. It is a sub-relationship of *knows*.

likes: Actor1 likes Actor2. Its inverse relationship is *isLikedBy*. It is a sub-relationship of *hasFeelingsFor*.

livesWith: Actor2 shares a residence with Actor1. Its inverse relationship is *livesWith*. It is a sub-relationship of *locatedNear*.

locatedNear: Entity1 is located near Entity2. Its inverse relationship is *locatedNear*. It is a sub-relationship of *objectRelationship*.

lostContactWith: Actor2 was once known by Actor1 but has subsequently become uncontactable. Its inverse relationship is *isNotInContactWith*. It is a sub-relationship of *knows*.

loves: Actor1 loves Actor2. Its inverse relationship is *isLovedBy*. It is a sub-relationship of *hasFeelingsFor*.

neighborOf: Entity1 is in the same locality as Entity2. Its inverse relationship is *neighborOf*. It is a sub-relationship of *locatedNear*.

occupiesOrPossesses: Actor physically occupies or possesses the thing. Its inverse relationship is *isOccupiedOrPossessedBy*. It is a sub-relationship of *hasInterestIn*.

owns: Actor owns or has some property rights to the thing. Its inverse relationship is *isOwnedBy*. It is a sub-relationship of *hasInterestIn*.

relatedToGoal: task is related to goal. Its inverse relationship is *relatedToTask*. It is a sub-relationship of *objectRelationship*.

relatedToSubgoal: subtask is related to subgoal. Its inverse relationship is *relatedToSubtask*. It is a sub-relationship of *objectRelationship*.

restrictsAction: entity restricts the action. Its inverse relationship is *isRestrictedBy*. It is a sub-relationship of *hasInfluenceOn*.

sendsMessagesTo: Actor1 sends messages to Actor2. Its inverse relationship is *receivesMessagesFrom*. It is a sub-relationship of *communicatesWith*.

talksWith: Actor 1 and Actor2 talk together. Its inverse relationship is *talksWith*. It is a sub-relationship of *communicatesWith*.

uses: Actor makes use of or derives benefit from the thing. Its inverse relationship is *isUsedBy*. It is a sub-relationship of *objectRelationship*.

worksWith: Actor2 works for the same employer as Actor1. Its inverse relationship is *worksWith*. It is a sub-relationship of *objectRelationship*.

wouldLikeToKnow: Actor1 would desire to know Actor2 more closely. Its inverse relationship is *isDesiredToBeKnownBy*. It is a sub-relationship of *knows*.

Possible Future Enhancements to Object Relationships

The use of the word "object" for this type of relationship is confusing. A new term may be created in the future. In addition, the form of the object relationship names is not consistent.

- Generally, the first part of the name is a verb, such as "is," "has," or "uses." However, sometimes this is not the case. For example, *acquaintanceOf* could be renamed to *isAcquaintanceOf* and would be in conformance with such a rule.
- Many relationships start with "has," and their inverse relationship starts with "is," such as *hasTrainer* and *isTrainerOf*. Yet others that might follow this rule don't, such as *isMentorOf* and *isMenteeOf*. Perhaps this rule should be implemented.
- There are other cases where different forms of the relationship name have been used. For example, *isInstanceOf* and *instantiates* are synonyms. Formal synonym definitions might be implemented.
- Various sub-relationships of *hasPart* have a restrictive, component meaning. That is, the list of parts is required to be complete. Yet other sub-relationships have a non-restrictive, aggregation meaning. That is, any or all of the list of parts may be present. In some cases, one meaning is used in one case, and the other is used in another case. The definitions might be separated, creating new relationships, and enforced.

13.5 Relationships Recapitulation

The basic relationships in the MCO are the object relationships that connect classes and instances. These relationships range from structural ontology relationships such as *is-a* and *hasInstance* to domain-specific relationships such as *hasTaskGoalPair* and *owns*. The data relationships are technical artifacts of the need to specify connections to data in an environment in which most of the elements refer to things, not numbers or rankings. The annotation relationships are artifacts of the OWL language that support connecting additional information to a central element such as a class or instance without having to include the information in the name. Possible future enhancements are suggested.

Chapter 14
Conclusion

This book is not the final word on what we can know about modern conflict. However, it is a distillation of what is currently known and what has been done toward creating an ontology of modern conflict at the theater level, the Modern Conflict Ontology (MCO). There is much that remains to be learned, particularly regarding the social theories that drive the estimates of outcomes of both conventional and unconventional conflict activities. This chapter includes a review of the ontology's contents and how they fit together, a discussion of the uses of the ontology, and concluding thoughts.

Chapter 2 provided a more detailed discussion of modern conflict. Multiple books can be and have been written on the various aspects of modern conflict. In comparison, this is a superficial description; however, it provided the background understanding of the domain that is the topic of the MCO. The chapter discussed modern conflict in detail. It defined the components of modern conflict, considered the relative likelihood of those components occurring, and examined the status of modern conflict as a complex adaptive system. In Chap. 2, we also reviewed the various theories of conflict that have been advanced and reviewed over the course of years and centuries. We presented several illustrations of different kinds of conflict. We concluded the chapter with a characterization of modern conflict, providing a solid basis for creating an ontology.

Chapter 3 provided an overview of ontologies and the concepts involved in creating and understanding them. That chapter only provided enough information to understand the application of ontologies to modern conflict in the MCO. We began by discussing the choice of the best foundational ontology for the creation of an ontology for modern conflict. In Chap. 3 we also presented an overview of the structure of the MCO. It is also useful to compare the MCO with a related ontology and we did so. The MCO is derived from the Unconventional Conflict Ontology (UCO) (Hartley, An Ontology for Unconventional Conflict 2018), so a discussion of the changes from the UCO to the MCO was in order.

D. S. Hartley III, *An Ontology of Modern Conflict*, Understanding Complex Systems, https://doi.org/10.1007/978-3-030-53214-7_14

14.1 The MCO: What We Can Know

The MCO defines the things we comprehend about a modern conflict, including the relationships among the pieces. As shown in the upper part of Fig. 14.1, it starts with the things that don't require a particular situation – the situation-independent parts – and continues with things that require a particular situation for full expression.

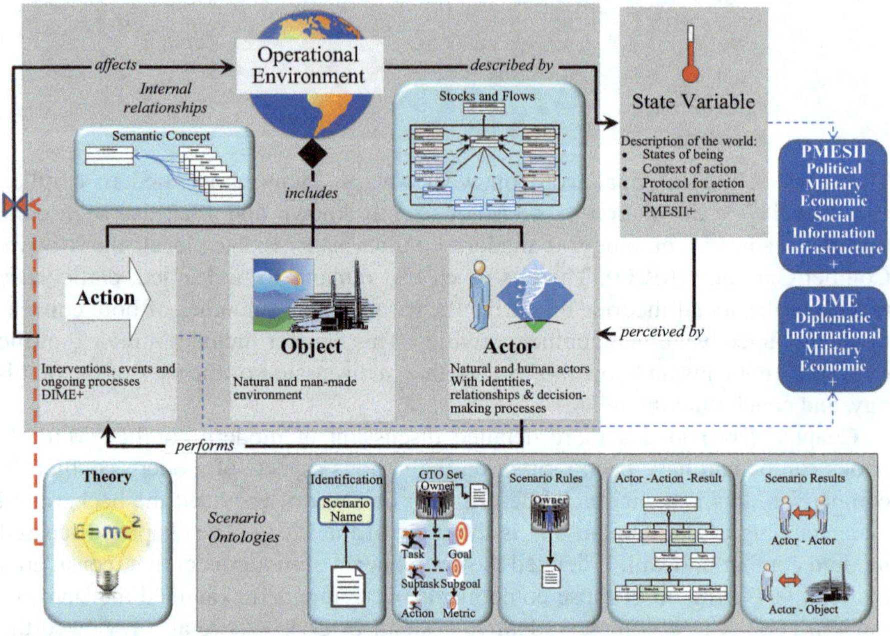

Fig. 14.1 Full context diagram

Situation-Independent Parts

The operational environment is divided into three parts, each with its own ontology:

- The active, motivating elements of the domain, the elements that do things, are called ***Actor*** classes. Generally, Actors are people, but not always. The Actor classes include individual humans, groups of humans, and non-human elements that act. The Actor Ontology performs the function of identifying the differences and commonalities among the Actor classes. Chapter 4 describes the Actor Ontology, a sub-ontology of the MCO. The chapter begins with an overview of Actors and insights into Actor classes. It continues with a discussion of each of the Actor categories and subcategories and a listing of all of the Actor classes in each. Chapter 7 discusses the new composite Actors and Actor plus Object classes. The multifaceted possibilities introduced by these classes are not immediately evident from their definitions in Chap. 4. The exposition is designed to build on the concepts, introducing one or two new facets at each stage, to make the complete picture more easily understood. Instances are used very liberally to ensure that the concepts are made concrete.

- Actors ***perform Actions***, as shown in the figure. Actions are the interventions, events, and ongoing processes that cause changes to the Objects and Actors in the MCO. Actions directly cause changes. As with the Actor Ontology, the Action Ontology performs the function of identifying the differences and commonalities among the Action classes. Chapter 5 describes the Action Ontology. The chapter begins with an overview of Actions and insights into Action classes. It continues with a discussion of each of the Action categories and subcategories and a listing of all of the Action classes and their definitions in each. Chapter 9 describes the DIME Ontology, which classifies the Actions according to the Diplomatic, Information, Military, and Economic (DIME) (plus an Other category), instruments of national power paradigm. Chapter 9 also illustrates how the Action classes are distributed across the DIME categories.
- Actions cause changes to or ***affect*** the entire Operational Environment, which consists of Actors, Actions, and ***Objects***. The Objects consist of natural environment elements, man-made environment elements, and conceptual elements, the passive elements that are only the recipients of Actions. As with the Actor and Action Ontologies, the Object Ontology identifies the differences and commonalities among the Object classes. Chapter 6 describes the Object Ontology. The chapter begins with an overview of Objects and insights into Object classes. It continues with a discussion of each of the Object categories and subcategories and a listing of all of the Object classes in each. Chapter 7 discusses the new composite Object and Object plus Actor classes. The multifaceted possibilities introduced by these classes are not immediately evident from their definitions in Chap. 6. The exposition is designed to build on the concepts, introducing one or two new facets at each stage, to make the complete picture more easily understood. Instances are used very liberally to ensure that the concepts are made concrete.

The elements in the Operational Environment are of interest because they represent entities that exist in the physical world; however, their state of being at any given time is also of interest. Each element is ***described by*** one or more ***State Variables*** or ***Metrics***. (Generally, each Metric will be a vector, with values expressing various significant aspects of the element.) These Metrics are ***perceived by*** the Actors, influencing their Actions, completing the cycle in Fig. 14.1. There are two Metric Ontologies.

- The MType Metric Ontology identifies the differences and commonalities among the Metrics by the type of information contained in the Metric. Chapter 8 describes the state variable or Metric Ontology, with particular emphasis on the meaning of Metrics. In that chapter we discuss the types of Metrics; however, we

only provide samples of the Metric classes because of their very large number (almost 6000).
- The PMESII Ontology identifies the differences and commonalities among the Metrics using the PMESII paradigm, which classifies the Metrics according to the Political, Military, Economic, Societal, Information, and Infrastructure (PMESII), plus kinetic and environmental categories and a null category. Chapter 9 describes the PMESII Ontology. Chapter 9 also illustrates how the Actor, Action, and Object classes are distributed across the PMESII categories through their Metric classes.

In addition to the relationships within each of the element ontologies and the Metric Ontologies, there are other relationships among the elements:

- The ***Stocks-and-Flows*** Ontology defines connections among the elements, such as Actions that increase or decrease the capacity of a particular class. The name is derived from the simplest operational concept, illustrated by a water tank (a stock of water) and two verbs, "fill" and "empty," that describe flows into and out of the water tank. Depending on the type of the class (Object, population, or organization), additional relationships are identified that link other Action classes, Actor classes, and Object classes to the particular class. Chapter 10 describes the Stocks-and-Flows Ontology and its operational connections. This ontology provides the explicit connections, of which a computer would be ignorant.
- The ***Semantic Concept*** Ontology describes connections among the elements that associate commonalities of meaning. A set of semantic concepts is defined, to which the elements are connected based on semantic similarity. These connections are obvious to a human because he or she knows the meanings of words, but not obvious to a computer. Chapter 10 describes the Semantic Concept Ontology.

The situation-independent parts of the ontology are constant for all situations in the sense that if a given element is part of a situation, the relationships that are specified in the situation-independent ontologies hold in the particular situation.

Scenario Parts

Chapter 12 describes the five Scenario Ontologies. These ontologies require instantiations for full understanding. That is, the structures are defined in the MCO; however, their value lies in describing a particular scenario and examples are required (and provided) to see how they work.

The scenario parts of the ontology consist of two components, a situation-independent structure and a situation-dependent set of relationships among the classes that describe a part of the situation.

- The ***Scenario Identification*** Ontology identifies a scenario.

- The Goal-Task-Owner ***GTO Sets*** Ontology defines the important parties and their agendas in a scenario. The agendas are the goals and tasks that are thought to be required for accomplishing the goals of each of the important parties (Owners) in a particular situation. The structure of the GTO sets can be defined independent of the situation; however, the particular owners, the Action classes that represent their tasks, and the Metric classes that represent their goals all depend on the situation. The owners' beliefs that the Action classes will actually accomplish the goals constitute their particular metric models.
- The ***Scenario Rules*** Ontology defines the restrictions the parties apply to their own actions in a scenario. These are simple situation-independent structures that specify the restrictions that owners place upon themselves in the choice of Actions that they may take. The actual relationships are situation-dependent.
- The Actor-Action-Result ***AAR Sets*** Ontology defines the connections of the parties' actions to results in a scenario. The AAR Sets describe the connections between particular Actions in a particular situation with the results that occur. Again, the structure of the AAR Sets can be defined independent of the situation, but the classes that are related are situation-dependent.
- The ***Scenario Relations*** Ontology defines the relationships between Actors [such as kinship], between Actors and Objects [such as ownership], between Objects [such as co-location], among entities in a group, and among entities in a network. The types of Actor relations, single pair, group, and network, have simple structures, independent of the situation. However, the actual relationships among the classes depend on the situation.

These five scenario parts are shown in the box at the lower right corner of Fig. 14.1, outside the cycle.

Theory

Chapter 11 described the ***Theories*** Ontology. In that chapter we described the structure of the ontology, its contents and their validity, and its uses with respect to model validity. The Theories Ontology is a work in progress. It contains a structure and a set of classes. However, the contents require review, emendation, and additions by experts in the various disciplines before it can be regarded as nearly complete.

The theory icon is also outside of the cycle in Fig. 14.1, but is also outside the situation-dependent box. The Theories Ontology focuses on answering three questions that relate to a real-world connection between the abstraction of the general ontology and the abstraction of the Theories Ontology.

- Why have things changed in the way they have changed? This reflects the change in a Metric value from one observation to the next.
- What is the meaning of the observed situation? This reflects the need to connect the set of Metric values to some larger understanding of the state of the situation

(a metric model). In detail, this is the inference that one Metric's value is based on the values of other Metrics.
- What will happen if I do some particular thing? This reflects the calculation of the "result" in an AAR based on the situation and an Action.

These theory connections are all represented in Fig. 14.1 by the dashed-arrow connection of the theory icon to a valve in the ***affects*** arrow. That is, while it is known that each Action affects other things, the particular effect is based on something, which we call a theory. This theory may be a detailed understanding of causal effects, or it could be a probabilistic statement that a set of effects is possible, each with a probability of occurrence.

Relationships

Chapter 13 explored the ***relationships*** that connect the Actors, Objects, and Actions into the complex structures that express the modern conflict domain. The basic relationships in the MCO are the object relationships that connect classes and instances. These relationships range from structural ontology relationships such as *is-a* and *hasInstance* to domain-specific relationships such as *hasTaskGoalPair* and *owns*. The data relationships are technical artifacts of the need to specify connections to data in an environment in which most of the elements refer to things, not numbers or rankings. The annotation relationships are artifacts of the OWL language that support connecting additional information to a central element such as a class or instance without having to include the information in the name.

14.2 Implementing the Ontologies

Constructing an ontology is a fascinating process. It requires knowledge of the fundamentals of ontologies and the choices available for their implementation. It also requires the willingness to delve into the details of the domain to be described. For domains of the size of modern conflict, this means taking pains to check the minutiae of descriptions of thousands of entries from multiple sources. It also requires the ability to create a vision of the overall picture. For this project, this vision started when Lee Lacy created the first version of the context diagram (Hartley and Lacy, Irregular Warfare (IW) Metrics Ontology Final Report, TRAC-H-TR-13-020 2011). This vision set the context for the entire ontology. Similar visions were required along the way in discovering symmetries and repeated structures in the domain.

As with any software process, organization is critical. However, an iterative or cyclical process is also critical, especially for larger ontologies. This means that you should be willing to rip out chunks of completed work in order to recast them with a different structure when a better understanding of the domain of knowledge leads to a revision of your thinking about the design.

Implementation History

The ontology that is described here is the result of two projects for the US Army Training and Doctrine Command (TRADOC) Analysis Center (TRAC) and several independent research and development projects by Hartley Consulting. As described in Chap. 1, the development of the MCO was one of incremental development, rather than a unified top-down effort. This had the disadvantage of gaps in the development process; however, it provided the advantage of time to reflect on what had been accomplished and to plan for and implement changes.

For example, ***version 1.0*** of the ontology, called the IW Metrics Ontology (IWO), used the concept of Lines of Effort (LOEs) to capture the agendas of major players (Hartley and Lacy, Irregular Warfare (IW) Metrics Ontology Final Report, TRAC-H-TR-13-020 2011; Hartley, Ontology Structures for Modeling Irregular Warfare 2012a). The LOEs for US forces were taken from the *Army Field Manual on Tactics in Counterinsurgency* (HQ Department of the Army 2009). However, in ***version 2.0***, called the IW Ontology 2, we introduced the GTO sets to broaden the concept (Hartley and Lacy, Creating the Foundations for Modeling Irregular Warfare 2013a, IW Ontology Final Report 2013b). Versions 1.0 and 2.0 were implemented in OWL Lite, with a Microsoft Access™ database holding the basic materials.

Version 2.0 listed properties as separate items; however, in ***version 3.0***, called the Unconventional Conflict Ontology (UCO), we merged properties into the Metrics, creating an MType Metric Ontology to differentiate the Metrics along the lines of the previous properties concept. Additionally, the Semantic Thesaurus of Version 2.0, which linked the Concepts to the Metrics, has been recast as the Semantic Concept Ontology, which links the Concepts to the Elements. The UCO and the associated Theories Ontology were implemented in a Microsoft Access™ database. Figure 14.2 illustrates the structure of the tables in the database.

The names of the tables are shown in the boxes; however, they are not important for this discussion. The figure includes the standard cardinality markings for the connections between tables (1 and ∞); however, these are also not important for this discussion. The groupings, colors, and connections illustrate the important points.

The situation-independent parts of the UCO occupy the top portion of the figure and consist of blue and orange boxes. The three element ontologies comprise the first three rows and connect to the orange **element** box in the top right corner. The next two rows represent the two Metric Ontologies, with the orange **metrics** box shown, connecting through a blue box, to the element box. The next row represents the Concepts Ontology, with the orange **concepts** box also connecting to the elements box. The last row (with a multipart blue box connection) represents the Stocks-and-Flows (SaF) Ontology, with the orange **SaFClasses** box also connected to the elements box.

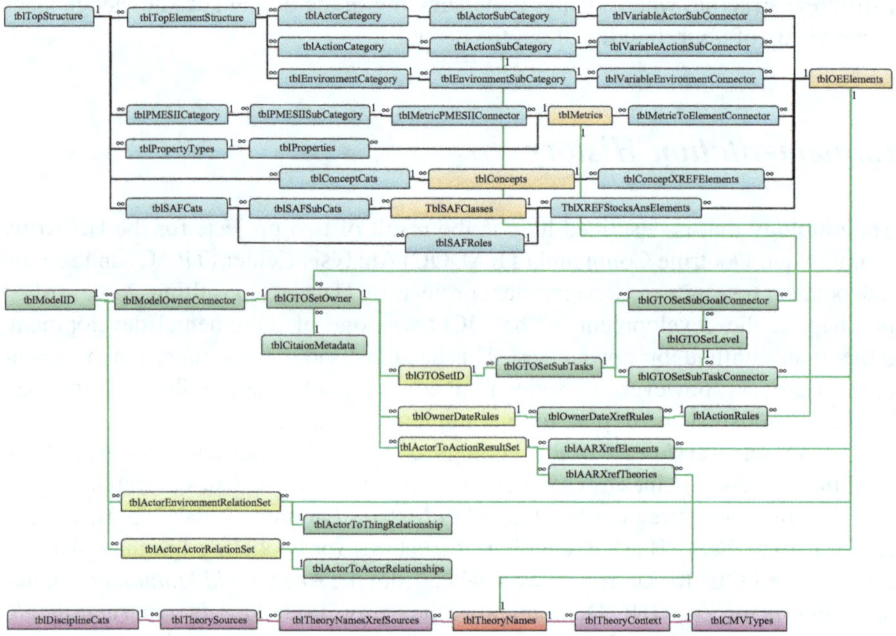

Fig. 14.2 UCO database structure

The situation-dependent parts of the UCO occupy the middle portion of the figure and consist of the green and yellow boxes. The complicated first rows represent the **GTO Sets** (yellow box), the **Owner Rules** (yellow box), and the **ActorActionResults Sets** (yellow box). These are connected (green lines) to ActionSubCategory box, the Metrics box, the element box, and the theories box. The next row represents the ActorObjectRelation structure, with the **ActorObjectRelations Set** (yellow box) and the list of possible relationships (green box). It is connected to the element box. The last row of this portion of the figure represents the ActorActorRelation structure, with the **ActorActorRelations Set** (yellow box) and list of possible relationships (green box). It is also connected to the element box.

The Theories Ontology is shown in the last row of the figure and consists of magenta and red boxes. The **TheoryNames** box is the red box and is connected (green line) to the ActorActionResults box.

The figure does not include the numerous queries that produce useful lists of related items in the database. However, these queries were based on the connections shown in the figure and can easily be imagined.

The Current Version

In an earlier book, I described the UCO (Hartley, An Ontology for Unconventional Conflict 2018). I also mentioned possible future enhancements aligning the ontology with a more formal ontology framework and adding combat to the domain. ***Version 4.0*** (currently version 4.26), the MCO, includes these enhancements.

It is not surprising that the addition of conventional combat required changes to the UCO. The changes included both major and minor changes. I have taken this opportunity to bring the ontology into closer agreement with the organizational conventions advocated by many professional ontologists. Further, the preparation of this book afforded the opportunity to review the details of the ontology and discover and correct wording and connection errors, as well as to discover a few missing Elements and Metrics.

General Changes from the UCO

The MCO is very similar to the UCO; however, there have been several major and many minor changes.

Table 14.1 displays a comparison of the basic counts between the MCO and its predecessor ontologies. The elements are divided into Actors, Actions, and Objects (using the newer terminology). The Metrics are listed separately. In the IWO (Hartley, Unconventional Conflict: A Modeling Perspective 2017) and the UCO (Hartley, An Ontology for Unconventional Conflict 2018), the sums of the three types of elements are larger than the displayed totals because several elements are repeated in two or three of the types. These elements have been separated in the MCO, so the sum equals the displayed total.

Table 14.1 Comparison of the MCO with predecessors

		IWO	UCO	MCO
Elements	Actors	102	101	114
	Actions	389	392	474
	Objects	209	179	191
	totals	682	653	779
Metrics		754	791	5930

The counts for the IWO and the UCO elements are similar, reflecting only minor changes: some elements were combined in the UCO and additional Metrics were created. However, the Metrics counts for the MCO are dramatically different; the changes are explained below.

A number of changes resulted from the choice of the Basic Formal Ontology (BFO) as a model for a foundational ontology. This influence resulted in the following changes:

- Assigned globally unique names to all classes, not just unique within their structure (e.g., no repeating of names in the Actor and Object Ontologies).
- Standardized the names of elements: start with small letter, noun phrase using the singular construction.
- Changed the terminology from passive environment or passive element to Object.
- Standardized the definition of elements: state the parent class name and then say how the thing is specialized from the parent class and what differentiates it from others with the same parent.
- Used a basic *is-a* taxonomy structure where possible, meaning that no multiple parents are allowed.
- Replaced the *is-a* relationships with other relationships where connections to multiple classes are required.
- Split some elements into the part that is an Actor, the part that is an Object, and the part that is an Action to avoid dual parents.
- Re-thought the PMESII connections for Metrics to avoid dual parents.
- Decided not to use processes.

A number of changes resulted because of adding conventional combat to the ontology:

- Added combat elements.
- Added composite elements to support the needs of combat concepts and improve the expression of non-combat concepts.
- Added relationships for military organizations, such as *hasTenant* and *hasMilSystemsGroup*.
- Added instances to help explain the concepts, leading to refinements of the structures.

Several structural changes were made to regularize the overall structure and to add new representations.

- Added the DIME Ontology.
- Refined the Concept Ontologies.
- Created the Scenario Ontology.
- Integrated the Theories Ontology.

Rethinking the state variables (metrics) led to several major changes:

- Realized the implications of the fact that the content of a metric is at the instance level of the metric and at the instance level of the element to which it is attached.
- Got rid of most common metrics. Only situation, identity, time, and location are common. For example, each element has an *identity* metric. The value of the identity applies to the instance of the element and is found in an instance of the *identity* metric. All metrics could have been implemented as common metrics except for the desire to connect to the PMESII paradigm. The particular PMESIISubcat class to which a particular metric should be attached depends, in part, on the element for which the metric provides state information. The four common metrics are not relevant to the PMESII paradigm; however, the rest

must be unique for each element. This, however, does create an explosion in the number of metrics.

- Reformed the metric structure. Make a metric *partOf* the PMESII class structure, not connected by *is-a*. This allows multiple connections to the PMESII class structure for each (non-common) metric and is a better representation for the relationship than that defined by the *is-a* relationship.
- Refined the metric structure of the UCO so that the set of metrics for a given entity form a metric vector containing one or more (or none) from each of the Metric Types in the MCO.

Although the BFO has processes, the MCO does not. A "process" is a thing that has a temporal starting point and a different temporal ending point. Whether a process should be included in an ontology depends on part on the granularity of the ontology. Meiosis and mitosis are cell division processes. Each takes a certain amount of time, requiring different temporal starting and ending points. Meiosis in human males takes about 74 hours and mitosis takes about 2 hours. If the time granularity of a particular ontology is measured in months, not hours, then it would generally not make sense to include these processes in that ontology.

Further, a "process" refers to a standard thing. That is, there is a standard starting state and a standard ending state. There are also generally standard sub-processes that occur in sequence. Variants are allowed, but must be specifiable. The random actions of a group of people at a beach, starting in the morning and ending in the evening, might be of interest for some ontology; however, they would not be identified as a process.

There are processes that occur in modern conflict: loading ammunition into a particular weapon system is a process and performing a pre-flight check of an aircraft is a process. However, the time granularity of the ontology militates against declaring these to be processes in the ontology. If they were significant enough to be included (a different granularity issue), they would be included as Actions, with essentially zero duration.

There are processes that might be candidates for "processes" in the MCO; however, they have a different problem. For example, a battle has a starting point and a different ending point and even has somewhat standardized starting and ending states. The starting state might be described as two separated forces, each with its own force strength. The ending state might be described as one of several discrete possible states: one side is victorious and the other side is defeated; both sides are rendered unfit for further combat, with neither truly victorious; the battle is inconclusive and the two sides separate; etc. For an ontology with very large granularity, defining this as a process might be useful. For the Modern Conflict Ontology, such a definition is not useful. The problem lies in the sub-processes: they are indeterminate. Here it is more useful to use Actions and admit that Theories are required to determine or explain the results, which are then followed by additional Actions.

Plans are documents of actors, actions, places, and timing. After-the-fact process descriptions are just chronicles of what happened, where, and when. Documents are just objects in the Object Ontology. Processes in the sense of industrial processes don't really exist in conflict, because the exceptions overwhelm the "standard" processes.

Changes to Actors from the UCO

Most of the general changes above were formal, e.g., changing spelling conventions and definition standards. For the most part, the structures of the Actor classes of the UCO describe the content adequately. However, the inanimate Actors category of the UCO has been replaced by the non-human Actors category in the MCO. The MCO also introduces an entirely new category, composite Actors.

Changes to Actions from the UCO

For the most part, the structures of the Action classes of the UCO describe the content adequately. However, the categories and subcategories have been changed, some Action classes have been revised, some new Action classes have been defined, and individual Action classes now appear in only one subcategory.

Changes to Objects from the UCO

The first change is that in the UCO terminology these things were environment elements, referring to their status as passive parts of the environment. That terminology was adequate, but clumsy. The term "object" is also perhaps only adequate, but less clumsy. This change is also reflected in the category and subcategory names. For the most part, the structures of the Environment Element classes of the UCO describe the content adequately. However, there have been changes at the category and subcategory level, which make cross-referencing difficult. For example, the *ShelterInfrastucture* subcategory has been eliminated, and the two classes within it have been moved to other subcategories. In addition, a new category, *compositeObject*, is introduced in the MCO.

Composite Classes Not in the UCO

Composite classes have been added to the Actor Ontology and the Object Ontology. The latter ontology includes composite classes containing both Object classes and Actor classes.

Changes to Metrics from the UCO

The Unconventional Conflict Ontology (UCO) positioned the PMESII Ontology as the principal Metric Ontology and the MType Metric Ontology as the subordinate ontology. The MCO reverses this ordering. The major change in the MType Metric Ontology is the creation of the *metricVectorKey* category and placing of three Metric subcategories in this category. The *flow* category has also been added. The *relationship* category has been renamed the *relational* category to reduce confusion between it and the relationships. The other category-subcategory relationships have been modified, and some of the subcategory names have been changed. Finally, the miscellaneous category has been eliminated. The major change in the PMESII Ontology is the connection of the Metric classes to the categories and subcategories of the ontology. In the UCO, this connection was an *is-a* relationship. In the MCO, this connection is an *isPartOf* relationship (inverse of *hasPart*), with an aggregation meaning, rather than a component meaning. This change allows a Metric class to be a part of multiple PMESII subcategories, without violating the one-parent rule for the *is-a* relationship.

Changes to the DIME/PMESII Ontologies from the UCO

I've been working with DIME/PMESII-type concepts since before the terms were invented. The concepts are good – better than what we had before. However, in my experience, they are better as concepts than as working tools. For example, Actions are what DIME is about, but they are better organized by their types than by DIME categories and subcategories. The reason is that Action classes are just that – classes. A particular Action class, such as one referring to building something, could be categorized as a Military Action because often the military performs construction or as an Economic Action because it is an example of economic power. However, the distinction can actually be made only at the instance level, where a given Action class is instantiated. The real point about DIME is that it isn't just military actions that are useful. Once you get past that, you are really just arguing about how many angels can dance on the head of a pin.

State Variables or Metrics are the same. The important point of PMESII is that you can't just measure damage and know what is going on. State Variables are really worse than actions because there are so many of them. Each element (Actor class, Object class, and Action class) has attached state variables or metrics. For the 779 element classes, there are about 6000 metric classes. "But there is more ..." Most metric classes are actually sub-vectors. If you want to discuss the capacity of a bridge, you might want to talk about its weight capacity and its throughput capacity (number of lanes for traffic) and its current values versus its ideal values (undamaged or expected values when the bridge is complete), etc. A full simulation would not have 6000 slots, but three to ten times that number. Then you have to remember that these numbers refer to classes, not instances. Each instance of a bridge has its own instances of the metrics.

With any metric ontology, we have an implicit conception of rolling up the details, for instance, metrics for the entire Infrastructure. The methodology chosen for such a roll-up is an explicit modeling decision. If we were to say that the infrastructure is 47% destroyed, what does that mean? What should it mean? How does a user understand what it means versus what the modeling decision really entails? Then there are cross-terms: the economy depends on the infrastructure. How? This requires another explicit modeling decision.

Changes to Semantic Concept Ontology from the UCO

Two new categories have been added, *militaryConcept* and *smallConcept*. This resulted in moving some of the *conceptNames*. Each *conceptName* (now called an *issue*) has the suffix "Issue" appended to the old name to emphasize its identity. For the most part, the structures of the Semantic Concepts of the UCO describe the content adequately. New classes have also been added, as appropriate. In this process, a few classes that should have been included in an issue, but weren't, have been added.

Changes to Stocks-and-Flows Ontology from the UCO

With the change of terminology from Environmental Element classes to Object classes, it would be proper to change terminology for the *environmentOriented* concept to *objectOriented* concept; however, the terminology of "object oriented" is already overused and applied to too many other things. For this reason, we have left this terminology as is and just remember that it refers to our Objects. For the most part, the structure of the SaF Concepts of the UCO describes the content adequately. Two new roles have been created, *nonhumanIncrease* and *nonhumanDecrease*, in the *organizationOriented* classes. This accounts for the use of dogs to detect buried people in disasters, to detect drugs by the police, and to detect hidden explosives by the military. Three additional SaF classes have been added, the *legitimacySAF* class, relating changes to legitimacy, the *populationGeographicSAF*, describing the relationships of a geographic population, and the *wholeOfGovernmentSAF* class, describing the relationships in the whole-of-government. These are environment-, population-, and organization-oriented, respectively. In addition, there have been new elements added to the ontology and further assignments of elements to the SaF classes.

Changes to Scenario Ontologies from the UCO

In the UCO, the Scenario Ontologies were referred to as situation-dependent ontologies. The concept is the same; only the name is different. In the MCO, the (very brief) Scenario Identification Ontology has been added to provide an externally defined scenario that can be used in the other scenario ontologies so that each can refer to a particular scenario. In the MCO, many of the diagrams have been changed to provide a better representation of the concepts involved and, in some cases, to represent the structural changes to the ontologies. Further, sample instantiations have been added to support a better understanding of the ontologies.

Changes to Theories from the UCO

The structures of the Theories of the UCO describe the content adequately. However, the treatment of validity has been modified in the MCO.

Changes to Relationships from the UCO

While the UCO certainly had definite relationships between ontology elements and structures, they were not explicitly described. The MCO explicitly names and defines its relationships and describes a structure for them.

Future Enhancements

Throughout the chapters, possible future enhancements have been described. They are collected here.

Possible Future Enhancements of AAR Sets

The current version of the AAR Ontology has minimal class structure, proceeding directly to the instance structure. A possible future version of the MCO might use an approach more like that of the GTO sets, providing a class structure instead of concealing the details in the instance structure.

Currently, the relationship between the action instance and the instance in a GTO set is implied, but not implemented, in the MCO. A future version might make this relationship explicit.

Currently, the relationship between the actor instances and the Scenario Relations groups is implied, not implemented, in the MCO. A future version might make this relationship explicit.

Possible Future Enhancements to Data Relationships

It is likely that an effort to fully populate the MCO for a particular conflict might motivate the creation of relationships between the Metrics and various data types (such as integer, floating-point, Boolean, or string), requiring new data relationships.

Possible Future Enhancements to Object Relationships

The use of the word "object" for this type of relationship is confusing. A new term may be created in the future. In addition, the form of the object relationships names is not consistent.

- Generally, the first part of the name is a verb, such as "is," "has," or "uses." However, sometimes this is not the case. For example, *acquaintanceOf* could be renamed to *isAcquaintanceOf* and would be in conformance with such a rule.
- Many relationships start with "has" and their inverse relationship starts with "is," such as *hasTrainer* and *isTrainerOf*. Yet others that might follow this rule don't, such as *isMentorOf* and *isMenteeOf*. Perhaps this rule should be implemented.
- There are other cases where different forms of the relationship name have been used. For example, *isInstanceOf* and *instantiates* are synonyms. Formal synonym definitions might be implemented.
- Various sub-relationships of *hasPart* have a restrictive, component meaning. That is, the list of parts is required to be complete. Yet other sub-relationships have a non-restrictive, aggregation meaning. That is, any or all of the list of parts may be present. In some cases, one meaning is used in one case and the other is used in another case. The definitions might be separated, creating new relationships, and enforced.

The MCO database also includes a rudimentary structure for modifying and using the ontology. A possible next step is to enhance this structure to support a robust set of uses for the ontology. For example, building the capacity to create scenarios with the full set of instantiations would be a relatively straightforward project.

14.3 Using the Ontology

Constructing the original IW Ontology (which led to the Unconventional Conflict Ontology and the Modern Conflict Ontology) was not a "science project," with its creation as the sole goal. It was constructed to be used for DIME/PMESII models. Two of the uses that were envisioned have already been implemented: developing existing models and performing VV&A on models. These and other uses are discussed here.

Tracking and Understanding the Situation

The ISSM was developed as a metric model of OOTWs, based on the work reported in *Doing Windows* (Hayes and Sands 1997). Figure 14.3 shows an ISSM tracking chart (Hartley, Interim Semi-static Stability Model 2006a). The chart tracks the values of several ISSM variables over time (horizontal axis), based on inputs taken from reporting of the situation. The vertical axis represents a scale from "horrible" at the bottom to "wonderful" at the top. The first variable in the legend, "Civil stability and durable peace exits," is the final output variable of the model (magenta line in the chart). The other seven variables are the penultimate variables that influence the final variable.

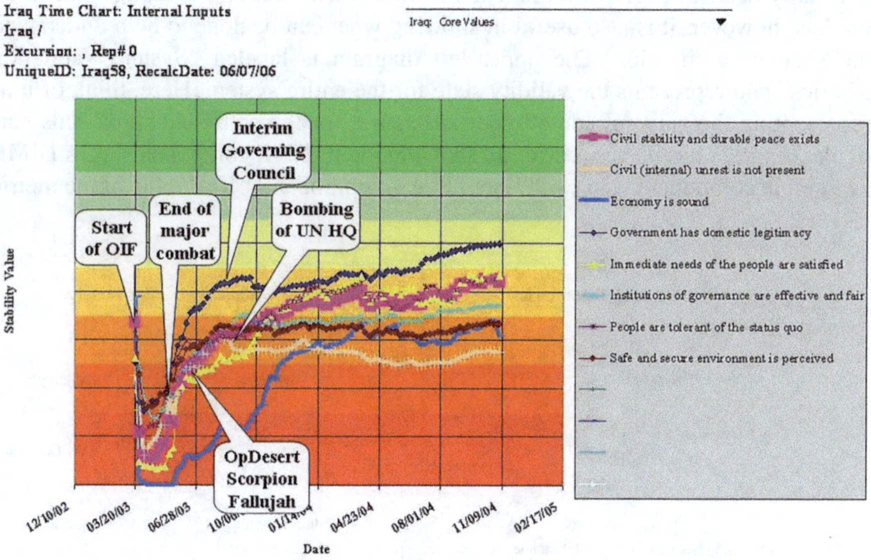

Fig. 14.3 The ISSM tracking model

The particular scenario being tracked in this case was Operation Iraqi Freedom (OIF) from its start on March 20, 2003, to November 9, 2004. The text boxes are added to indicate significant events on the timeline. The inferences from this chart are that the situation was improving as of the end date, with some sectors doing better than others. The lowest sector, "Civil (internal) unrest is not present," might have been regarded as a warning. In fact, later events in this sector led to a reversal of the gains shown in this chart.

The UCO provided the opportunity to create a more complete method for tracking the course of an unconventional conflict, a method that is based on a more complete understanding of its context and components.

Consider the impact of a hurricane on a small country. The hurricane winds, storm surge and flooding, and spawned tornados damage the infrastructure. The damage leads to reductions in the business economy, including jobs. Damage clearance takes weeks, competing with the normal economy. During this time there is increased criminal activity (e.g., looting) and increased civil unrest. Support by police and national armed forces are required. The financial costs of the clean-up cause economic dislocation, but fund some temporary jobs. The long-term recovery includes gradual improvement to the economy and jobs. All of this is reflected in the changes in values of the state variables. However, some kind of metric model will be required to provide a clear picture of the damage and progress in recovering from it.

Although a metric model can provide a single number that represents the state of the situation, perhaps on a scale from "horrible" to "wonderful," a little more detail is usually needed. Figure 14.4 is taken from a discussion of validating systems of models; however, it is also useful in showing what can be done to help understand the state of a situation. The upper left diagram is labeled "System Validation Metrics" and represents the validity state for the entire system. Here, think of it as representing the value of some important metric for the entire situation. This particular diagram has 9 axes (currently showing the PMESII components, plus DIME as a single component, and two "User Issues" components). The value of the metric

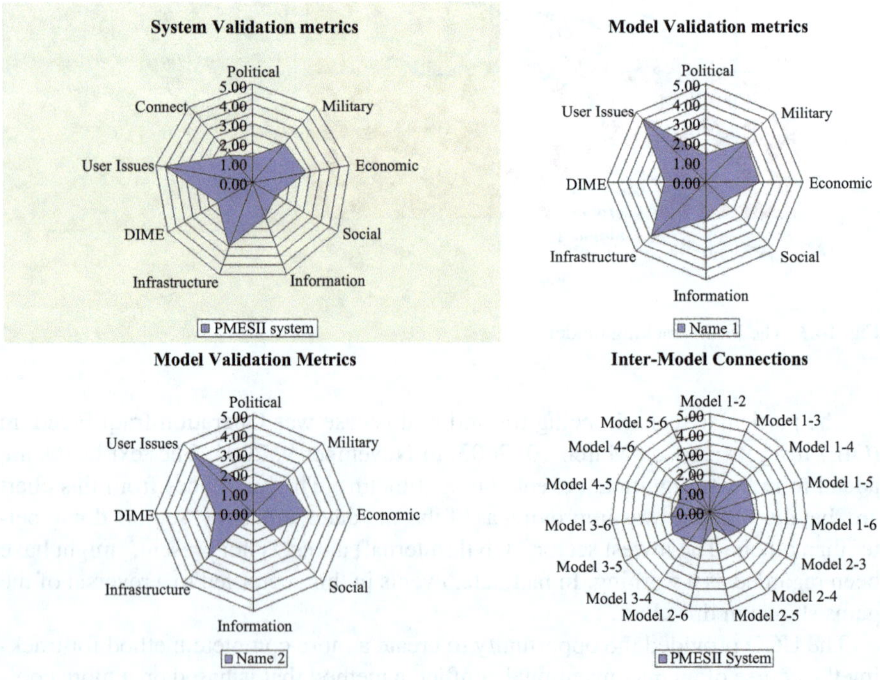

Fig. 14.4 Spider diagrams

within each of the desired components is shown by its distance from the center. When the values on each axis are connected and the polygon is colored, the relative strengths on each axis become obvious. In our use for understanding a situation, the axes might be provinces within the country. The other diagrams in the figure could then represent further breakdowns for each province. Alternatively, the main chart could represent some composite metric, and the sub-charts could represent the component values.

The point is that there are multiple ways of tracking and understanding the relevant information about a situation.

Similarly, the MCO provides the opportunity to create a more complete method for tracking the course of a modern conflict, including combat operations. A more comprehensive metric model would be required. This model would be based on a more complete understanding of modern conflict, provided by the MCO.

I have tried to explain the MCO as simply as possible. However, it is not simple. I have provided counts of the various classes, but did not emphasize that in an actual situation, each important individual, object, or action has its own instantiation, multiplying the entities to be tracked. Each instantiation has its own set of Metrics to be tracked and its own connections to other instantiations. Each time something changes, new values of the Metrics obtain. The volume of information required is staggering. Looked at from the viewpoint of gathering information to feed the ontology, you could be forgiven for thinking it a fool's errand.

However, there is a confluence of events outside the scope of the ontology that changes the state of affairs. As Ken Jobson and I say in another book, we are approaching the Panopticon – the surveillance state. Already almost everything you do on your smartphone is recorded and the information is disseminated. Grocery stores record how much, how frequently, and what you buy. Credit card companies do the same thing. The coming Internet of Things (cognified objects such as smart TVs, refrigerators, and cars) will increase the information that is recorded and disseminated about you. Currently that data are stored in separate databases by various companies and the government (except in China, where the government gets it all (Lee 2018)). As these databases are combined centrally, the volume of data will be much larger than the staggering amount required for the ontology. Understanding it will require an ontology – and the MCO or its equivalent will be a sub-ontology. Filling in the information needed to track a situation will be automated. Even choosing the instances – which individuals, actions, and objects need to be included – will be matter that AI can handle (Hartley and Jobson, Cognitive Superiority: Information to Power, the Road to Winning in the Sixth Domain 2020).

Building or Improving Models

TRAC used the list of Metrics in the IW Metrics Ontology (Hartley and Lacy, Irregular Warfare (IW) Metrics Ontology Final Report, TRAC-H-TR-13-020 2011) to plan for improvements to the IW Tactical War Game (TWG). TRAC personnel reviewed each Metric (and its definition) to determine whether the concept it represented was already included in the model. Those Metrics that were not already included were binned into three categories: those that should be included in the next year's model improvements, those that should be included in future improvements, and those that were not germane to the model's intended uses. The TRAC users passed these results to the IW TWG modelers. Other model users can employ the Ontology in the same manner for any existing model in the unconventional conflict domain.

The UCO can be used in building a model of unconventional conflict (Hartley, Unconventional Conflict: A Modeling Perspective 2017). The Unconventional Conflict Ontology supports holistic modeling. Holistic modeling means that we can both see the entire picture of what needs to be modeled and make informed decisions about what to model and what to omit. The UCO also separates the things we understand best from the things we understand least. That is, it separates the easy (relatively speaking) things from the hard things and allows us to put more effort into deciding how to address the hard parts. These "hard parts" include adjudication (the choice of theories and their implementations), the GTO sets (metric models), the tactical decision rules, and the modeling of intelligence operations (ground truth versus perception). This separation also means that we can perform verification, validation, and accreditation more efficiently and can describe the competence of the model more accurately.

Similarly, the MCO can be used in building a model of modern conflict. The MCO retains the advantages described for unconventional conflict and extends them to modern conflict. The complexity of the MCO that was described in tracking a situation also holds for building a model; however, it may be somewhat mitigated by the fact that a model does not have to include everything, depending on the purpose of the model. On the other hand, all of the data will have to be supplied for the model, and, given that the scenario will likely not be a current real-world scenario, much of the data will have to be gathered manually. Despite this complexity drawback, if the MCO accurately portrays the elements of the real world that will need to be modeled, the volume of data required is a function of the real world, not the MCO.

During the middle 1970s, some of the most difficult data to find for combat models consisted of what was called "weapons' effects data." One example would be data to define the effectiveness of the main gun of one tank on another tank. Depending on the granularity of the model, this might be:

- A single probability of kill number
- A probability of hit and a probability of kill-given-a-hit pair of numbers
- A probability of hit and a set of probabilities for mobility kill (the tank can't move), firepower kill (the tank can't shoot), a mission kill (the tank can't perform its mission), and a catastrophic kill (all of the preceding and the tank can't be repaired later)
- One of the above types, but from different angles (such as front, side, or rear) or different aim points (such as hull or turret)

In the MCO, *effectiveness* is a metric for actors, not for actions such as strikes. The name for the equivalent metric in the MCO is *damage*. While the difference in names is not significant, the fact that there is only a single metric available might appear to be a problem. However, remember that each metric will generally be a vector, not a single value. Given the possibilities above and the existence of multiple other weapons systems and targets, the question arises as to whether a single vector will be adequate.

The answer lies in the MCO philosophy of separating out theories and making calls to them obvious. The first point to be taken from the possibilities above is that each involves a separate theory of vulnerability (or at least different applications of the theory). Thus, the requirements for the damage metric vector depend on the choice of theories to be used in the model. The second point is to notice that a very specific vignette is implied by the statement of metric needs, one in which one tank fires on another as if they were alone on the battlefield. The MCO is based on modern conflict in which damage to military assets is not the only, and sometimes not the major, factor in success. Suppose the vignette is placed within an urban environment with culturally significant buildings and civilian housing surrounding the tank battle. A round that misses the target tank may cause collateral damage to the surrounding infrastructure and may cause injuries or deaths among the civilian population, which must also be assessed. Further, there may be sociological and cultural repercussions.

Such examples are the reason for the creation in the ontology of Actor-Action-Result (AAR) sets. These are pre-defined or situationally created vignettes with explicit calls to theories to assess the results of the vignettes. Their complexity will define the nature of the damage metric vectors. Generally, larger-grained models will require lower complexity damage metric vectors, and finer-grained model will require higher complexity metrics. As always, there is a price to be paid for detail. This is true no matter what the source of the model. At least with a model based on the MCO, the needs are explicit.

Supporting VV&A of Models

VV&A is an integral part of modeling. Frequently it is thought of as an expensive add-on to the modeling process; however, any good modeler is checking and testing his or her model during the design and implementation process and endeavors to have the results accepted when the work is finished. This is the core of VV&A.

The underpinnings of the ontological VV&A approach ensure that the problem of testing for coverage is relatively simple. The things that should be covered are exposed to see if they are explicitly modeled (or not modeled). This makes it relatively simple to verify the implementation of the conceptual model and perform validation on the more mundane aspects of the model. The explicit exposure of the use of theories makes the estimation of the validity of the more difficult aspects of the model easier.

Hartley Consulting has also used the Irregular War Ontology to improve its VV&A tool (Hartley, DIME/PMESII VV&A Tool 2008), which was used to perform verification and validation (V&V) of the HSCB Testbed (Hartley, I2WD HSCB Testbed: 2009 V&V Strategy & Plan 2009) and the Deployable Exercise Support system (DEXES) (Hartley, Validation Process for the DEXES II Conceptual Model 2012b).

The increased coverage and reduced dependence on the DIME and PMESII decompositions of the MCO will require a rewrite of parts of the DIME/PMESII VV&A Tool. The parts that support the active testing and record-keeping parts will be largely unchanged.

Inferring Agendas

In creating a model, the agendas of all parties are defined by the inputs. Within the model, one party may or may not know the agendas of the other parties. A party might know public aspects of other parties' agendas, but not private aspects. External to the model, the modelers will have done their best to produce credible guesses for the various agendas.

In dealing with real life, each party will have an agenda; however, it may not be clear, even to the owner, exactly what that agenda is. And that agenda may not be internally consistent. The GTO set structure provides a tool for estimating a party's agenda by examining its actions. Figure 14.5 shows a version of the GTO set structure. Each agenda has an owner (or each owner has an agenda), and each agenda

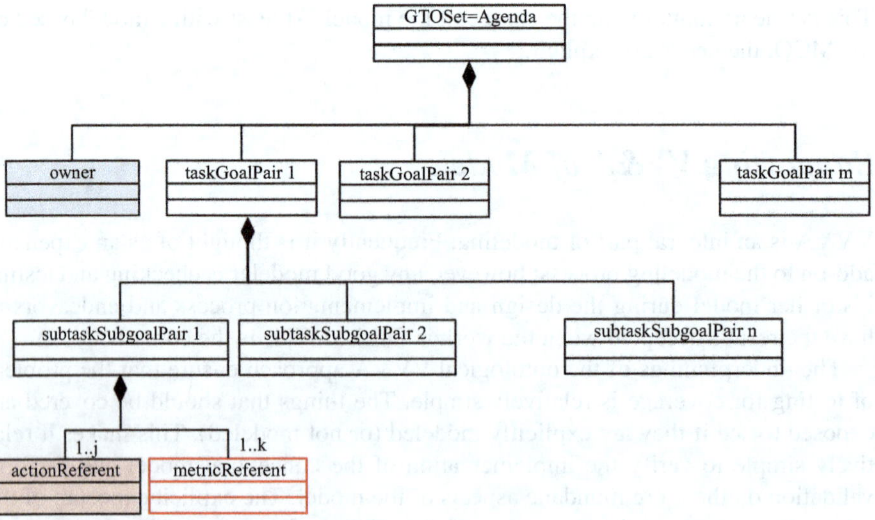

Fig. 14.5 Agenda structure

consists of from one to *m taskGoalPairs*. Each *taskGoalPair* has from one to *n subtaskSubgoalPairs*. Finally, each *subtaskSubgoalPair* has from one to *j actionReferents* and from one to *k metricReferents*.

From the top-down perspective, you think of defining the *taskGoalPairs* to match the agenda (and similarly down the tree). However, from the bottom-up perspective, you could say that the *taskGoalPairs* characterize the agenda. If you only knew them, you could infer the agenda. Similarly, if you only knew the *subtaskSubgoalPairs*, you could infer the *taskGoalPair*. And, if you only knew the *actionReferents*, you could infer the *subtaskSubgoalPair*.

The key lies in noting that each party does things; that is, it implements Actions. From our work in implementing ontologies, we have a fair number of instantiations of varied agendas. Thus, we have a set of candidate *subtaskSubgoalPairs* and their *actionReferents*. Figure 14.6 illustrates a notional matrix of all of these subtask instantiations against all possible Actions, showing which ones are referents for the particular subtask.

Subtask	Action 1	Action 2	Action 3	Action 4	Action 5		Action i
1	X		X	X			X
2	X	X			X		
3		X	X	X			X
4		X		X	X		
5			X	X			X
6	X				X		
7			X	X			X
8				X	X		
9				X	X		X
j					X		

Fig. 14.6 Subtask – Action matrix

Now we consider a set of observations of Actions. In this matrix (Fig. 14.7), the observation consists of the identity of the owner of the Actions and the Actions that occurred. By checking against the characterizations of the subtasks in Fig. 14.6, we have determined that Owner 1 has employed subtask 1 three times and subtask 4 and 6 once each. If these three subtasks were part of a single task with a supposed goal, then we would be well on the way to identifying the agenda of Owner 1.

Unfortunately, the real world is not quite as simple as shown in Fig. 14.7. Sometimes the owner of an Action is not clear. And, in general, the entire set of Actions that characterize a particular subtask will not be implemented at the same time (or may not be obviously related). Thus, the real matrix would be very sparse with question marks in the owner column, and no inferred subtask column would be present.

Observation	Owner	Action 1	Action 2	Action 3	Action 4	Action 5		Action i	Subtask
1	1	X		X	X			X	1
2	1	X		X	X			X	1
3	2		X	X	X			X	3
4	1		X		X	X			4
5	3			X	X			X	5
6	1	X				X			6
7	2		X	X	X			X	3
8	3			X	X			X	5
9	1	X		X	X			X	1
m	3			X	X			X	5

Fig. 14.7 Observation – Action matrix

We may make a statement about the probability of an Action (or set of Actions) taking place, given an agenda; however, we want to know what the probability is that an agenda is held, given the occurrence of an Action (or set of Actions). They are not the same. Bayes' theorem gives the connection (where P(X) means the probability of X and P(X | Y) means the probability of X given Y):

$$P(\text{Agenda} \mid \text{Action}) = \left[P(\text{Action} \mid \text{Agenda}) * P(\text{Agenda})\right] / P(\text{Action}).$$

Thus, statistical analyses of sufficient numbers of observations will yield probabilistic inferences of subtasks and tasks and, hence, subgoals, goals, and agendas.

14.4 Concluding Thoughts

Modern conflict is a complex, messy thing. It normally involves multiple Actors, with their own conflicting agendas and differing concepts of legitimate actions. This complexity means that understanding modern conflict is very difficult. For example, answering the simple question "who won?" is no longer simple for modern conflict. Because modern conflict contains unconventional conflict as the norm now, understanding it is a necessity, despite the difficulties.

The Modern Conflict Ontology supports understanding the entire situation. It also supports understanding how the parts of the situation fit into the entire situation. Perhaps most important, it defines what we do not know or at least do not know well – the thing we have called the Theories Ontology – and how Actions create effects.

The point of the ontology is to describe what we can know about conflicts in general and to support recording what we know about a particular conflict. To those ends, we need to be able to describe the actors, objects, and actions of the conflict. We need to talk about the relations among them. We need to talk about what drives them to do what they do. The core message of this book is that understanding modern conflict is now easier than it had been because of the development of an ontology that describes the entire modern conflict domain.

Bibliography

W. Anderson, B. Peterson, An ontology of modern military organizations and their structure. *Workshop on the IEEE Standard Upper Ontology (IJCAI01)* (IEEE, 2001)

R. Arp, B. Smith, A.D. Spear, *Building Ontologies with Basic Formal Ontology* (The MIT Press, Cambridge, 2015)

I. Berlin, *The Hedgehog and the Fox.* (H. Hardy, Ed.) (Princeton University Press, Princeton, 2013)

B.J. Bernardoni, R.F. Deckro, M.J. Robbins, Using social network analysis to inform stabilization efforts. Mil. Oper. Res. **18**(4), 37–60 (2013)

R.A. Bitzinger, *Assessing the Conventional Balance in Europe, 1945–1975, N-2859-FF/RC* (Rand, Santa Monica, 1989)

Chairman of the Joint Chiefs of Staff, *Joint Operations, Joint Publication 3-0* (Chairman of the Joint Chiefs of Staff, Washington, DC, 2011)

Chairman of the Joint Chiefs of Staff, *Joint Planning, Joint Publications 5-0* (Chairman of the Joint Chiefs of Staff, Washington, DC, 2017)

Chinese Academy of Military Science, *The Science of Military Strategy (2013).* (4. W. Co, Trans.) (4th Watch Publishing Co, Lexington, 2018)

R.A. Clarke, R.K. Knake, *The Fifth Domain: Defending Our Country, Our Companies, and Ourselves in the Age of Cyber Threats* (Penguin Press, New York, 2019)

C.V. Clausewitz, *On War.* (M. Howard, & P. Paret, Trans.) (Alfred A. Knopf, Inc, New York, 1993)

J. Coles, J. Zhuang, Introducing terrorist archetypes: Using terrorist objectives and behavior to predict new, complex, and changing threats. Mil. Oper. Res. **21**(4), 47–62 (2016)

Department of Defense, dtic.mil, 3000.05 (2009, September 16). Retrieved April 24, 2016, from DoD Issuances: http://www.dtic.mil/whs/directives/corres/pdf/300005p.pdf

Department of Defense, dtic.mil, 3000.07 (2014, August 28). Retrieved April 24, 2016, from DoD Issuances: http://www.dtic.mil/whs/directives/corres/pdf/300007p.pdf

DoD Chief Information Officer, DM2-DoDAF meta-model (2010). Retrieved August 23, 2017, from DoDAF: http://dodcio.defense.gov/Library/DoD-Architecture-Framework/dodaf20_conceptual4/

Dublin Core Metadata Initiative, Dublin core metadata element set, version 1.1 (2017). Retrieved August 23, 2017, from Metadata Innovation: http://dublincore.org/documents/dces/

E.L. DuBois, W.P. Hughes, L.J. Low, *A Concise Theory of Combat, NPS-IJWA-97-001* (Institute for Joint Warfare Analysis, Naval Postgraduate School, Monterey, 1997)

T.N. Dupuy, *Numbers, Predictions & War: The Use of History to Evaluate and Predict the Outcome of Armed Conflict* (HERO Books, Fairfax, 1985)

D. S. Hartley III, *An Ontology of Modern Conflict*, Understanding Complex Systems, https://doi.org/10.1007/978-3-030-53214-7

P. V. Fellman, Y. Bar-Yam, A. A. Minai (eds.), *Conflict and Complexity: Countering Terrorism, Insurgency, Ethnic and Regional Violence* (Springer, New York, 2015)

M.T. Flynn, M. Ledeen, *The Field of Fight* (St. Martin's Press, New York, 2016)

J.L. Gaddis, *On Grand Strategy* (Penguin Press, New York, 2018)

M.A. Gallagher, D.J. Caswell, B. Hanlon, J.M. Hill, Rethinking the hierarchy of analytic models and simulations for conflicts. Mil. Oper. Res. **19**(4), 15–24 (2014)

V. Gerasimov, The value of science in prediction. *Military-Industrial Kurier* (2013, February 27). Retrieved March 1, 2019, from: https://inmoscowsshadows.wordpress.com/2014/07/06/the-gerasimov-doctrine-and-russian-non-linear-war/

T.R. Gruber, A translation approach to portable ontology specifications. Knowl. Acquis. **5**(2), 199–220 (1993)

N. Haken, J. Burbank, P.H. Baker, Casting globally: Using content analysis for conflict assessment and forecasting. Mil. Oper. Res. **15**(2), 5–19 (2010)

B.H. Liddell Hart, *Strategy*, 2nd rev edn. (Meridian, New York, 1967)

D.S. Hartley, *Predicting Combat Effects* (INFORMS, Linthicum, 2001)

D.S. Hartley, Interim semi-static stability model (2006a). Retrieved May 4, 2016, from Hartley Consulting: http://drdeanhartley.com/HartleyConsulting/TOOLBOX/issm.htm

D.S. Hartley, *Operations Other Than War (OOTW) Flexible Asymmetric Simulation Technologies (FAST) Prototype Toolbox: ISSM v4.00 Analysts' Guide* (Dynamics Research Corporation, Orlando, 2006b). Retrieved from: http://drdeanhartley.com/HartleyConsulting/TOOLBOX/ISSM%20AG%2012Dec06.pdf

D.S. Hartley, DIME/PMESII VV&A tool (2008). Retrieved May 4, 2016, from Hartley Consulting: http://drdeanhartley.com/HartleyConsulting/VVATool/VVA.htm

D.S. Hartley, *I2WD HSCB Testbed: 2009 V&V Strategy & Plan* (US Army RDECOM CERDEC I2WD, Ft Monmouth, 2009)

D.S. Hartley, Ontology structures for modeling irregular warfare. Mil. Oper. Res. **17**(2), 5–18 (2012a)

D.S. Hartley, *Validation Process for the DEXES II Conceptual Model* (Hartley Consulting, Oak Ridge, 2012b)

D.S. Hartley, DIME/PMESII models, in *Conflict and Complexity: Countering Terrorism, Insurgency, Ethnic and Regional Violence*, ed. by P. V. Fellman, Y. Bar-Yam, A. A. Minai, (Springer, New York, 2015), pp. 111–136

D.S. Hartley, *Unconventional Conflict: A Modeling Perspective* (Springer, New York, 2017)

D.S. Hartley, *An Ontology for Unconventional Conflict* (Springer, New York, 2018)

D.S. Hartley, K.O. Jobson, *Cognitive Superiority: Information to Power* (Springer, New York, 2020)

D.S. Hartley, L.W. Lacy, *Irregular Warfare (IW) Metrics Ontology Final Report, TRAC-H-TR-13-020* (US Army TRAC, Ft Leavenworth, 2011)

D.S. Hartley, L.W. Lacy, Creating the foundations for modeling irregular warfare, in *Advances in Design for Cross-Cultural Activities, Part II*, ed. by D. M. Nicholson, D. D. Schmorrow, (CRC Press, Boca Raton, 2013a), pp. 13–23

D.S. Hartley, L.W. Lacy, *IW Ontology Final Report* (US Army TRAC, Ft Leavenworth, 2013b)

D.G. Haut, *OOTW* (D.S. Hartley, Interviewer, 1994)

D.G. Haut, Progress in computer-assisted training simulations for lesser regional contingencies. *JUORS VII, 28 November – 1 December 1995.* Tokyo, Japan (1995)

D.G. Haut, M.L. McCurdy, Modeling and simulation for the new Pacific community: A USPACOM perspective. Mil. Oper. Res. **1**(1), 19–25 (1994)

B.C. Hayes, J.I. Sands, *Doing Windows: Non-Traditional Military Responses to Complex Emergencies* (CCRP, Washington, DC, 1997)

R.L. Helmbold, *The Advantage Parameter: A Compilation of Phalanx Articles Dealing with the Motivation and Empirical Data Supporting Use of the Advantage Parameter as a General Measure of Combat Power* (US Army Concepts Analysis Agency, Washington, DC, 1997)

Homer, *The Odyssey.* (E.V. Rieu, Trans.) (Penguin Classics, Baltimore, 1946)

Homer, *The Iliad.* (W.H. Rouse, Trans.) (Signet Classic, New York, 1950)

V. Horbulin, The "hybrid warfare" ontology. Strateg. Prior. **38**(1), 4–13 (2018). Retrieved from: https://niss-priority.com/index.php/journal/article/view/183

HQ Department of the Army, *Army Field Manual on Tactics in Counterinsurgency, FM 3-24.2* (Department of the Army, Washington, DC, 2009)

W.P. Hughes, *Fleet Tactics and Coastal Combat*, 2nd edn. (Naval Institute Press, Annapolis, 2000)

W.P. Hughes, *Re: A Concise Theory of Combat* (D.S. Hartley, Interviewer, 2018, April 18)

IDEAS Group, DM2_HTML_130326 (n.d.). Retrieved August 23, 2017, from International Defence Enterprise Architecture Specification Group: http://www.ideasgroup.org/dm2/

IFOMIS, BFO basic formal ontology (2017, March 31). Retrieved August 24, 2017, from BFO Basic Formal Ontology: http://ifomis.uni-saarland.de/bfo/

Joint Force Development, *Competition Continuum, Joint Doctrine Note 1-19* (Chairman of the Joint Chiefs of Staff, Washington, DC, 2019)

J. Keegan, *A History of Warfare* (Vintage Books, New York, 1994)

L. Kello, The meaning of the cyber revolution. Int. Secur. **38**(2), 7–40 (2013)

L. Kello, *The Virtual Weapon and International Order* (Yale University Press, New Haven, 2017)

M.L. King, A.S. Hering, A.M. Newman, Evaluating counterinsurgency classification schemes. Mil. Oper. Res. **19**(3), 5–25 (2014)

P. Kucik, E. Paté-Cornell, Counterinsurgency: A utility-based analysis of different strategies. Mil. Oper. Res. **17**(4), 5–23 (2012)

V. Kuikka, A combat equation derived from stochastic modeling of attrition data. Mil. Oper. Res. **20**(3), 49–69 (2015)

L.W. Lacy, *OWL: Representing Information Using the Web Ontology Language* (Trafford, Victoria, 2005)

F.W. Lanchester, Mathematics in warfare, in *The World of Mathematics*, ed. by J. R. Newman, vol. 4, (Simon and Schuster, New York, 1956), pp. 2138–2157

K.-F. Lee, *AI Superpowers: China, Silicon Valley, and the New World Order* (Houghton Mifflin Harcourt, Boston, 2018)

G. Lubold, D. Volz, Chinese hackers breach navy data. Wall Street J., A1 (2018, December 15–16)

A. Maan, *Narrative Warfare* (Narrative Strategies, Ink, 2018)

N. Machiavelli, *The Prince.* (D. Donno, Trans.) (Bantam Books, New York, 1966)

A. Macias, US intel report says mysterious Russian explosion was triggered by recovery mission of nuclear-powered missile, not a test (2019, August 29). Retrieved November 16, 2019, from www.cnbc.com: https://www.cnbc.com/2019/08/29/intel-says-russian-explosion-was-not-from-nuclear-powered-missile-test.html

P.R. Mansoor, Introduction: Hybrid warfare in history, in *Hybrid Warfare: Fighting Complex Opponents from the Ancient World to the Present*, ed. by W. Murray, P. R. Mansoor, (Cambridge University Press, Cambridge, 2012)

M.E. Mateski, T.A. Mazzuchi, S. Sarkani, The hypergame perception model: A diagramatic approach to modeling perception, misperception, and deception. Mil. Oper. Res. **15**(2), 21–37 (2010)

S. McFate, *The New Rules of War: Victory in the Age of Disorder* (William Morrow, New York, 2019)

W. Murray, P. R. Mansoor (eds.), *Hybrid Warfare: Fighting Complex Opponents from the Ancient World to the Present* (Cambridge University Press, Cambridge, 2012)

B. Olson, U.S. becomes net exporter of oil, fuels for first time in decades (2018, December 6). Retrieved December 9, 2018, from The Wall Street Journal: https://www.wsj.com/articles/u-s-becomes-net-exporter-of-oil-fuels-for-first-time-in-decades-1544128404?mod=searchresults&page=1&pos=1

A. Oltramari, L.F. Cranor, R.J. Walls, P. McDaniel, Building an ontology of cyber security. STIDS, 54–61 (2014)

Onto-Med, Ontologies (2010). Retrieved August 23, 2017, from Ontologies in Medicine: http://www.onto-med.de/ontologies/index.jsp

T.J. Paciencia, D.J. Richmond, J.J. Schumacher, W.L. Troy, A framework and system for theater air attrition modeling. Mil. Oper. Res. **23**(2), 41–59 (2018)

G.S. Patton, *The Probable Characteristics of the Next War and the Organization, Tactics, and Equipment Necessary to Meet Them* (Army War College, Washington, DC, 1932)

C. Paul, C.P. Clarke, B. Grill, Qualitative comparative analysis of 30 insurgencies, 1978-2008. Mil. Oper. Res. **17**(2), 19–40 (2012)

S. Perlo-Freeman, Arms race (2019, July 4). Retrieved September 21, 2019, from Encyclopedia Britannica: https://www.britannica.com/topic/arms-race

M. Pillsbury, *The Hundred-Year Marathon: China's Secret Strategy to Replace America as the Global Superpower* (St. Martin's Press, New York, 2015)

Protégé. (n.d.). Retrieved July 16, 2016, from http://protege.stanford.edu/

L. Qiao, X. Wang, *Unrestricted Warfare*. (unknown, Trans.) (Echo Point Books & Media, Brattleboro, 1999)

J. Schroden, R. Thomasson, R. Foster, M. Lukens, R. Bell, A new paradigm for assessment in counterinsurgency. Mil. Oper. Res. **18**(3), 5–20 (2013)

R. Shearer, B. Marvin, Recognizing patterns of nation-state instability that lead to conflict. Mil. Oper. Res. **15**(3), 17–30 (2010)

P.W. Singer, E.T. Brooking, *Like War: The Weaponization of Social Media* (Houghton Mifflin Harcourt, Boston, 2018)

B. Smith et al., Basic formal ontology 2.0: Specification and user's guide (n.d.). Retrieved August 24, 2017, from: https://github.com/bfo-ontology/BFO/wiki

Sun-Tzu, *The Art of War*. (S.B. Griffith, Trans.) (Oxford University Press, New York, 1963)

R. Taylor, S. Germano, Spy chiefs agreed to contain Huawei. Wall Street J., A8 (2018, December 15–16)

The White House. (2018, September). Retrieved February 2, 2019, from WhiteHouse.gov: https://www.whitehouse.gov/wp-content/uploads/2018/09/National-Cyber-Strategy.pdf

Thucydides, *History of the Peloponnesian War*. (R. Warner, Trans.) (Penguin Classics, London, 1954/1972)

L. Tolstoy, *War and Peace*. (R. Pevear, & L. Volokhonsky, Trans.) (Knopf, New York, 2007)

J. Trindle, E. Groll, J. Hudson, U.S. freezes assets of Russian businessmen and bank close to Putin (2014, March 20). Retrieved September 21, 2019, from Foreign Policy: https://foreignpolicy.com/2014/03/20/u-s-freezes-assets-of-russian-businessmen-and-bank-close-to-putin/

U.S. Marine Corps, *Warfighting, MCDP 1* (U.S. Marine Corps, Washington, DC, 1997)

M.C. Űnal, Counterinsurgency and military strategy: An analysis of the Turkish Army's COIN strategies/doctrines. Mil. Oper. Res. **21**(1), 55–88 (2016)

US Army TRADOC, TRADOC-G2/oe-ontology (2017, January 31). Retrieved April 13, 2018, from GitHub: https://github.com/TRADOC-G2/oe-ontology

Virgil, The Trojan Horse, in *Great Spy Stories from Fiction*, ed. by A. Dulles, (Harper & Row, New York, 1969), pp. 289–293

Wall Street Journal Editorial Staff, How America Broke OPEC. Wall Street J., A12 (2018, December 15–16)

Wikipedia, Ontology (information science) (2016, July 6). Retrieved July 6, 2016, from Wikipedia: https://en.wikipedia.org/wiki/Ontology_(information_science)

Wikipedia, NATO bombing of Yugoslavia (2018a, April 28). Retrieved April 29, 2018, from Wikipedia: https://en.wikipedia.org/wiki/NATO_bombing_of_Yugoslavia

Wikipedia, Siege (2018b, April 22). Retrieved April 29, 2018, from Wikipedia: https://en.wikipedia.org/wiki/Siege

Wikipedia, Syrian civil war (2018c, April 28). Retrieved April 29, 2018, from Wikipedia: https://en.wikipedia.org/wiki/Syrian_Civil_War

Wikipedia, Bosnian war (2019a, March 21). Retrieved May 2, 2019, from Wikipedia: https://en.wikipedia.org/wiki/Bosnian_War

Wikipedia, Chengdu J-20 (2019b, November 12). Retrieved November 16, 2019, from Wikipedia: https://en.wikipedia.org/wiki/Chengdu_J-20

Wikipedia, China-United States trade war (2019c, September 20). Retrieved September 21, 2019, from Wikipedia: https://en.wikipedia.org/wiki/China%E2%80%93United_States_trade_war

Wikipedia, Datei:Operation Desert Storm.jpg. (n.d.). Retrieved September 4, 2017, from Wikipedia: https://de.wikipedia.org/wiki/Datei:Operation_Desert_Storm.jpg

J.C. Wylie, *Military Strategy: A General Theory of Power Control* (Naval Institute Press, Annapolis, 2014)

Index

D. S. Hartley III, *An Ontology of Modern Conflict*, Understanding Complex Systems, https://doi.org/10.1007/978-3-030-53214-7